Badiou, Infinity, and Subjectivity

Continental Philosophy and the History of Thought

Series Editors: Christian Lotz, Michigan State University, and Antonio Calcagno, King's University College at Western University

The *Continental Philosophy and the History of Thought* series seeks to augment and amplify scholarship in continental philosophy by exploring its rich and complex relationships to figures, schools of thought, and philosophical movements that are crucial for its evolution and development. A historical focus allows potential authors to uncover important but understudied thinkers and ideas that were nonetheless foundational for various continental schools of thought. Furthermore, critical scholarship on the histories of continental philosophy will also help reposition, challenge, and even overturn dominant interpretations of established, well-known philosophical views while refining and reinterpreting them in light of new historical discoveries and textual analyses. The series seeks to publish carefully edited collections and high-quality monographs that present the best of scholarship in continental philosophy and its histories.

Titles in series:

Badiou, Infinity, and Subjectivity: Reading Hegel and Lacan after Badiou, by Mohammad Reza Naderi

Foucault's Critical Philosophy of History: Unfolding the Present, by Adam Takács

Reading Continental Philosophy and the History of Thought, edited by Christian Lotz and Antonio Calcagno

Reframing Twentieth-Century French Philosophy: The Roots of Desire, edited by Elodie Boublil

Hegel and Heidegger on World and Nature: Reconciliation or Alienation, by Raoni Padui

The Idea of Beginning in Jules Lequier's Philosophy, by Ghislain Deslandes

Hannah Arendt and the History of Thought, edited by Daniel Brennan and Marguerite La Caze

Rethinking Philosophy with Borges, Zambrano, Paz, and Plato, by Hugo Moreno

The Ontological Roots of Phenomenology: Rethinking the History of Phenomenology and Its Religious Turn, by Anna Jani

Negative Dialectics and Event: Nonidentity, Culture, and the Historical Adequacy of Consciousness, by Vangelis Giannakakis

Marxism and Phenomenology: The Dialectical Horizons of Critique, edited by Bryan Smyth and Richard Westerman

Groundwork of Phenomenological Marxism: Crisis, Body, World, by Ian H. Angus

Max Stirner on the Path of Doubt, by Lawrence S. Stepelevich

Badiou, Infinity, and Subjectivity

Reading Hegel and Lacan after Badiou

Mohammad Reza Naderi

LEXINGTON BOOKS
Lanham • Boulder • New York • London

Published by Lexington Books
An imprint of The Rowman & Littlefield Publishing Group, Inc.
4501 Forbes Boulevard, Suite 200, Lanham, Maryland 20706
www.rowman.com

86-90 Paul Street, London EC2A 4NE

British Library Cataloguing in Publication Information Available

Library of Congress Cataloging-in-Publication Data Available

ISBN 978-1-66693-104-4 (cloth)
ISBN 978-1-66693-105-1 (ebook)

I dedicate this work to my loving and supportive wife, Fariba Vaezi, for her constant care and support, not just through my studies but also throughout our lives together. I want her to know that this was impossible without her standing by my side and holding my hand throughout this journey.

Contents

List of Tables and Figures

TABLES

FIGURES

Acknowledgments

I would like to thank Alain Badiou, for his guidance and encouragement over the years I spent developing this book. I would also like to acknowledge the contribution of Gabriel Tupinambá to what I have been able to produce here. Our numerous meetings and hours of discussion shaped this work. Gabriel diligently reviewed the draft materials and provided crucial interventions and comments, many of which were incorporated into the text. I thank him for his brilliance, diligence, and dedication. I've truly enjoyed our conversations and learned immensely from them. I would also like to thank Dennis Yao. Dennis read the entire work word by word and provided crucial suggestions to make the work more readable and fluent. I also would like to thank Jelica Šumič Riha and Jana Berankova for reviewing and making many suggestions to improve the arguments. I thank Lexington publishing editors Jana Hodges-Kluck, Deanna Biondi, and Emilia Rivera for their assistance in taking this book into production.

Previous versions of parts of this book have been published as follows: parts of chapter 1 with the heading "Mark and Lack: Formalism and Fidelity" were published with the same title in *Crisis and Critique: Philosophy and Science*, Volume 5, issue 1, 15-03-2018, edited by Agon Hamza & Frank Ruda, and reproduced with the kind permission of Crisis and Critique. Parts of chapter 8 were published in *Filozofski vestnik*, Vol. 43 No. 3 (2023): Reconfigurations of Limits, Varieties of Perversion, with the kind permission of Rowman & Littlefield.

Introduction

The twentieth century ends without having mastered infinity. The infinite, a powerful category of thought during romanticism, posed a particular challenge to twentieth-century thought. The more this century sought a decisive break with romanticism, the more it struggled with a sense in which it could claim infinity as its own. We can name at least two ways in which the infinite was intertwined in the thought of romanticism: first, in God, the infinity of the ideal that was represented and expressed in all aspects of thinking, in art, science, religion, politics, and especially in the highest exaltation of amorous relationship wherein it is God who beckons from behind the veil of the beloved. In this sense, the work of art, the edifice of political and religious hierarchy, the intoxicating prose, and the divine poem were all mediums for the ideal Itself—the finite had captured the infinite, the infinite that was incarnated in the finite. Second, in the artist, the scientist, the poet, the minister, and the lover, we find the medium between reality and the ideal—they possess the power to bring the infinite into the finite arrangement of literary expressions, artistic creations, scientific theories and the canon of laws. This second aspect of romanticism is the extraordinary subjectivity capable of mediating the transcription of the infinite into the finite.

It is then clear why romanticism captured the dialectic of infinite and finite so succinctly by suturing thinking to art, by which, since Kant, the infinite is primarily thinkable as the sublime and the enigma of its articulation with the finite is made sensible in the problem of taste and the question of the artistic project. To bring the infinite to the finite is also to produce a sensible form of the sublime, after all. It is no wonder, then, that the same thinker who sought to reconfigure the dialectics of the finite and the infinite from the standpoint of a new reading of the logic of incarnation, in which the infinite is always already within the finite, also claimed that art was no longer the medium of subjectivity. If Hegel brings to closure this romantic recourse to the sublime, opening the possibility of new, immanent thinking of the infinite, Marxism, then, could be seen as the closure of this possibility, as it interprets Hegel's salvage of the infinite as the remainder of the romantic period—an attachment to the ahistorical. This is a procedure that all the great "thinkers of suspicion"—Marx, Nietzsche, and Freud—seem to share to some degree: the infinite as a defense against the real of the finite, asserting that the incarnation of the infinite into finite was finally a step toward banishing the infinite altogether from the thought of the twentieth century.

Therefore, in what ways did this disruptive measure of the twentieth century sever the task of thinking from romanticism take shape? Primarily in prohibiting the incarnation of the infinite in finite—its dialectic ended with the banishing from thought the category of infinite as a myth, belonging to the era when metaphysics shackled thinking, and in pursuit of all dominant doctrines of the subject took finitude as the defining and foundational category—to the detriment of the concrete existence of infinities and the capacity of directly engaging with them—instead replacing this dialectic with the dialectic of the one and the other.

This sequence allows us to understand the profound challenge of what Badiou has called the "passion of the real" in the twentieth century: How to orient ourselves towards something beyond the finite without giving in to the seduction of the sublime "beyond" of the romantics? This seems to be the locus of the dialectics of the one and the other: to find in the finite its point of irreducible otherness without "substantializing" it into a new sublime One. The subject is no longer thought, thus, as a mediation between the finite and the infinite, but rather as a procedure-oriented by the margins of alterity in a finite—thus unifiable—situation. The "real" becomes the name for this point of contact between finitude and otherness—a point that is isomorphic not with the infinite but with the inaccessible. However, the reduction of the infinite to the search for the immanent point of inaccessibility also re-establishes a new suture: it might be political, psychoanalytic, artistic, or even scientific, but at heart, it is an ethical suture—it becomes the underlying ethical concern of avoiding the closure of the Other back onto the One.

The twentieth century could only think of a determinate theory of the subject by excluding infinity. This orientation is what Badiou refuses to accept!

This study will attempt to reconstruct the journey through which infinity comes to reclaim a defining position in the thought of our times: a journey marked by the development of the philosophy of Alain Badiou. This philosophy, which is in close dialogue with politics, psychoanalysis, art, and mathematics, has constructed a potent formulation of how the thinking of infinity underlies not only the development of thought itself from Greeks up to the modern era but also, more importantly, the hope for the future of humanity, society, and even the destiny of this planet.

The focus of this study is the role of infinity, as an axiom of thought, on the thinking of being in general, on ontology as the science of being, and particularly on what it means to be a subject. That is the study of rigorously conceiving the solution to the following problems:

1. What does it mean that being is infinite?
2. What sort of thinking is adequate to this proposition?
3. What theory can recuperate the thinking of the infinite?
4. What sort of impact does this have on what we think and do today?

Concerning the questions above, the overall approach taken by this study is not to tackle these inquiries head-on using Badiou's mature work but to choose a particular itinerary within Badiou's vast philosophical oeuvre that started in the mid-1960s when he was among the editors of *Cahiers pour l'Analyse*, and build up to his magnum opus of 1988, *Being and Event*. In particular, we choose three works from three different eras of Badiou's intellectual development that best represent his commitments during those eras: "Mark and Lack" from his logicist and structuralist period, *Theory of the Subject* from his political and revolutionary period, and *Being and Event* from his mature period. In reading these three works, we can trace three significant concerns that run through, more or less, the entirety of Badiou's career so far: the questions of beginning, interiority, and novelty (and historicity concerning novelty). The real concerns to which these questions are addressed differed in each era, but the problems remained the same. What does it mean for a structure to have interiority? What does it mean that a structure has closure? Wherefrom is the being of structure endowed? How does novelty emerge in a structure? These questions motivated Badiou's rejoinder to Jacques-Alain Miller in 1967, which bears the title "Mark and Lack." This seminal work in mathematical and structuralist logic was Badiou's debut, at the age of thirty, in the larger scene of the heated intellectual debates in France at the time.

Moreover, roughly a decade since, and after the events of May 1968 in France, the questions turned to a different domain but remained the same. What do we mean by politics exists? When we speak of a political subject, from what position are we talking? What precisely changes when the revolution happens, and what happens after?

The importance of Badiou's thought regarding interiority, novelty, and beginning was his commitment to defining these terms through non-metaphysical, particularly non-essentialist, propositions. While he disagreed with structuralists, logicists, analytical philosophers, and Althusserians, he agreed with them on this point that the ensuing theory should not fall back into an anthropological or psychological theory of genesis. This led him to critique structuralism and logicism to seek a formulation that allowed for structured interiority, whose beginning could be thought, wherein novelty would not be grounded in any trans-structural providence. We attempt to show how this critique led him to ontology and mathematics and, finally, the formula that ontology = mathematics.

In pursuing this itinerary, two things should become clear. The first is that despite the way Badiou is usually read as being divided into episodes—before May 1968 and after May 1968 and before the ontological turn and after—the above thread of concerns runs through his entire philosophical development. We read these works as different chapters of the same inquiry. That, in itself, is a significant advance in Badiou's still young scholarship. In addition, we also encounter a central concept in his thought that, like an encompassing surface, is interwoven by the threads of his interests. These chapters, in some sense, tell the story of the clarifying, polishing, and deepening of this conceptual surface that we call "discipline." The concept of discipline for us acts as a marker to navigate through our itinerary—to track the development of a unique concept that at each stage becomes the central pivot for a confrontation: the debate in *Cahiers* with Miller regarding subject and structure, the debate in *Theory of the Subject* with Lacan (among others) regarding the subject and the real, and the debate in *Being and Event* with Heidegger (among others) regarding the subject and infinity. In each case, we have extracted the elements of this concept from the ways that Badiou conducts his responses to his opponents, in his unique and usual way of not rejecting them outright but showing that what they aspire to do is a particular case of a more general theory. In our view, this strategy—a critique by way of regionalizing a previous system of thought—seems to be a veritable new form of critique following the concept of discipline.

By constructing this concept, we then track Badiou's intellectual development and reconstruct this development as the unveiling of a consistent and crucial theoretical position that we believe has eventually enabled Badiou to complete his critique of structuralism, dialectic, and the modern version of idealism in the forms of phenomenology and linguistic philosophy, as well as developing new ontological commitments that answer in a robust and yet non-essentialist way the question of beginning, interiority, and novelty.

As the last chapter of this work shows, discipline is our way of coming to terms with Badiou's much-avowed Platonism. In other words, the position from which Badiou can critique both structuralism and dialectics, develop a new ontology, come up with the theory of event and procedures of fidelity, reconstruct the concept of the subject, and, finally, resurrect the much-abandoned category of truth, is a position whose central tenet resembles Plato's theory of methexis, participation. Discipline, as a reworking of Plato's theory, allows Badiou to strip away the reference to objects and builds a theory of participation whose sole touching point is the subject.

While discipline is a central concept in Badiou's thought from his very early days, its actual robustness is derived from another principle, first declared by another Greek philosopher, Parmenides: being and thinking are the same. Discipline is how the sameness of being and thinking is "carved out" and made thinkable by not making prior commitments to the One, the Substance, or a cosmic physical reality. The way that Badiou solves the identity of being and thinking is by declaring being as infinite and thinking as axiomatic. If being is infinite, the only mode of thinking adequate to the infinity of being is axiomatic. In that mode of thinking, thought can progress with all objective references stripped away. The concept of discipline is thus that of thinking which can live up to these two poles or conditions: infinite being and axiomatic form.

Therefore, the axiom of Infinity—the axiom that combines a declaration about being with a modality of thinking in one condensed form—is the central thought in Badiou's philosophy and, in our view, the marker of a new era in thought. This era will mark itself as a rupture with our era characterized by finitude as the founding principle of thought. Badiou's central insight was to discover in Cantor's axiom of Infinity the proper form of a declaration that has defied philosophy for millennia.

The axiom of Infinity is, of course, one of many axioms in the set theory. However, this study proposes that the axiomatic method's power is only genuinely tested by the axiom of Infinity axiom rather than other axioms of the set theory. In more general philosophical terms, this means that the infinite genuinely challenges the "linguistic turn" of philosophy and the reduction of mathematics to language. This is further reinforced by the fact that all the debates in the philosophy of mathematics concerning the validity of Cantor's work turn around the validity of "actual" infinities as a mathematical category: intuitionism, constructivism, Wittgensteinian pragmatism to be particular.

Finally, an essential facet of Badiou's philosophy in this new engagement with infinity is its instrumental subjective aspect. Thinking infinity is not a speculative thought but is the thought that organizes situations of being, configures the organization of human and nonhuman elements in a construct Badiou calls the subject, underwrites the fidelity of the subject to the construction of truths, and enriches its world with those truths. It draws new possibilities in light of the truths: to question, discover, pursue, and produce new things and worlds. To what a great living master of the East, Hossein Elahi Ghomshei, calls creating new realms of gold—whose standard of measure is eternal truths. Like what Ghomshei alludes to, truths are about bringing new realms to existence. In this sense, the thought of infinity is very much a category of the subject—its formation, matter, procedure, and, finally, what its being is consumed for, are all backed by the axiom of Infinity. In this sense, we say the subject is infinite—infinity is the backbone of the subject, what it is and does. The twentieth century has realized that the void is a condition of the subject. However, Badiou concludes that the void is necessary but still not a sufficient condition for the thinking of the subject: to protect the alterity of the pure multiple under the void is an ethics, but to advance toward its infinity is a discipline.

This study is divided into three parts, each reading a different work by Badiou and his interlocutors. Part I is the exploration of the subject and the structure. Chapter 1 is dedicated to the great debate in *Cahiers pour l'Analyse* between Miller and Badiou. Miller delivered a lecture titled "Suture" as an intervention during Lacan's *Seminar XI*, later published in *Cahiers*. In this lecture, Miller used Frege's logical construction of cardinal numbers as the entry point to a more general but repressed logic which he called the logic of signifier. The tenets of this logic, which he had laid out in another text called "Action of the Structure," also published in *Cahiers* sometime before, is that the stability of any structure requires a repressed element he called the subject. He used Frege's construction as evidence for this theory and concluded that the logic of the signifier is the logic of all logic. The debate proper began when Badiou, as we mentioned, wrote a rebuttal to Miller to defend the independence, or the interiority, of logic and, in particular mathematical logic, from epistemological interventions such as Miller's. However, the way Badiou develops this interiority, combined with the thought of production of novelty in logic through the concept of stratification, for the first time sowed the seeds of the thought of discipline, which, as we said, matured in Badiou's works and thoughts.

Part II of the book is titled "Subject and Real" and is dedicated to reading Badiou's *Theory of the Subject*. It picks up from what was developed previously: now we have the concept of discipline and the centrality of the domain of politics for Badiou, and the question is to show in what sense politics is a discipline (similar to the sense in which logic/science was shown to be a discipline). This new task is not as straightforward. For reasons described in this part, Badiou realized that to establish the place of politics as a discipline, he must first achieve a critique of structuralism, and to do this, he needed to "step out" of structuralism but not into metaphysics. He discovers that the dialectic provides such a foothold. He then turned to Hegel and the dialectic of "something and other" in the *Science of Logic*. Badiou was also aware that Lacan had made a similar critique of structuralism. However, the main issue with Hegel and Lacan is their idealist commitments, which prevented them from taking the dialectic and critique of structuralism far enough. Badiou's reading of these two great masters is an incredible display of reading philosophical texts with fidelity, intensity, and charity. He aims to tap into the riches of their thoughts and illustrate that they possess more robustness than they may have realized, possibly due to their prior commitments. The outcome of the interlocution with Hegel and Lacan in *Theory of the Subject* allows Badiou to carve a space of interiority and novelty for politics. Now we have two domains of human endeavor that are both disciplinary: science and politics. How can we account for their difference? How do we justify the being of multiple disciplines? What is the place from which this principle operates? These are among some of the questions with which we end part II. At the end of part II, we show that we need to step out of the historical sequence of the disciplines, that is, their dialecticism, to think of their beginning. This is the first time the question of beginning as a programmatic question shows itself. This is the question of where we can think of the being of disciplines.

Part III of the work focuses on the reading of *Being and Event* and the attempt to build a framework to answer the questions raised at the end of part II. In this part, we show Badiou's complete working through of the concept of discipline based on the axiom of Infinity. Badiou's question of the beginning of disciplines that we raised at the end of the previous part leads us to examine how the prior thinking of this question either sutures the question of beginning to the thought of the One or the thought of the ground, both of which caused the overdetermination of this question in the form of religious or transcendental metaphysics characterized by Hegel and Heidegger. We examine the conditions by which the notion of beginning can be de-sutured from these two terms and find a solution in Albert Lautman's philosophy of mathematics. Lautman was a French mathematical philosopher who the Nazis executed at age thirty-six. His writings, gathered posthumously, significantly influenced Badiou (as he admitted in *Being and Event*). We review some of his main ideas through various sections of *Mathematics, Ideas and the Physical Real*. His interpretation of axioms and axiomatic thinking is the base according to which we read Badiou's references. Also, Lautman's reconstruction of the axiom theory using Platonist operators of one and dyad remains a source of inspiration for us to understand the kinship of axiomatic thinking with the Platonist paradigm of participation. Using the axiomatic framework that allows thinking of the question of beginning without overdetermination, we return to two primary categories of *Being and Event*: the void and the event. We show different senses in which Badiou makes use of the category of the void, each of which watches over different parts of his overall framework: the void as the suture to being qua being, the void as the operator of the count, the void as a particular gap, and the void as a singular term in ontology. However, the essential sense of the void lies in the subtractive nature of Badiou's ontology.

In the discussion of events, we begin to see the programmatic value-add of the category of discipline. We show a rupture in the flow of *Being and Event*, which we can now make sense of by assuming two performative domains within the same work: ontology and discipline. While ontology looks after the categories of being whose governance is through the axioms of the set theory, discipline encompasses a theory for historical situations. This division makes sense of the rupture in the structure of *Being and Event*. It provides a theory to understand the rupture within the discipline of ontology as well: a rupture that is precisely around the axiom of Infinity and the "Continuum Hypothesis."

The last chapter of this part takes on the examination of the category of the subject. It discusses what could be thought of in this category in light of the idea of infinite, event, and generic. This chapter captures another encounter with Lacan in two episodes: according to what Badiou presents as the last meditation of *Being and Event*, and another critical text by Badiou, "Subject and Infinity," where Badiou goes over Lacan's formula of sexuation. In the first encounter, we highlight the difference between the two thinkers, which should make complete sense now. Instead of the Lacanian assertion of "it thinks," following Badiou, we posited the Platonist dictum of "being thinks"—but stated from within an ontology which has already turned the Lacanian thinking of the real into a conditioned region of being, absorbing its properties and effects. This decentering of the function of thinking watches over two crucial elements. The first is the placement of the category of truth. The decentering of thinking from the subject also decenters the category of truth outside of the subject to the being itself. It also extroverts the subject: the subject that finds its primary cause in an unspeakable being finds a secondary cause in the fidelity to truth and begins to compose consistency. In the second encounter, Badiou examines Lacan's account of the field of the infinite feminine jouissance and shows how Lacan, due to prior commitments again, mutilates the thought of infinity by converting it to the all familiar infinity as inaccessibility to conclude that feminine jouissance has the structure of fiction—it is infinite (as what the psychoanalytical theory needs to assert) only as fiction—so he can maintain the thought of the subject as finite. For Badiou, this result is not acceptable. The subject is of the order of infinite, if not in its cause, then in its consistency. The generic procedure that underwrites the possibility of the subject as the composer of consistency deploys the infinity of the situation of being in the capacity of its guarantor. In his usual genius, Lacan detected the need for infinity in the question of the differences between sexes. However, for reasons that this chapter presents, he refrains from committing to actual infinity, so he recoils to the realm of fiction. Two, the two in love, for Badiou is of the order of infinite. So, from one side, Badiou seems to strip away something from the Lacanian subject—the category of truth—but from the other side, this very fact stretches the subject to infinity—to infinity that is now causing the subject as the locus of operations to which Badiou refers as generic procedures. This decentering of the subject, the fact that it is now determined by something that is not inside it, changes the model of the subject from one that is orbiting around a center to a shooting star fixated on a destination in the void.

In the last chapter of this work, "Thinking Discipline," we offer a shortened summary of the main theses that this work has to offer. It discusses the consequences of the thought of discipline on general logic and, in particular, dialectic, Badiou's Platonism, and the new discourse of infinity and finite that, in our view, should gradually replace the dominant twentieth-century discourse of the one and the other.

This work attempts to build a retroactive construction of Badiou's intellectual endeavor charted within three major works spread over four decades. Accordingly, the claim of this book is that, although Badiou's thought can be divided into different periods within this spread, with distinct priorities in each, his overall thought has been defined and shaped according to the categories of infinity and subjectivity. Furthermore, we have identified three themes that have dominated his discourse during the development of Badiou's thought since the late 1960s: interiority, novelty, and beginning. These three themes each identify the different periods of Badiou's intellectual engagement. While all three themes are present in all periods, in each period, one of these themes was central to Badiou's involvement and acted as the portal through which his intellectual and theoretical intervention came forth. The first period, marked by his participation in *Cahiers*, is dominated by the question of interiority. The main concern in his debate with Miller is to show, contrary to Miller's logic of signifier, how the sciences can stay clear of ideology without needing external discourse. Sciences have the means to break away from ideologies built inside them, with no help from metaphysics and epistemology. In the second period and during the political sequence, the thought of novelty was the dominant thought for Badiou. The centrality of this thought guided Badiou's division of Hegel and Lacan to establish a logic for periodization and regionalization. This was when Badiou first committed himself to the interiority requirement of a discipline whose grammar he developed during the first period, which he carried over to the region of politics. In the third period, during his writing of *Being and Event*, the dominant thought for Badiou was the question of beginning—what is the proper beginning for thought that would fit those commitments (interiority and novelty) that he had developed in the previous two sequences? It is also crucial to point out that regardless of whether this is a question of interiority, novelty or beginning, Badiou's interventions into these questions surround the category of infinity. In all his interventions, the point of view that he supports is powered by an implicit or explicit commitment to some form of actual infinity—infinite stratification in the discussion of interiority, infinite as that which grounds consistency versus cause in the discussion of novelty, and axiomatic thinking as the only form of thinking adequate to the infinity of being, without the One and ground in the discussion of beginning. In addition, in all three themes, this surge of infinity into the theory sought and supported by Badiou is always taking place around the category of the subject: Badiou's rejoinder to the logic of signifier and the subject as the marker of the void in Miller's presentation, the proposition of subjective process as powered by the consistency of the real, and the axiomatic intervention of a subjective figure in the pursuit of truth, are all subjective forms around which the eruptions of the thought of the actual infinities blaze through. This clearly shows that the thinking of infinity matters the most for Badiou when subjective formation, intervention, and process are at stake. Therefore, this study aims to show how the thought of infinity determines the figure of the subject whose being is consumed by a truth.

Further to the interplay and compossibility of the themes of interiority, novelty, and beginning during each period of this development, this study proposes discipline as the logic of these themes in general, such that the being of truth can be instantiated within the region that is carved by the logic of discipline in support of the figure of the subject. Discipline manifests how a region of being (infinitely) could be thought (axiomatically) such that the properties of interiority, novelty, and beginning are maintained. From Badiou's works, it is abundantly clear that such regions are conditions of truth: politics, art, science, and love. From our standpoint, these conditions are circumscribed by the logic of discipline: they are disciplinary regions or simply disciplines. Following Badiou's development, we've only shown how science and politics possess these disciplinary properties. However, we believe that a similar endeavor will be able to demonstrate that art and love are also regions of being marked by this logic, that they indeed are disciplines too.

THE RELATION OF THIS STUDY TO THE EXISTING SECONDARY LITERATURE

What makes the current work distinct is that we elaborate the trajectory of Badiou's philosophy by following a leading thread that has been mostly latent not only in the secondary literature about Badiou but, as this work shows, even in Badiou's primary texts. We can summarize this leading thread as follows: the dominance of axiomatic thought and the category of mathematical infinity.

It is in light of this leading thread that we demonstrate two other major propositions about Badiou:

1. The coherence of his intellectual development from the early interventions to the publication of *Being and Event*, in each episode of which he works out the new categories of interiority, novelty, and beginning.
2. The formation of a theory we dubbed discipline that makes Badiou the next great dialectician after Hegel and Lacan.

The distinctiveness of this work is thus not so much its primary engagement with Badiou's political, philosophical, or even mathematical commitments but to the commitment of what he later himself called "the principle of 'maximality,'"[1] which roughly states: whatever is not forbidden is permitted. It is a choice of thought to maximize what it can think of being. We should be hopeful that thinking, which is adequate to the task of discovering all the useful properties of numbers, is already endowed with this capability if it functions in the axiomatic register. It is not by providence that we are able to think maximally of being; it is by the method. The maximality principle is the direct outcome of axiomatic thought, particularly the axiom of Infinity.

This is what makes this work different from the current secondary literature on Badiou. We elaborate on the method of Badiou's thinking and try to show that philosophical categories are the consequences of this method.

As a brief survey, we can categorize secondary literature on Badiou into the following three categories:

1. The early reaction among the French intellectuals, most of whom were Badiou's friends and colleagues.
2. Early reaction in the English-speaking world to Badiou's growing corpus that was still not fully translated into English.
3. The later reaction to his work, which focuses on various philosophical commitments Badiou has pronounced.

In the first category, we see the reception of his thought among other French philosophers of his time, some of his close friends or colleagues who spent time with him at *Ecole Normale Supérieure*. Much of the articles and criticism published in France by well-established philosophers primarily mark the points of distinction between their projects and Badiou's system, rarely engaging with the internal logic of his thinking. For example, we can mention Lyotard's review of *Being and Event*,[2] which accuses Badiou of a Schmitt-like theory of sovereignty based on the decisionism inherent in the declaration of an event. Also, the famous section in *What Is Philosophy?*[3] by Deleuze and Guattari briefly discusses key concepts of *Being and Event*.

They try to show that Badiou is a victim of the confusion between concept and function, which, in the end, results in him conflating intensive and extensive dimensions of multiplicities. Also, Daniel Bensäid's article criticizes Badiou's return to philosophy, which is unable to differentiate Marx as a scientist from Marx as a philosopher.[4] Among these encounters with Badiou, a short text by Jean-Toussaint Desanti stands out.[5] For us, this work stands out because of Desanti's attention to Badiou's "method." In relation to Badiou's choice of naming the chapters of his seminal work "Meditations," Desanti makes the following remark:

> Three closely related requirements can be discerned in this word: the fidelity to an initial decision (the choice in favour of the intrinsic); the deployment of a movement of thought orienting itself within its own discourse; and the "exhibition" of the conceptual connections which render that discourse consistent. That these requirements are closely related means, at the very least, that the initial decision will always be taken up again, reproduced, its validity ceaselessly renewed in the coordinated unfolding of the discourse itself.[6]

This is very close to what we attempt to establish in this work. The difference is that while Desanti makes that judgment about the meditations, we see this movement of thought, based on an initial decision and the fidelity to it, throughout Badiou's entire career from the late 1960s to the publication of *Being and Event.*

But, more importantly, Desanti's short essay directly influenced the author's later research. After reconstructing Badiou's main meta-ontological theses, Desanti demonstrates that the category theory meets the exact requirements but makes room for another approach to multiplicities which is relational and extrinsic. In a way, the whole project of *Logics of Worlds* responds to this intervention by Desanti.

A second group of commentators, still in France, are the students and disciples of Badiou, such as Mehdi Kacem and Quentin Meillassoux. Kacem, whose initial books were inspired by his reading of *Being and Event* in 2000,[7] attempted "unprecedented synthesis between my mathematized ontology and an original theory of affect and jouissance, which he extracts and reorganizes on the basis of Deleuze and Lacan."[8] However, as intense as his proximity to Badiou's project was, his later publications marked his later distancing,[9] and with the breakout of the pandemic, this distancing became even more severe.[10] The other famous disciple of Badiou is the philosopher Quentin Meillassoux.[11] It is a partnership preserved until today owing to their shared conviction of the need to go beyond finitude, their common praise for mathematics, and the possibility of a rational ontology. From his first engagement with Badiou's philosophy, Meillassoux demarcated his points of disagreement with Badiou, for instance, his articles on the relationship between the event and history[12] or Badiou's theory of appearance.[13]

In the second category, we have the secondary literature that effectively introduced Badiou to the English-speaking world even before any of his major works were translated into English. In fact, Badiou's reception outside of France is due almost exclusively to the intervention of Slavoj Žižek, who dedicated a long article to him in *Atlantic Quarterly*[14] and almost a third of *The Ticklish Subject.*[15] Due to this early influence of Žižek, an essential part of this first wave of international interpretations came through psychoanalysis and discussion of the concept of a political subject in Badiou and Lacan. In this context, commentators such as Lorenzo Chiesa,[16] Sam Gillespie,[17] Adrian Johnston,[18] Jodi Dean,[19] Frank Ruda,[20] and Geoff Pfeiffer[21] were generally interested in the points of contact and divergence between Badiou, Lacan, Hegel, Žižek, and Althusser. Among these, Gillespie's and Ruda's work is worth mentioning as they picked up two themes that are close to our inquiry. By focusing on the proclamation of mathematics is ontology, Gillespie came close to unveiling the essence of this formula, which contains the following two assertions:

1. Being is not one; being is beings, and beings are multiplicities.
2. The science of beings is axiomatic.

As we have mentioned earlier in this introduction, this is what we have named the category of beginning under Badiou's theory of the discipline, and we have dedicated part III of this work to analyze it. Ruda's work has also been a source of inspiration for our work. Ruda's attention to an early text of Badiou that unveils the two dialectical matrixes in the *Science of Logic* is also important for us in constructing Badiou's transition from logic (science) to politics as two examples of disciplines that our work constructs. But what makes this work distinct from interesting and important remarks in Gillespie's and Ruda's is that we argue that these two themes are part of the same methodological commitment within a larger theory: the theory of the discipline.

A significant and early contribution to Badiou's young English canon was Peter Hallward's book in 2003,[22] which was published before the English translation of *Being and Event*. Hallward's work was the first attempt to systematically absorb and examine Badiou's work. It remained, for decades, the only attempt to connect the main theses of *Being and Event* with the rest of the Badiouian bibliography, devoted to the different political, amoral, artistic, and scientific conditions of his philosophy. Another important contribution made by Hallward was the edition of a volume we quoted from before, *Think Again*. This volume, as discussed above, contains crucial and somewhat critical essays on Badiou's "new" philosophical program after the publication of *Being and Event*, especially in France.

The third group of contributions belongs to the secondary literature after the publication of *Being and Event* and *Theory of the Subject* in English. The publication of *Theory of the Subject* was very influential, providing a view into Badiou's political years. Two important works that ensued from this soon after are the works of Bruno Bosteels and Christopher Norris. In his *Badiou and Politics*,[23] Bruno Bosteels was interested in demonstrating the continuity between Badiou's "youthful" work, explicitly Marxist, and his mature philosophical work. Christopher Norris also published a book[24] commenting chapter by chapter on *Being and the Event*. Despite the significant effort, he still does not describe the mathematical constructions of the book's second half, like most commentators.

During this period, a distinct group of commentators was formed in Australia. Justin Clemens,[25] A.J. Bartlett,[26] and Oliver Feltham[27] (the latter, translator of *Being and Event* into English), along with many others, such as Hallward and Bosteels, produced exegetical systematization and reconstruction of Badiou's philosophical concepts. An important distinguishing feature of these commentators is that they worked in groups for a considerable time, sharing the same university environment and organizing reading groups of Badiou's works and mathematics study groups. Among many contributions that this group of authors has produced, Bartlett's work is of a distinct status. Bartlett uses Badiou's key notions such as truth, event, and subject to retroactively ready and reconstructs the Platonic corpus. In light of what Bartlett argues that the Socratic "not-knowing" is a state in which a militant of truth always dwells. The state of "ignorance" for Socrates is the denial of the Sophist's claim to knowledge. Bartlett's work is the first single work that tries to "apply" Badiou's categories in a discipline not directly addressed by Badiou.

Another group of commentators that have also built working ties around Badiou's work is the one that, for a while, was organized around the Jan Van Eyck Academy in the Netherlands and the philosophy department at Middlesex University. Unlike the commentators linked to psychoanalysis and the Hegelian heritage, thinkers such as Nina Power,[28] Lucca Frazer,[29] Ray Brassier,[30] and Alberto Toscano[31] found in Badiou a way of connecting Marxist politics to a doctrine of the inhuman, recovering important aspects of Deleuzian philosophy in a context more suited to scientific treatment. However, at the same time that it is possible to unify them through the projects they shared, the differences between these authors—who soon established themselves as original thinkers with no direct relation to Badiou—are striking: Nina Power's Feuerbachian feminism, Frazer's recovery of Sartrean themes through the theory of computation, Brassier's combinations between Sellarsian philosophy and Marxism, or Toscano's political philosophy of infrastructures share almost nothing, except, perhaps, the return to a rationalism of the potency of the inhuman.

In a similar key, we could group Deleuzian reception of Badiou with authors such as Clayton Crockett,[32] Roland Faber,[33] Jon Roffe,[34] and, in a different way, Paul Livingston.[35] The first three, resuming the tendency of the first Badiou's reception, are concerned, above all, with demonstrating that the main Badiouian theses are already contained in Deleuze's work. On the other hand, the latter proposes an innovative reading of different philosophical currents in the twentieth century—represented especially by Deleuze and Badiou—and suggests that both positions are consistent but depart from different interpretations of the effects of the theorems of Gödel for the philosophy of language.

One last important and notable work is Burhanunddin Baki's book.[36] Baki's book is the only one that directly involves Badiou's mathematical ontology and the mathematical framework mobilized in *Being and Event*. Another important and notable author that has recognized the importance of advanced mathematics in Badiou's meditations is Fernando Zalema.[37] Although his work is not related directly to Badiou, his recognition of Badiou and also his edition of the only English language book on Lautman[38] have been important resources to our work.

The point of this quick survey is to show that, since Hallward's book in 2003, no study on Badiou has taken place that seeks to examine in detail the order of reason internal to Badiou's thought. As rich and diverse as it is, the current secondary literature on Badiou treats his thought in fragments: dwelling on Badiou's relation to Lacan or Deleuze, debating his alleged political or art theory, and so on, and even fewer engage directly with his mathematical ontology. As we have said, the current work is not only to show the integrity of Badiou's mature thought (as of the first volume of *Being and Event*) but also to show that this thought had been in slow and painstaking progress since the late 1960s. Consequently, we aim to show the philosophical system in Badiou follows an inner logic, the theory of discipline, which in turn has a correlation to a certain set of commitments that Badiou kept since the beginning of his career, which we have phrased as the "maximality principle."

As the result, this work has rarely been able to benefit from or engage with the secondary literature. With few exceptions (Jacques-Alain Miller, Albert Lautman, Sylvan Lazarus, Slavoj Žižek, and a few others), this study almost exclusively engages with primary texts of Badiou and his main interlocutors (primarily Descartes, Spinoza, Hegel, Heidegger, Lacan, and Althusser). Deleuze is a notable absence from this list, and this was not by design but a consequence of what has led our trajectory. Badiou's primary texts we used for our aim did not engage with Deleuze. That being said, there could be many points of contact between the theory of discipline and Deleuze that can explore and enrich both philosophies. This, however, has to remain a separate and hopefully consequent effort.

NOTES

1. Alain Badiou, *The Immanence of Truths, Being and Event III*, 39–40. London: Bloomsbury Academic 2018.
2. Jean-François Lyotard, "Alain Badiou" in *Cahiers du Collège International de Philosophie n.*8, 1989, 211–225; also see Badiou, A., *Infinite Thought, Truth and the Return to Philosophy*, 129-130. London: Continuum 2004.
3. Gilles Deleuze, *What Is Philosophy?*, with Felix Guattari, 150–153. New York: Columbia University Press 1994.
4. Daniel Bensaid, "Alain Badiou and the Miracle of the Event" in *Think Again, Alain Badiou and the Future of Philosophy*, 94–105. London, New York: Continuum 2004.
5. Jean-Toussaint Desanti, "Some Remarks on Intrinsic Ontology of Alain Badiou" in *Think Again, Alain Badiou and the Future of Philosophy*, Continuum, London, 2004, 59–66.
6. Desanti, "Some Remarks on Intrinsic Ontology of Alain Badiou," 61–62.
7. Mehdi Belhaj Kacem *Événement et répétition*, Auch: Tristram, 2004. Also Belhaj Kacem, M., *L'Affect*, Auch: Tristram, 2004.
8. Alain Badiou, *Logics of Worlds*, 524. London, New York: Continuum 2009.
9. Mehdi Belhaj Kacem, *Transgression and the Inexistent, A Philosophical Vocabulary*, London, New York: Bloomsbury Academic 2014.
10. See Belhaj Kacem's interview with Daniel Tutt here: https://danieltutt.com/2021/12/13/the-concept-is-a-weapon-interview-with-french-philosopher-mehdi-belhaj-kacem%EF%BF%BC/.
11. Quentin Meillassoux, *After Finitude, An Essay on the Necessity of Contingency*, London, New York: Continuum 2008.
12. Charles Ramond, *Alain Badiou: Penser le multiple*, L'Hartman, 2002.
13. Fabian Tarby & Isabel Vodoz (eds.), *Autour de Badiou*, Germina, 2011.
14. Slavoj Žižek, "The case of Alain Badiou" in *South Atlantic Quarterly*, 1998—available here: https://www.lacan.com/zizekbadiou.htm.
15. Slavoj Žižek, *The Ticklish Subject, The Absent Centre of Political Ontology*, London, New York: Verso 2000.
16. For example, "The Body of Structural Dialectic: Badiou, Lacan, and the 'Human Animal'" (2014)—available here: https://badioustudies.files.wordpress.com/2016/11/1-lorenzo-chiesa-the-body-of-structural-dialectic-badiou-lacan-and-the-e28098human-animal_-pp-1-23.pdf.
17. Sam Gillespie, *The Mathematics of Novelty*, Melbourne: re-press 2008.
18. Adrian Johnston, *Badiou, Žižek and Political Transformations, The Cadence of Change*, Evanston: Northwestern University Press 2009.
19. Jodi Dean, *The Communist Horizon*, London, New York: Verso 2012.

20. Frank Ruda, *For Badiou, Idealism without Idealism*, Evanston: Northwestern University Press 2015.

21. Geoff Pfeiffer, *The New Materialism: Althusser, Badiou, Žižek*, Oxfordshire: Routledge 2015.

22. Peter Hallward, *Badiou, a Subject to Truth*, Minneapolis: the University of Minnesota Press 2003.

23. Bruno Bosteels, *Badiou and Poloitics*, Durham&London: Duke University Press 2011.

24. Christopher Norris, *Being and Event: A Reader's Guide*, London, New York: Continuum 2009.

25. Justin Clemens & A.J. Bartlett (eds.), *Badiou: Key Concepts*, Oxfordshire: Routledge 2014.

26. A.J. Bartlett, *Badiou and Plato, an Education*, Edinburgh: Edinburgh University Press 2015.

27. Oliver Feltham, *Alain Badiou: Live Theory*, London, New York: Continuum 2008.

28. Nina Power & Alberto Toscano (eds.), *On Beckett*, Manchester: Clinamen Press 2004.

29. The translator of *The Concept of Model*, Melbourne: re-press 2007.

30. Ray Brassier (ed.), *Badiou: Theoretical Writings*, London, New York: Continuum 2004.

31. The translator of *Logics of Worlds*.

32. Clayton Crockett, *Deleuze Beyond Badiou*, New York: Columbia Press 2013.

33. Rolad Faber & Henry Krips & Daniel Pettus (eds.), *Event and Decision: Ontology and Politics in Badiou, Deleuze and Whitehead*, Cambridge: Cambridge University Press 2010.

34. Jon Roffe, *Badiou's Deleuze*, Oxfordshire: Routledge 2014.

35. Paul M. Livingston, *The Politics of Logic, Badiou, Wittgenstein, and the Consequences of Formalism*, Oxfordshire: Routledge 2014.

36. Burhanunddin Baki, *Badiou's Being and Event and the Mathematics of Set Theory*, London, New York: Bloomsbury Academic 2015.

37. Fernando Zalamea, *Synthetic Philosophy of Contemporary Mathematics*, New York: Sequence Press 2009.

38. Albert Lautman, *Mathematics, Ideas, and the Physical Real*, London, New York: Continuum 2011.

Part I

SUBJECT AND STRUCTURE

INTRODUCTION

One way to triangulate the central thesis of the current work is by referring to the relation of the subject to structure and the different ways in which this relation can be thought of. The compossibility of the subject and structure has a strong philosophical pedigree, but it found a particular interest in the French philosophical milieu of the 1960s, especially ones gathered around Lacan and Althusser. Badiou was one of those young philosophers and participated in the new ways in which this relationship was being discussed and challenged.

The claim of this work is that Badiou has hardly ever left the general arena of this debate whose boundaries are drawn by the concepts of subject, structure and science—the latter being the element that sutures the two together, through its rigor and the activity of scientists: only fidelity to the rigor of scientific formalism can unleash the unpredictable results of the sciences, and such a practice is thoroughly axiomatic, in contrast to its commonly perceived empirical nature. The terms of this debate permeate Badiou's entire intellectual career, which is marked by an oscillation between politics and mathematics and one that finds its dialectical unity in the axiom of Infinity.

This relationship therefore goes to the heart of our proposal: it is only through the axiom that actual infinities exist that subject and structure can be thought as compossible, through an immanent causality of truths that the subject enables. But this relationship had equal importance among Marxist theorists, for whom structure was the primary factor, and for psychoanalysts whose practice puts forward the figure of the subject as the primary concern.

This era of French philosophy, whose heated debates are the focus of the present chapter, took place on the heels of the demise of the existentialist capture of this relationship. For Sartre, it was the praxis and the subject that defined the structured ensembles, and for reasons that are beyond the scope of this work, Sartre's formulation of this relationship had fallen short of attracting the young philosophers of the 1960s. Fortunately, an alternate proposal became available from an unlikely source: psychoanalysis. Jacques Lacan, a towering intellectual source with a commitment to scientific formalism as well as a great deal of philosophical influence, opened up new, exciting ways that this relationship could be thought of and rehabilitated from the rubbles of existential doctrines.

As Badiou has remarked in a number of places, the works of Jacques-Alain Miller, who was the foremost student of Lacan as well as the first to programmatically recapture the essence of Lacan's thought with regard to the subject, science and structure, play a key role in this debate. Badiou, in fact, develops his own distinct voice in *Cahiers pour l'Analyse* (*Cahiers*) largely in relation to these works. So, while Lacan's thought as the great master loom behind these two and a generation of young philosophers in the 1960s, it is Miller's particular recapture that sparked Badiou's attention to this project, as Miller was able to uniquely weave together the strands of thought scattered throughout Lacan's oeuvre into a philosophical program that would otherwise be difficult for anyone to recognize.

But the most important influence is that, in fact, the terms through which Miller put this program together have had a lasting impact on Badiou, that as we said, has remained with Badiou until the present time—and perhaps the reason why Badiou has remarked on Miller and *Cahiers* as often as he does in his works and interviews.

In part I, we are reconstructing these terms from their historical source, so we can view the major aporia they opened in the thought of Alain Badiou, an aporia whose final resolution in our time heralds a major shift in thought. For our purpose, we will closely read two major works in *Cahiers*, "Suture" by Miller and "Mark and Lack" by Badiou.

Chapter 1

The Debate in *Cahiers pour l'Analyse*

SUTURE: THE SUBJECT AND NUMBER

Jacques-Alain Miller delivered the lecture named "Suture" as an intervention at Lacan's seminar Critical Problems for Psychoanalysis on February 24,1965, a few days after his twenty-first birthday.[1]

"Suture" is a word picked from ordinary language and is used in reference to a very specific field in Lacanian psychoanalysis—the logic of the signifier. Here is how Miller defines the meaning of this word:

> Suture names the relation of the subject to the chain of its discourse; we shall see that it figures there as the element which is lacking, in the form of a stand-in [tenant-lieu]. For while there lacking it is not purely and simply absent. Suture, by extension—the general relation of lack to the structure—of which it is an element, inasmuch as it implies the position of a taking-the-place-of [tenant-lieu].[2]

Central to the efforts of Miller was Gottlob Frege's conception of natural numbers. Frege believed that numbers are logical constructs generated by pure thought: what has been referred to as Frege's logicism, which is to build numbers as well as the rest of arithmetic on a logical ground that leaves nothing to intuition or any preconceived notion of number or any predisposition to perceptive laws of arithmetic. The construction of number, in contrast to the intuitive conception, in Frege's system, is a purely logical operation—as Badiou remarks, in Frege's system, the number is neither empirical nor transcendental nor, in other words, the number is an epistemological trait with no ontology.[3] For Badiou, the key text for Frege's conception of numbers is *The Foundations of Arithmetic*.[4] The fundamental argument, extremely dense, occupies §55 to §86 (from pages 67 to 99 in the cited edition).

Frege: Cardinal Number and the Number Sequence

Frege's ideas about numbers start with the recognition that cardinal numbers, numbers that are used to answer the question: "how many are there," bear a striking kinship to natural numbers. Frege's insights concerning this relationship start with the observation that a statement of cardinal number (e.g., "There are eight planets") is an assertion about a concept. To explain this, Frege notes that one and the same external phenomenon could be counted in different ways; for example, a certain external phenomenon, say a large modern company, could be counted as one business entity, 3 divisions, 19 departments, 120 organization units, or 50,000 employees. Each way of counting the external phenomenon corresponds to the manner of its conception. This leads Frege to the assertion that when we talk about "how many," and in response, we provide a cardinal number, we are, in fact, referring to the size of the set corresponding to all the objects in the world that fall under a given concept. Frege called this set the extension of a given concept: the extension of a given concept C is the set of all objects to which the attribution of the concept is a true statement, and cardinality N is the size of this set. If S is the extension of C, then cardinality N, denoted by $\#S$, is a number corresponding to the size of S.

In a more formal language, let C be the function from the domain W of all objects in the world to the range B (truth values) comprising two values, T and V (true and false).

$$C\,(W \rightarrow B)(x, y){:}\ x \in \mathrm{W},\ y \in B$$

We define S (the extension of C) as a subset of C:

$$S \subset C\,(W \rightarrow B)\,(x, y){:}\ x \in W,\ y = T$$

We define N as the cardinality (denoted by #) of the function S:

$$N = \#S$$

It can be noted that for every function C_i there is a corresponding S_i and N_i.

Therefore, the question "How many are there?" can be formulated as the question "How many truth cases of a concept C are there?" In Frege's view, the statements of number, which answer such questions (e.g., "There are x number of truth-cases for the concept C"), tell us something about the concept involved. For example, the statement, "There are eight planets in the solar system" tells us that the ordinary concept "planet in the solar system" falls under the second-level numerical concept being exemplified by eight objects.

The next step in Frege's formalization of cardinal numbers is to introduce the concept of equinumerosity. First, assume that N. gives the numerosity of concept C Then the equinumerosity of any given concept with another concept, say C_i and $C_{j,}$ with corresponding numerosities of N_i and N_j, is ascertained by showing $N_i = N_j$. Accordingly, the concept of equinumerosity between C_i and C_j is defined as such: C_i and C_j are equinumerous if there is a relation *R* such that:

1. Every object in S_i is *R*-related to a unique object in S_j.
2. Every object in S_j is such that there is a unique object in S_i which is *R*-related to it.

In other words, C_i and C_j are equinumerous if and only if there is a relation that establishes a bijective (one-to-one and onto) correspondence between S_i and S_j. If we let $C_i \approx C_j$ stand for equinumerosity, then the definition of this notion can be rendered formally as follows:

$$C_i \approx C_j = \exists R[\forall x(x \in S_i \rightarrow \exists! y(y \in S_j \,\&\, Rxy)) \,\&\, \forall x(x \in S_j \rightarrow \exists! y(y \in S_i \,\&\, Ryx))]$$

Using the above, the following facts can be shown about equinumerosity:

1. $\forall x(x \in S_i \leftrightarrow x \in S_j) \rightarrow C_i \approx C_j$
2. $C_i \approx C_i$
3. $C_i \approx C_j \rightarrow C_j \approx C_i$
4. $C_i \approx C_j \,\&\, C_j \approx C_k \rightarrow C_i \approx C_k$

Based on this, Frege defined the cardinality *N* of *C* in terms of the principle now known as Hume's Principle:[5]

Hume's Principle: The cardinality of C_i denoted by N_i is identical to the cardinality of C_j denoted by N_j if and only if C_i and C_j are equinumerous.

$$N_i = N_j \leftrightarrow C_i \approx C_j$$

Using Hume's Principle, Frege defines a cardinal number in the following way:[6]

Definition of cardinal number: *x* is a cardinal number if and only if $\exists C$: $x = N$

The definition of cardinal numbers proposed by Frege is a purely logical definition with no appeal to the intuitive concept of numbers. For any cardinal number, there exists at least one concept whose cardinality can be expressed as (or is equal to) that cardinal number.

The above logical definition of cardinal numbers acts as a stepping-stone toward the understanding and generating of natural numbers. The first insight into this relationship is the recognition of the following list (list 1):

$C_0 = [\lambda x\ x \neq x]$

$C_1 = [\lambda x\ x = N_0]$

$C_2 = [\lambda x\ x = N_0 \vee x = N_1]$

$C_3 = [\lambda x\ x = N_0 \vee x = N_1 \vee x = N_2]$

etc.

The first concept in the list, C_0, is the concept of all objects that are not identical to themselves, which, according to Leibniz's identity principle, has an empty extension; that is, there is no object that sustains a truth case for this concept. N_0 is the cardinality of the extension of this concept, and we name 0 the cardinal number equal to N_0, which is the cardinality of an empty extension.

The second concept in the list, C_1, is the concept of N_0, the cardinality of an empty extension. It can be observed intuitively that the cardinality of this concept is 1; there is only one N_0.[7] We call this cardinality N_1, and we name 1 the cardinal number equal to N_1.

The third concept in the list, C_2, is the concept of all those objects that are either equal to 0 or equal to 1 (or N_0 and N_1 respectively). It is obvious that the cardinality of this concept is 2. We call this cardinality N_2, and we name 2 the cardinal number equal to N_2. It is evident that the rest flows in a similar fashion:

$C_i = [\lambda x\ x = N_0 \vee x = N_1 \vee \ldots \vee x = N_{i-1}]$ and $N_i = \#C_i$

Why should this construction be important to Frege? Because it resembles the sequence of natural numbers starting from 0 with the striking property that for each concept C_i only the numbers corresponding to the concepts in the sequence preceding C_i fall under C_i. The definition suggests that all the natural numbers could be primarily construed on the basis of cardinal numbers in a sequence that maintains precedence and antecedence. This furthermore supports Frege's conviction that natural numbers are also logical constructs with no basis in the intuitive or realist conception of numbers. In addition, Frege was able to use this definition of natural numbers to provide a logical analysis of mathematical induction and prove that mathematical induction can be used validly to demonstrate the properties of natural numbers, an extremely important result for making good on his logicist ambitions. Frege could then use mathematical induction to prove some of the basic laws of natural numbers.

But Frege ultimately realized that, although we may identify this sequence of cardinal numbers with natural numbers, such a sequence is simply a list: it does not constitute a definition of all natural numbers, but only the numbers defined in the sequence. Such a concept was required since Frege, unlike Dedekind and Peano, wanted to derive the axioms of number theory from Hume's Principle and the definition of cardinal numbers. Again, this is important for Frege to do so since he believes the basis for the axioms is an intuitive conception of numbers, which he has set out to reprimand.

Frege proceeded to show, through a theorem later on known as Frege's Theorem, that the five Dedekind/Peano axioms for number theory can be derived from Hume's Principle in second-order logic.

Miller: The Logic of Signifier

Miller articulates his objective as follows:

> What I am aiming to restore, piecing together indications dispersed through the work of Jacques Lacan, is to be designated the logic of the signifier—it is a general logic in that its functioning is formal in relation to all fields of knowledge including that of psychoanalysis which, in acquiring a specificity there, it governs; it is a minimal logic in that within it are given those pieces only which are necessary to assure it a progression reduced to a linear movement, uniformally generated at each point of its necessary sequence. That this logic should be called the logic of the signifier avoids the partiality of the conception which would limit its validity to the field in which it was first produced as a category; to correct its linguistic declension is to prepare the way for its importation into other discourses, an importation which we will not fail to carry out once we have grasped its essentials here.[8]

In other words, Miller attempts to show that the general field of logic, used by Frege in order to generate the concept of cardinal number and resituate that of natural numbers, is itself rooted in a more "primordial" logic, the logic of signifier.

Miller describes his procedure as "archeological" through which he "excavates" the logic of signifier from under the visible and familiar landscape of general logic. The relationship between these two logics, however, is not a relationship that has been forgotten—the excavation is not digging up a lost origin—as Derrida has done with the origin of geometry—but a relation of repression and misrecognition. That is, the logic of the signifier is repressed while, and precisely because, it lays the ground for general logic.

Miller stages this excavation of the logic of signifier by honing into Frege's definition and generation of natural numbers. Miller wants to show Frege's steps involve the above-cited misrecognition of the function of the subject whereby Frege silently and unconsciously grounds the logical generation of numbers on a theory of the subject—something Frege in fact, intended to avoid, as for him, any "psychological" account of number is too empirical and thereby too contingent. While Miller agrees with Frege that an empiricist and psychological account should be excluded, he thinks that the subject as found in the logic of signifier is different.

The first instance of the repression is when Frege transforms "things" into "objects." As previously mentioned, Frege speaks of objects as instances of a concept. The underlying assumption that Miller elucidates is the supposition that an object, which for Frege is a countable unit—since in order for an object to fall under a concept, it has to be considered a unit—stands for a thing. For a thing to be an object, that is, a countable unit, one must already assume that a thing is an identity, and cannot be more than one. So, before we can advance to the operation of the count, we must commit to the identity principle.

This means that each operation of subsumption (of objects under concepts) makes use of another concept: the concept of "identical to itself." It is important to note that Frege openly commits to this concept. Hence, he should not have an issue with admitting that identity underlies the operation of subsumption. But therein lies the problem.

Frege's evocation of the concept "not identical to itself," defined formerly as $C_0 = [\lambda x\ x \neq x]$, yields a contradiction: the subsumption operation assumes $\neg C_0$ in order to show that the extension of C_0 is empty. If every subsumption leverages the assumption that each thing is an object, that it is identical to itself, then how may one use such an operation to count each object that is not identical to itself? When Frege posits the concept C_0, he at the same time disrupts the operation of subsumption, which will evidently make the construction of any extension for C_0 an impossibility.

As already mentioned, 0 is the cardinal number Frege assigned to the extension of C_0. It is supposed to stand for the size of an empty set. In Frege's system, the (logical) derivation of 0 is the first step in generating the other natural numbers.

Assuming C_0, however, will simultaneously make the construction of the C_0 extension an impossibility. It is obviously clear that the assertion "C_0 extension is impossible" is totally different from the assertion "it is an empty set."

Unlike the dialectic of evocation and revocation employed by Miller to categorize this antinomy as a misrecognition, the foregoing remarks seem to justify that this antinomy is, in fact, a paradox that threatens the very foundation of the orientation according to which concepts are given logical priority over collections of objects. It may be possible to show that this antinomy has a similar status to that of Russell's paradox, for which the axiom of Separation in ZFC is specifically devised. This axiom does the reverse: it assumes that collections (or sets) have logical and ontological priority over concepts. According to this axiom, any set can only be constructed using the elements that are contained in other sets. In this case, the operation of subsumption must therefore pick members from other sets—sets which may have no formulaic conception, yet have cardinalities. The axiom establishes that the concept of cardinality is unrelated and is logically prior to the Fregean concept, thereby grounding the claim that numbers are not logical artifacts.

By adopting the principle of identity as the gold standard for truth, Frege has immediately concluded that the extension for C_0 is empty. In this case, the emptiness of S_0 is not attained through the operation of subsumption but is a direct result of the identity principle. If that were to be true, then the logical orientation is guilty of having a double standard, utilizing different ways to arrive at conceptual extensions—one for C_0 and one for all other concepts. Miller evidently subscribes to such an interpretation. This is evident in the following passage:

> In effect, let there be the concept "not identical with itself." This concept, by virtue of being a concept, has an extension, subsumes an object. Which object? None. Since truth is, no object falls into the place of the subsumed of this concept, and the number which qualifies its extension is zero.[9]

The logic of signifier in Lacan revolves around two key concepts, which not only speak to the position and formation of the unconscious but also speak, in a broader sense, to the conditions for any system of exchange (e.g., language, the markets). Lacan names these two concepts alienation and separation—both of which leverage a third concept, that of the subject. Alienation is the process through which some entity from outside (from the register of real) comes to occupy a position within the system, during which it loses something of itself. This loss is due that the entity must pay in order to be "part of the system," but it also continues to function as a mark that the entity carries with it: even if the entity is now counted as a member of the system and subjected to its structure, the loss will forever "determine" it. It is important to emphasize that it is not what the subjected entity retains of itself that officiates its inscription within the system, but the fact that it loses part of itself. In other words, the subjective position is not determined by what the thing becomes by being subjected to the system but by what was lost in that subjection.

The logic of the signifier is what thematizes the logic of this process—it is the thematization of the loss and the ensuing determination by the objectified loss. Central to this logic is the notion of the subject: it is a role or position within the structure that, while holding the structure together, captures the relation of the system to something that does not exist but yet "determines" the system. The subject is thus a conceptual apparatus that performs two crucial roles in any structure:

1. It inscribes what does not exist as something that is registered within the structure.
2. It allows for what does not exist to cause the structure to exist and to expand.

In short, the claim here is that any structure is built around a subjective core—it is this subjective core that is the "essence" of any structure. No structure exists without a subjective core. In this sense, the subject is a metalogical or even ontological apparatus, and by no means refers to an experiential or substantial entity, such as human individuals.

The last feature/component refers to the dynamism that allows circulation within a given structure—what Miller and the rest of the editorial board of *Cahiers* call a discourse. A simple analogy for this is a word or shape puzzle that, within its square box, contains one empty space that allows the other boxes to move around, one space at a time, thus allowing for the puzzle to be solved. The "emptiness" of the subjective position is metaphorically similar to the empty space within the puzzle. Similarly, the systemization at work in symbolic life is marked by an essential incompleteness, which elicits a subjective effect. This empty space allows for discourse in structure and the very movement of jouissance, whether within language, the market, or a kinship structure.[10]

It is not surprising, thus, that Miller finds, within the Fregean construction of number, the prime example of the initial loss and the construction of the ensuing circulation in the structure thematized by what he calls the logic of signifier. Miller emphasizes that he makes explicit what, already in Frege's text, exists in an implicit form. It is evident in list 1 under the definition of cardinal numbers that there are two important moments in the generation of numbers. We've already explored C_0, the concept that underlies the number 0. The other fundamental moment is, of course, C_1, which in Miller's words, is the "proper name" for 0. So, while 0 is summoned to name the cardinality of a concept that is discarded for its nonexistence (or its impossibility), 1 counts the name of this impossible object and marks its initiation into the chain of succession. Thus 0 names the exclusion, and 1 counts this exclusion: the operation of counting grounds the entire succession of numbers, as the rest of numbers is the repetition of this operation. However, the consistency that underlies this repetition is the marker of an initial exclusion.

"Suture" is thus the term that refers to the naming (counting) of a non-being (metaphor) that, at the same time, ties this non-being to a chain of repetition (metonymy): the insistence of what was originally excluded makes the structure come to life. Miller's name for non-being is the subject, and for structure is discourse, and thus suture marks the relation of the subject to its discourse: subject figures as the element which is lacking in the form of a placeholder.

This also makes clear the Lacanian definition of the subject as what presents a signifier to another signifier in the chain: every signifier is based on some specific construction (i.e., repetition) made possible by the same annulled entity which Miller calls the subject.

The logic of the signifier is the logic of all logics, and the signifying chain is the structure of all structures. This bold pronouncement is likely the reason Badiou entered this debate. The editors of *Cahiers* believed they had discovered a logic that renders all other logics consistent.

As a student of Althusser, Miller is keen to bring out the philosophical question latent in Lacan's enterprise. In "Suture," we find him laying out the foundation of an answer to his own question: "What is your ontology Dr. Lacan?"

The answer is a pre-ontological discourse. Miller sees Frege, Marx, and Freud as each building discourses, which, despite belonging to different fields of human knowledge, all share a common characteristic of being discourses of overdetermination. Every such discourse is marked by repetition, a sign of impossibility. So, these discourses can be read as a response or consequence of a latent, but grounding, impossibility, or inconsistency. Frege's service to Miller was to direct his attention toward the identity principle of Leibniz. From that principle, Miller grasped that the principle of inconsistency/impossibility is non-identity, for which the psychoanalytic term is subject. Therefore, every discourse takes place with a subject sutured to a repetitive chain that marks the return of the original impossibility.

It is important here to emphasize that Miller's usage of Frege's brand of logicism is a programmatic usage, which sees Frege's construction as a formal proof of the primordiality of the logic of signifier over, not only logic but the entirety of mathematics—and possibly all scientific discourse.

The examination of the validity of this claim is the topic of the next section, where Badiou sees Miller's construction (as well as Frege's logicism) to be fraught with a fundamental misunderstanding of what axiomatic thinking is, resulting in an ideological subversion in which a regional field of study (psychoanalysis) sutures itself to an imaginary Truth and claims to ground the entire field of knowledge globally.

MARK AND LACK: FORMALISM AS FIDELITY

From the outset, in the very first paragraph of his seminal essay "Mark and Lack," Alain Badiou lays his cards on the table. It is in the spirit of the epistemological rupture that he speaks, the rupture that breaks epistemology away from ideology (and common sense alike, one might add) as it evacuates from science any notion of Truth, and replaces it with "a mechanism of production."[11]

What does this rupture say about logic? Or rather, the question here is, what logic does Badiou have in mind?

Badiou asserts that there is an ideological representation of logic in which there exists a presupposition of the positing of Truth rather than the construction of an object. In this representation, which Frege is in part responsible for by abruptly likening "a proposition to a proper name whose reference, or denotation, is the True or the False," "logic incessantly coordinates as many linked inscriptions as necessary in order for it to pass from one invariable name-of-the-True to another."[12]

Classical logic focuses on the forms and categories of statements in order to sort through valid and invalid conclusions. It involves the study of the truth-involving relations between sentences; it is interested in giving a general answer to the question: When does the truth of one set of sentences guarantee the truth of some other sentence, or what does it take for one sentence to follow from some others? Doing this involves giving some kind of analysis of sentences into their parts since whether one sentence follows from another is typically a matter of relations between such parts. The prime example of such analyses is the Aristotelian logic, which breaks down sentences into subject and predicate and determines how the most general distinctions between classes of sentences are distinctions between modes of predication, a movement Aristotle called a syllogism. Using syllogism, we can make valid inferences by suitably categorizing the subject and the predicate of sentences within a given argument.

In contrast, modern logic, that is, the logic which Badiou is talking about, and which is usually assigned to Frege, is worried about the truth-values of statements. In other words, Frege is worried about the instances of a statement, and whether two statements that have the same instances are equal statements. So, all statements for Frege are (complex) denoting terms: they are terms that denote truth values. This is an important difference between classical and modern logic. In classical logic, the quantifiers do not play a significant role. With Frege, we have the transcription of the old statements of categorical logic in a language employing variables, quantifiers, and truth functions. Modern logic, mostly invented by Frege, presents the first complete axiomatization of propositional and first-order "predicate" logic (the latter of which Frege invented outright).

This is an important clarification for Badiou. The revolution inaugurated by Frege in logic intended to reduce mathematics to logic and logic to a conceptual construction of truth functions. This not only makes logic a system of conceptual construction foreign to the real, but it also makes mathematical objects and theory an exercise in tautology wherein concept comes first and number thereafter. We know from the later Badiou that this orientation is in direct opposition to the role of mathematics that he has in mind—mathematics is not only not tautological but it is also the science of being qua presentation. What is important here is that, even at this stage of his intellectual career, Badiou notices that the orientation which equates logic and mathematics to complete conceptual construction is a sort of metaphysics in disguise, one that is at the service of producing ideology of science and not science itself.

It is the latter interpretation of logic formalized by Frege that is at the center of theorization in Miller (and Lacan). This interpretation, as we saw, starts from the conceptual formalization of zero based on Leibniz's principle of identity, and propagates through the number system such that each number is the recounting of the same non-identity principle. Miller has used this conceptualization of zero and number in Frege's system as the basis of his concept of suture, central to the logic of the signifier. The subject stands for the nonidentity whose implication by its nonreflective part in the structure is masked by its reflective part, yet a causal connection binds both. The reflective reduplicates reality in the imaginary, and the non-reflective registers itself as an absence, a lack, but the two parts feed off each other in a repetitive entanglement such that the more that lack presses on, the more the imaginary will prevail, in turn intensifying the force of the lack. But in this relationship, the imaginary is a reduplication, that is, a recounting of the lack every time. This is why Badiou claims that, for Miller, True is another name for the lack.[13]

Within this logic that contains a twofold process of reduction to lack (True) and to nonidentity, Badiou wants to posit a different logic, the logic of stratification, in order to show the true closure and foreclosure of science (particularly mathematics) such that it does not lack anything that it doesn't produce elsewhere. This effectively demonstrates that "The logic of the Signifier is a metaphysics: a representation of representation, an intra-ideological process and progression."[14]

What is at stake in our view, however, is not just whether two views of science, one more or less attributed to Althusser and one being worked out by Miller and other *Cahiers*' editors, can converge. Rather, the question for Badiou is really whether the logic of the signifier can commit to the epistemological rupture or not. Badiou, as we will witness, is committed to the original Bachelardian project, in which the objective process of science lacks nothing it cannot produce within itself. The issue is that for Badiou, Miller et al. have attempted to bypass this notion of scientificity by succumbing to Frege, whose method allows them to identify a repressed element in science: the non-self-identical element that is repressed and then sutured to the entire process. This allows Miller et al. to import a primitive, the subject, into the foundational theory. The real problem for Badiou in accepting the theory of the signifier revolves around the same point of the subject within the structure. For Lacan and Miller, the entire process of language is marred by an imaginary process. This process is necessary for the dynamism embedded in the speech. For this dynamism to hold, it is necessary to assume in the clinical setting the role of a reflective element. Miller's ambition is to use the same reflective element on a much broader stage, in science itself. He generalizes something operative to a specific field of human sciences, beyond its legitimate applicability.[15]

Therefore, Badiou's project in "Mark and Lack" contains three components. First, he shows that in science (including mathematics), which the epistemological rupture demarcates, there is no lack, and that nothing within science, is motivated or dependent on the functioning of a lack. Second, any deliberation on the foundation of mathematics and science should first consider requirements from within those fields and not from psychoanalysis or historical materialism; it is up to mathematics to define what is required for the foundation of mathematics, and if the lack does not appear anywhere in the theory we should take that hint very seriously. Third, in conclusion, he shows that there is an inversion which underlies the way in which the theory of the signifier is formulated. The theory of the signifier, with its roots in psychoanalysis, is extending a requirement specific to a particular discourse to the rest of the discipline of science, which to Badiou is a clear sign of an ideological recapture, "in which every science comes to mime its own reflection."[16]

In what follows, we will follow Badiou's argument closely to show each of these points that may be scattered through his remarkably concise and dense essay.

The first thing to note is Badiou's definition of the theory of logic: "The theory of logic pertains to the modes of production of a division in linear writing or inscription."[17] According to this concise and preliminary definition, logic comprises two main components:

1. Linear writing—a mechanism that produces strings of signs or syntagms
2. Modes of production of a division—a mechanism that takes the above syntagms and divides them into a dichotomy, or two sets

At the two ends of the spectrum, we have finitely many individual marks, which like atoms, are indivisible and independent, which we call alphabets, and we have two disjoint classes of strings of marks, which we call derivable and non-derivable.

The above two operations can also be presented as three: concatenation as putting marks together in syntagms, formation as dividing syntagms into syntactically correct and incorrect groups, and division as further dividing the syntactically correct group of syntagms into derivable and non-derivable classes.

The way Badiou uses the latter categorization and his repeated reference to mechanism is reminiscent of the Leibnizian logic machine or its modern reincarnation, the Turing machine.[18] The above processes then resemble a set of algorithms that execute based on its raw materials, producing outputs consumable by the next operation in line.

We define concatenation as the operation that draws from a set of alphabets and produces finite sequences of marks in a linear order, which may include repeating marks. The machine can produce these sequences of letters in whatever order, where each sequence will be fed into a subsequent algorithm. Badiou calls the output of the machine from this first phase of the operation (concatenation) the set *S*.[19]

The second operation, formation, starts reading from *S*, and decides whether the sequence of marks is valid syntactically or not—the rejected clauses are usually labeled as "non-sense." Furthermore, this split into valid and invalid subsets is a dichotomy, there is no remainder. This fact is a by-product of Gödel's proof. As Badiou points out, the very possibility of Gödel undecidability presupposes the existence of a dichotomic mechanism with its raw material. This is an important assertion for Badiou. First, it is only based on a perfect dichotomy that we could proceed to the operation of derivation, which is the next operation in line. But second, it is based on an entirely decidable closed mechanism with no compulsion to repeat that we can even recognize the existence of what is known to be un-closable, and thus internally limited. So, the second point Badiou wants to make with this remark is this: "The exhibition of a suture presupposes the existence of a foreclosure."[20] Foreclosure comes before suture and is logically prior to it. Therefore, to establish any claims regarding the existence of a structure whose integrity is preserved by the operation of suture, we must have a system that is closed but, at the same time, is not caught in an endless loop.

The word "algorithm" in computer science usually refers to a set of instructions that, for a given input, are deterministic and will halt. A loop is a part of the algorithm that may cause non-halting behavior for certain inputs. The essence of computability theory asks:

1. Whether for a certain problem, there exists an algorithm that can provide the answers.

2. Whether the algorithm is complete, that is, for some certain given conditions (e.g., inputs), the algorithm is deterministic and halts.

In computability theory, there are countless such algorithms. The very existence of these algorithms should be a counterexample to the *generality* of the logic of signifiers and the structural dynamics that it intends. The algorithmic nature of the formation operation is guaranteed to result in the set of well-formed expressions, which Badiou names *E*.

The third operation, derivation, uses the well-formed expressions from the set *E* and determines whether the expression is derivable (it is a thesis in the system) or it is non-derivable (it is a non-thesis in the system). If *e* is an expression in *E* and *e* is a thesis, then *e* belongs to the set of theses *T*, and if *e* is a non-thesis, it belongs to the set of non-theses *NT*.

Furthermore, there is an operator ¬ (negation) such that:

$e \in T \Leftrightarrow \neg e \in NT$, and vice versa,

$e \in NT \Leftrightarrow \neg e \in T$.

So, if we have two expressions, e_1 and e_2, both belonging to *E* and $e_1 \in T$ and $e_2 \in NT$, we can also write $\neg e_2 \in T$ and $\neg e_1 \in NT$.

If the above were true, then the derivation operation, like the operation before it, would create a perfect dichotomy: *T* and *NT*, and the relation between each expression and its negation, such that they never belong to the same set. If we use the symbol … to denote the relation between an expression and its negation, that is, *e* … ¬*e* means the relation between *e* and ¬*e*, then, following Badiou's lead, we can say that the mechanism of derivation, in case of a perfect dichotomy, cuts right through the middle of all such relationships according to which each expression and its negation belong to opposite sides:

Mechanism of derivation

e ……..|……… ¬*e*

If a perfect symmetry such as this existed, David Hilbert's dream of having an effective procedure (an algorithm) capable of proving all truths about axiomatic systems would be possible.

Hilbert, at the great mathematical congress held in Paris in 1900, posed ten problems to the world of mathematics. In 1928, he rearticulated them in three major categories of problems, which Stephen Hawking summarizes as follows:[21]

1. To prove that all true mathematical statements could be proven, that is, the completeness of mathematics.
2. To prove that only true mathematical statements could be proven, that is, the consistency of mathematics.
3. To prove the decidability of mathematics, that is, the existence of a decision procedure to decide the truth or falsity of any given mathematical decision.

Gödel's incompleteness proof in 1931 dashed Hilbert's hope, at least in part, since it disproves the first of these challenges.[22] Gödel proved that, in a consistent system, it is possible to have expressions such that neither it nor its negation is provable, that is, $e \in NT$ and $\neg e \in NT$ at the same time. Such an expression is termed undecidable. The undecidability of a well-formed expression, however, does not disturb the fact that the derivation mechanism cuts E into a dichotomy. The dichotomy exists even after Gödel. The issue is that the relation between the two parts of this dichotomy, T and NT, is no longer a perfectly symmetrical relationship, such that when an expression belongs to one set, its negation always belongs to the other set because sometimes an expression and its negation are both un-provable. So the meaning of incompleteness is that one of the undecidable expressions (e or $\neg e$) must be a thesis, but in a consistent system, we cannot prove either of them, so, therefore, T is not a complete set of all theses.

The existence of undecidable expressions does not speak to the undecidability of systems, which corresponds to Hilbert's third problem. However, after Gödel, it is no longer possible to prove the truth of all expressions in a given system; we have to suffice with determining whether they are derivable or non-derivable (while they may still be true).

It may be useful to provide some remarks regarding the third problem, whose aim is not to prove or disprove an individual statement, but to come up with a procedure to determine any statement in the system (whether it is derivable or not). After Alan Turing, we now call this the "computability problem," that is, a problem that can be solved using a Turing machine, and in computer science, we express this problem as the ability to articulate the solution through an algorithm that a computer can execute. If all axiomatic systems were decidable, what Hilbert had hoped for, then for every system, there could be finitely many algorithms that can decide the status of every correctly formed statement in that system, whether they are true, false, or undecidable.

First, Alonzo Church and then, almost simultaneously and independently, Alan Turing disproved Hilbert's decidability problem (using lambda calculus in the case of Church and in the case of Turing using a completely novel method, which we call the Turing machine today).

The intriguing way in which Turing solved this problem gave rise to another problem, which is almost as important as the three problems of Hilbert.

Given the Existence of an Algorithm to Solve a Problem, Is the Solution Practical?

The informal term "practical" implies the existence of an algorithm for solving the task that runs on a Turing machine, such that the time to complete the task varies as a polynomial function on the size of the input to the algorithm (as opposed to, say, exponential time). Computer scientists call this the complexity problem, based on which they divide questions into two classes: first, the general class of questions for which some algorithm can provide an answer in polynomial time, called "class P" or just "P." In contrast, for some questions, there is no known practical way to find an answer, but if one is provided with the answer, it is practical to verify it. The class of questions for which an answer cannot be found but can be verified in polynomial time is called "NP," which stands for "nondeterministic polynomial time." It is important to note that both P and NP classes of problems are solvable problems; that is, there exist algorithms that can find the answer to their questions, but in the case of NP, this answer cannot be given in any "practical" way.

Despite this categorization, it was not possible, before 1971, to prove that a problem is NP. In order to prove that a problem is NP, we must prove that there is no algorithm that can find answers in polynomial time. In 1971, Stephen Cook found the first NP problem. He proved that the Boolean satisfiability problem is an NP problem. This result is now known as the Cook-Levin theorem. Using this theorem, it was now possible to show certain problems are at least as hard as the Boolean satisfiability problem, and, therefore must also be NP. This led to a new subset of NP problems we now call NP-complete problems, the set of problems to each of which any other NP problem can be reduced in polynomial time, and whose solution may still be verified in polynomial time. That is, any NP problem can be transformed into any of the NP-complete problems. Informally, an NP-complete problem is an NP problem that is at least as "tough" as any other problem in NP.

A number of important and useful problems are proven to be NP-complete. For example, the prime factorization problem is an NP-complete problem, something that mathematician John Nash hinted at in 1955. Prime factorization is the basis for encryption because when the key is known, its verification is P, but when the key is unknown, the answer to the algorithm runs in exponential (non-polynomial) time, relative to the length of the key.

This little excursion may be useful because it accentuates a stark contrast of ultimate importance to Badiou as well as to our purpose. The early to the mid-twentieth-century discoveries regarding what is provable, what is computable, and what is complex, Gödel, Church, and Cook's results, reveal the consequences and limitations of axiomatic thinking.[23] They are not signs of a repressed lack in science—it was not the case that scientists are re-experiencing an impossibility that keeps repeating itself—but the affirmation of the scientific method: the fact that science continues to think beyond its determination set out by a particular discourse—in this case, Hilbert's program—is what the generation of French philosophers at the time referred to as the epistemological rupture. One of the central aims of our work is to show that the culmination of this "rupture" is what informs and underlies Badiou's mature project: the excess of being over language, underwritten by the axiom of Infinity. This, of course, was not yet present at the time Badiou was writing his audacious rejoinder to Miller, but from the way Badiou is troubled by Frege's and Miller's appropriation of logical laws to reappropriate certain metaphysical imports ideologically, it is clear how Badiou is on the path to discover what will eventually inform his entire project.

What Are These Metaphysical Imports?

The identity of marks is an intra-symbolic law and has no import to the realm of physical objects.[24] When we say a symbol x is identical to x wherever we write x, we are simply stating a self-evident fact, whose negation is unthinkable. That is, the lack of self-identity of symbols will not be symbolized: there is a lack of marks for such an absurdity.

Badiou makes the latter point clear by devising a function to represent the equal sign: instead of writing $x = y$, we write $I(x,y)$. In that case, we always have the following two formulae:

$I(x,x)$, every variable is equal to itself

$$I(x,y) \leftrightarrow [A(x) \leftrightarrow A(y)]$$

The latter formula, written in first-order logic, means that for any given function A if the value of A is the same for x and y, it is because x and y are equal and vice versa.

What about $\neg I(x,x)$, a completely permissible or well-formed expression in the system? Is it not the mark of non-self-identity we are seeking? Absolutely not! The expression $\neg I(x,x)$, the formula for self-inequality, is permissible solely on the basis that the first x in the function is the same as the second: that the same mark written in two different places are not two different things is the meaning of the self-identity of marks. As Badiou puts it "The production of the logical concepts of equality and self-inequality presupposes the foreclosure of what is scripturally non-self-identical. The lack of the equal is built upon the absolute absence of the non-identical."[25]

Here one can see the power of stratification that logic is capable of applying to itself. The production of $I(x,x)$ and $\neg I(x,x)$ is the outcome of the function of concatenation we saw above. The function of formation puts both of these expressions in E since they are both well formed. The function of derivation puts them into two different categories of expressions T and NT. If the set of true expressions in logic is the outcome of the last operation (derivation), then in that set, there is no presence of $\neg I(x,x)$. The identity of marks or graphemes is the law of the first operation (concatenation), whereas the expression of self-equality as the truth is the outcome of the third operation (derivation).[26]

By mistakenly assigning the law of self-identity to objects, a metaphysical assumption on its own, a space is opened for a mark within the realm of symbols for the relation of self-difference that exists in the realm of physical objects. The mistake is the exportation of this law to a domain where it has no jurisdiction. This metaphysical move opens up the symbolic/logical order to the registration of an impossible relation among objects, which by itself may not be a wrong move: it is possible in a language to name an impossibility, un-think-ability of a relation between objects in a separate domain, but language cannot name something that is unthinkable.

The second metaphysical move ironically committed by Frege, a celebrated pioneer in the axiomatization of logic and mathematics, was to think that logic actually provides a stratification of objective reality, that it is a language which speaks of a domain other than itself, and accordingly, he thought it quite legitimate to name this impossible relationship of self-difference by a mark inside logic, that merely indicates this impossibility in the other domain.

So, what about zero, then? If there is nothing that sutures logic to the empirical domain, then how can logic produce the notion of zero?

Let's consider the formulae we defined above for the equal sign:

$$I(x,x),$$

$$I(x,y) \Leftrightarrow [A(x) \leftrightarrow A(y)]$$

And let's say that in a theory based on first-order-logic-with-equality, the above formulae are part of the axioms of that logic.[27] Now consider the first part of this formulae: $I(x,x)$ and its negation $\neg I(x,x)$, which, as we've seen, are produced by the function of concatenation in S and are slated as well-formed by the function of formation in E. Finally, the function of derivation dispenses them separately into T and NT, respectively.[28]

Now we include a new mechanism M_4 that adds to M_3 a predicate 0 defined as follows.

Let the function $R(x,y)$ express that variables x and y are related within our system. Let us also assume that R can be reflexive; that is, $R(x,x)$ states that variable x has a certain relationship R with itself. Now, let us name a new function, which says whether a variable is linked to itself through relation R, as $Ar.R$, and accordingly, $Ar.R(x)$ means that x has the relation R to itself.[29]

We are now operating in second-order logic, in which we can select not just variables but also functions. In that sense, we can express the above in the following way:

$$\forall R,x\text{: } Ar.R \leftrightarrow R(x,x)$$

Given this definition, it is easy to see how I is such an R according to the axioms. That is, the axiom of equality (identity) will permit us to derive $Ar.I$ and not derive $Ar.\neg I$ within M_3. Let's define the 0 predicate as the following:

$$0 = Ar.\neg I$$

Or in other words:

$$\forall x\text{: } 0(x) \leftrightarrow \neg I(x,x)$$

Now, the predicate 0 is an inscription that is accepted by M_2, due to its well-formed-ness, and therefore it is in E, rejected by M_3 because it cannot be derived, and added back again to T by M_4. What M_4 did was add a mark to T for a non-derivable relation. It is important to note, as Badiou says, "The zero marks in M_4 (in predicative form) not the *lack of a term* satisfying a relation but rather *a relation lacking* in M_3, the relation $\neg I(x,x)$. We must nevertheless add: if the relation can be lacking in M_3, it is *only insofar as it figures in* M_2."[30]

How is the 0-predicate derived here by Badiou different from the 0-term derived by Frege (and used by Miller)? The difference is that the latter marks the lack of a term, and the former marks the lack of a predicate. Frege's version of zero states that there is no term x that can satisfy the negation of Leibniz's identity principle, which, as we saw, is something fraught with metaphysics. 0 as predicate, however, implies that an expression that exists in one stratum (M_2) is erased from another stratum (M_3), for which we devise a mechanism (M_4) to add a mark as a trace of this erasure. This is not just a "zero-sum game." Lack of a term and lack of a predicate are totally different things for one important reason: term (especially in Frege's use) is a non-logical artifact whereas the predicate $Ar.\neg I$ is an artifact of logic produced by M_1 and ratified by M_2. We're still well within logic: 0 is not a mark of what logic lacks but what logic produces to trace the lacking of a mark within its extensible stratification. That is why we could say science is stratified in such a way that no lack is marked in it that does not refer to another mark in a subjacent order differentiated from the first.[31]

Here we should emphasize two points, very important to the overall argument that Badiou is producing. In a number of places in this short essay, Badiou notes that we should differentiate logic from the discourse about logic. The discourse about logic, which is usually used for pedagogical reasons, provides intuitive or commonsensical conceptions that are foreign to logic itself.[32] This criticism, which goes to the heart of Bachelardian theory of epistemology as a non-empirical endeavor, tells us that we should bracket common sense and empirical concepts from the scientific domain, logic included, and differentiate the discourse about a discipline from the discipline itself. There is a convenient way that logicians speak about their theories, but when it comes to logic as a discipline, they abandon the niceties of the discourse and stick to what the discipline itself works with. Practitioners and people whose professions do not involve the direct treatment of a particular discipline (logic or science) are the usual victims of the discourse and mistake discourse for discipline. Philosophers are prime examples of such victims, and the ideological recapture/representation is what this error produces.

Gödel's theorem is a very famous case for such confusion between discourse and discipline. As we've seen, there is nothing in the incompleteness theorem that speaks about the lack of predicate logic, first-order logic, or arithmetic. It instead shows that language, in a predictable way, falls short of calling out or deriving all true statements. The logical result of this incompleteness theorem is, in fact, Cohen's generic procedure, which will be discussed at length in the following chapters and which embraces Gödel's incompleteness results to show how we can constantly extend a consistent/semi-veridical ground model (an initial denumerable set) by forcing an indiscernible (or generic set). In contrast to Cohen's use of Gödel's theorem, which we may categorize as a legitimate use by a logician who remains within the discipline, Lacan's usage of Gödel, in Badiou's eyes, is illegitimate and influenced by misconception of what the incompleteness theorem really means. Lacan's misconception states that this theorem proves the lack in the Other, which is the language or the overall battery of signifiers, and thus there is a need for a (reflective) element that has to suture the Other constantly, thence the role of the duped (and foreclosed) subject(s). The proper recapture of Gödel is by Cohen, whose underlying and enabling thought was backed up by the axiom of Infinity. The recapture of this in ontological terms is the excess of being over language. We will speak at length in subsequent chapters about the difference between this recapture versus the ideological recapture of the theory of lack and suture. For now, however, it has been made quite clear that logic lacks nothing, and using Gödel's incompleteness theorem as the proof that the Other is lacking (because logic or arithmetic is lacking) is an ideological representation of the theorem.

The very concept of suture, which has motivated this response by Badiou, is itself an ideological representation caused by the conflation of discourse and discipline:

> To deploy the concept of suture in the very place where it is inadequate (mathematics), and to conclude that this concept enjoys a universal legitimacy over discourses by exploiting scientists' conflation of their own activity (science) with its (ideological) re-presentation, is to reflect science in ideology: it is to de-stratify it so as to prescribe to it its lack.[33]

The other noteworthy point pertains to the stratification of logic. We saw how, with stratification, logic could produce conceptual tools to address its own needs. But the question is whether there is a way to produce a logic of stratification itself. In the footnotes and in passing, Badiou names two logicians who have attempted to answer this question: Wilfred Quine and Hao Wang.[34] Quine attempted an axiom of Reducibility to flatten out the strata to a single stratum, and conversely, Wang created an "expansive" system Σ to traverse the strata. But, according to Badiou, both attempts have failed. This failure means that a single meta-logic for logical stratification does not exist. He captures the meaning of this failure as follows:

> For our part, we are convinced that the stratified multiplicity of the scientific signifier, which is inherent to the process of scientific production, is irreducible to any of its orders. The space of marks does not allow itself to be projected onto a plane. And this is a resistance (or limitation) only from the viewpoint of a metaphysical want. Science wants the transformation-traversal of a stratified space, not its reduction.[35]

The effort to create a single theory that rules over the stratification of science is itself emitted from a metaphysical want—the desire to totalize. The discipline of science is of the order of the infinite, whose totalization is an impossibility. Even in his early career, Badiou's theoretical conception is imbued by the thought of infinity.

We began this inquiry by reviewing the dashing intellectual thrust made by the editors of *Cahiers* announcing a new unified theory of discourse. At the heart of this new unified theory was the announcement that we should look at the action of the structure in the presence of a reflective element. As seen earlier in this chapter, the abrupt introduction of a reflective element into the language of structuralism could be only justified when this element is seen as primitive, as nowhere does the theory elaborated in "Action of the Structure" assume the presence, let alone the action or impact, of such an element.[36] Seen as a primitive, Miller derived a number of intriguing properties not present otherwise in the traditional theory of structuralism: the function of misrecognition, the imaginary and real registers, the lack and the suture, and metonymical causation, to name a few—in the center of which there is a reflective element implicated by the structure and framed in such a way that its principle of existence is of nonidentity. However, "Action of the Structure" can only be understood, in our view, by reading "Suture," as it is only in the latter work that we find a derivation of the concept of subject in this theory. In Suture, Miller appropriates Frege's arguments to establish the role of the subject in number theory, but we can understand the true scope of this derivation only by looking back at "Action of the Structure." So, while "Suture" derives the subject, "Action of the Structure" utilizes it for building the overall scope of the program.

It is only when we look at both works by Miller that we can also properly understand the scope of Badiou's rebuttal. First, he has the task of countering the misconception inherent to both Frege's and Miller's derivations. Badiou establishes the fact that there is nothing within the foundation of mathematics that requires an element of non-identity. Second, neither mathematics nor logic (nor science in the way demarcated by Bachelard) is sutured to anything outside of it—like the way Frege's conception of numbers required bootstrapping by a recourse to the metaphysical version of the Leibniz's principle of identity. Science is foreclosed to anything outside it, it has a lack of lack, and this lack is without a trace. Third, science is infinitely stratified, which allows the production of abstractions inside its realm, without recourse to any theory or discourse outside of it—whose strata are subject to a law that derives their depth and breadth formulaically.

The immediate conclusion of this is that there is no subject of science:

> Foreclosure, but of nothing, science may be called the psychosis of no subject, and hence of all: universal by right, shared delirium, one has only to maintain oneself within it in order to be no-one, anonymously dispersed in the hierarchy of orders.[37]

As Tom Eyers has also remarked, it is ironic that Badiou should choose "foreclosure," a psychoanalytical term, to describe science as something about which "psychoanalysis has nothing to say."[38] But what looms under this term goes well beyond the psychoanalytical concept of foreclosure. The founding role of this term clearly illuminates the traces of the axiomatic orientation of thought, and while the term "foreclosure" implies closedness and protectiveness, logic, science, and mathematics enjoy much openness and bountifulness. That is precisely the sense of creativity that axiomatic thinking provokes. It frames the thought around a finite set of decisions or ideas, not to restrict thinking, but to free it and to explore possibilities in ways not otherwise possible. Axiomatic thought is what allows science to grow on its own merits alone: foreclosure, in this case, is not prohibitive in any sense. On the contrary, it is the founding principle of something productive and affirmative: the mechanism of production, partly exhibited in "Mark and Lack," upon which science can produce its signifiers, expressions, and abstractions according to its internal laws and funding decisions. In that sense, although Badiou does not yet make any note of the axiomatic thought, we think a retroactive reading of this work, under the light of his mature oeuvre, leaves no room for doubt that he is embarking on a project to juxtapose the axiomatic orientation of thought against the theory of discourse whose roots are in structuralist humanities: linguistics, anthropology, psychoanalysis, and last, but not least, historical materialism.[39]

As the unified theory of discourse claimed to recover the repressed Truth in science and give a unifying voice to the discourses of overdetermination, it is evidently tormented by the same traumatic core as philosophy itself—it attempts to bestow to science what it does not need and what it does not want: the Subject: "We can claim in all rigour that *science is the Subject of philosophy*, and this precisely because there is no Subject of science."[40]

Ontology of Discipline and Epistemology of Discourse

In "Mark and Lack," Badiou uses the consequences of Gödel's theorem to clarify a significant difference that exists between two "positions" ideology could occupy in relation to science. The position that has been discussed so far pertains to discourses that are outside of science proper. Badiou captures the intent of this category of discourse as such:

> The (metaphysical) project which, following Hilbert, enjoins every formal system to seal itself around the internal statement of its own consistency.[41]

These discourses go by different names (e.g., metaphysics), but in the context of the discussion, Badiou chooses to call them epistemological discourses. Epistemology, therefore, refers to those discourses that have science as their subject but do not produce, nor claim to produce, scientific theories themselves. In all their forms, epistemological discourses attempt to find from outside of a given discipline the unifying principle according to which the discipline can be defined and organized. Epistemological discourses, in that sense, are transcendental to the disciplines they study. Badiou uses Husserl's treatment of mathematics as nomology as an example of such a discourse. Being nomological means that the domain of mathematical objects can be exhaustively defined. As a formal system, mathematics is closed and saturated. In addition, it means that the progression of the system by establishing different levels can take place without contradiction.[42] The nomological definition implies that the technique establishing something like meta-mathematics can be stopped at any time, once the increased facility permits statements about reality to be obtained. These are assertions, none of which belongs to mathematics proper. They belong to a discourse that occupies a position outside mathematics. They are discursive assertions about a disciplinary practice.

In contrast to this, there are certain assertions about a given discipline that are part of the discipline itself. The assertions are not part of the disciplinary theoretical body but nevertheless are considered part of the discipline. Two famous examples of these in mathematics are Hilbert's program and *Principia Mathematica*, both of which are programmatic projects that make assertions about the structure and nature of mathematical theory in general. Badiou captures the nature of these programs as such:

> The project which, by means of the completely controlled reconstruction of a logistical system, claims to exhaust what otherwise presents itself according to the opacity that results from a history: let us call this "intuitive" arithmetic.[43]

We can also say that programs like Hilbert's or *Principia Mathematica* are representations of mathematical discipline, but representations that are intradisciplinary, representations that are immanent to a given discipline—which Badiou calls the "intuitive" part of a given discipline.

Therefore, according to Badiou, there are two representational domains, outside and inside a discipline, and they correspond to two aforementioned positions of ideology.

Now, one way that Badiou's critique of the project of a unified theory of discourse can be understood is as follows. Badiou does agree with the overall intent of the program to identify and reveal ideology, but both the position he identifies for ideology and the method he thinks science itself takes against ideological representation are different from those in Miller's proposal. The unified theory of discourse targets the transcendental position of ideology, and its method is to devise a theory that aspires to science in order to remain non-ideological. Badiou is critiquing both the target and the method of this theory. He is critiquing the target because the ideological representation of science outside of science is not what affects the practice of science. What matters to the practice of science are ideological representations that are immanent to it, the intradisciplinary representations.[44] What has a positive influence on science, and its advancement, for instance, is when science breaks with its own representative designations, that is, with what science thinks about itself (as opposed to what philosophy thinks about it).

In addition to this, Badiou also critiques the method of the unified theory of discourse. Not only because Badiou has been able to unwind the metaphysical core of its notion of subject that is the linchpin of this theory, but because, in the end, this theory is just another discursive tool, and discourse is not how science plows through its ideological obstacles. How has science been historically so effective in constantly removing from its path its own ideological designations? The answer is disciplinary engagement: it is the practice of constantly breaking away with how the discipline is represented inside the discipline itself, whose term for Badiou is "Formalism" in contrast to the representation of the discipline, or "Intuition." Therefore, what matters for science is the face-off between formalism, "an entirely coded scriptural artifice," and intuition, "the immanence of an historico-institutional discourse living off the abbreviations, equivocations, and univocal smoothing of an inoffensive mass of 'normal' signifiers legitimated by custom and practice,"[45] a wholly intradisciplinary conflict.

Yet again, Gödel's theorem plays a key role in demonstrating this dialectic between intuition in science, which involves "certain ambiguities produced in language by the (ideological) concept of Truth," and "formalism's fidelity to the stratifications and connectivities at work in the history of the science, insofar as they expel from the latter every employment of the True as (unlimited) principle."[46] Gödel's proof also had consequences for nomologist conception of mathematics as well. It proved that mathematical content is not nomological, that is, mathematics is not tautology. Mathematics can provide material; that is not the result of analytical manipulation of axioms. As Cavaillès mentions,[47] Gödel's statement, albeit undecidable, still represents an increase in knowledge. That the undecidable statement is legitimate implies, for Cavaillès, that formal systems possess their own proper content, different from experiential content. In addition, the expansion of formal systems does not take place in a predictable way; rather, based on its own sort of content, formal systems exhibit their own sort of necessity. Together, these consequences proved that formal systems such as mathematics are not closed systems (as Husserl had postulated). Nomology can thus be perceived as the epistemological principle around which Husserl could conceptualize mathematics as such. By prescribing or uncovering an inherent limitation in the ability to completely identify true statements in a formal system, Gödel proved that mathematics is not a nomological system. Unlike Hilbert's program, mathematicians have never been concerned about whether mathematics is a nomological system.

Badiou thereby interprets formalism as engagement, an operator of fidelity operative inside science. The general theory of discourse describes how the discourses of overdetermination break away from ideology. Badiou's rejoinder here is to show that the proclamation of the latter still leaves us in the realm of ideology because it still speaks of the discourse of science as opposed to the science itself. Therefore, it still posits the discourse as something separate from the discipline. With Miller, we start with ideology until science allows us to spot ideological misrecognition. For Badiou, on the other hand, we start with scientific thinking, and only when we try to think why or how a science is a science do we fall into ideology. Therefore, the development of the sciences involves the critique of the philosophical idea of science. For Badiou, given the way he criticizes philosophy and the alternating chain of science and ideology, the point is that we don't need to look beyond science to see the oscillation between science and ideology. Within science, we have both scientific thinking and scientific ideology: the case in point is the debate between Gödel and *Principia Mathematica*, the latter standing for an ideological recapture of mathematics inside mathematics itself. The Church/Turing and Cook/Levine theorems are also examples of such a break from Hilbert's program, which equally stands for an ideological recapture of mathematics, again within mathematics proper. That is why Badiou says that, through science, we learn that there is something un-sutured.[48] In contrast to the claim that science is the science of suture, we learn that through scientific practice, we get something that is not a hiatus between ideology and science. Scientific practice is a constant separation between formal means of thinking and formal means of representing. In this sense, science continually breaks with its own designation in the representational space. Therefore, rather than staging the debate between science and ideology at the level of discourse, Badiou brings back this debate to the level of discipline: the level of production and practice of true science. In "Mark and Lack," Badiou has shown us that logic is a practical endeavor: the practice of recognizing, concatenating, and deriving marks, traces and expressions. With this, he brings the debate over logic from the level of discourse to the level of discipline. It is within the discipline itself that a division with ideology takes place. This division then concerns the being of science; it is not a division between science and something else.

Now, why is this important? This distinction anticipates three significant themes in Badiou's later work. The move from discourse to discipline means the move to the very entailment of things, to the level of their being, and thereby designates a move from epistemology to ontology. From this point of view, the second theme is motivated by the recognition of the scientific process as a process of fidelity. And finally, the fact that the hiatus does not take place between two realms, but between a realm and itself, gives rise to the theory of event. In addition, the way Badiou presents Gödel's theorem as a nexus of interconnected commitments to logical construction as well as the confrontation with tacit or declared ideological positions, suggests, in an implicit form, a sort of engaged theory of subjectivity. This is what we will take up in the next part, to show how Badiou, starting from here, will undertake the project of the subject.

In an interview with Peter Hallward in 2007, Badiou mentions that an understanding of structuralism inspired the *Cahiers* project as a certain Lacanian interpretation of scientism.

> They sought to find in scientism itself, in extreme forms of formal thought, something to support the Lacanian theory of the subject. In my view that is why Miller's text "Suture" is programmatic. It is a fundamental text in this regard, because this is the text that manifests the synthetic genius for which Miller must undeniably be recognized: he shows that for Frege the logicist reconstruction of the theory of numbers conceals an operation which can only be interpreted as the operation of a subject. I would say that this was the general orientation.[49]

In the early 1960s, the general problem of the relation between structure and subject was raised anew among the generation of young philosophers in France. Previously, in the model upheld by Sartre's existentialism, it was subject and consciousness that were considered as the primary in this relationship, and all forms of the structure were engendered on the basis of an absolutely simple and initial determination, which was practice. However, Sartre's generic philosophy had lost its grip: theory of structure outgrew its initial grounding "on the basis of the simple intentionalities of consciousness."[50]

The alternate approach was a commitment to structure first but to enable the structure to harbor the element of the subject. As Badiou mentions, Lacan was the one who proposed the alternative to Sartre. Lacan's major philosophical influence was the ability to bring together, in an unusual way, a theory of formal structures, developed as the logical theory of signifiers, and a theory of subjective adventures.

This is a very important assertion, and for reasons that cannot be developed in this work, none of the French structuralists, and in particular Althusser, were able to make such a proposal, mainly due to their lack of commitment to the category of the subject. The project that *Cahiers* took up was then to elicit what was already present in the Lacanian theory of the subject, what *Cahiers* called the unified theory of discourse. This theory started by engaging in the most extreme formal rigor and by taking up the intellectual power of mathematics and logic. At the same time, this theory described how an element of indeterminism must be sutured and present for the proper functioning of the structure, at the center of which are the operators of metaphor and metonymy. Badiou, around this time, was a member of *Cahiers* and, in his own words, completely committed to the agenda that brought this group together—to raise anew the question of the relation between structure and subject post-Sartre, and although he undermines *Cahiers*' manifesto, by no means he ever abandoned the original project.

Badiou's work in this era showed one thing: *Cahiers*' and psychoanalysis's commitment to science is not thorough enough. The introduction of indeterminism and non-identity to science, perceived as the realm of self-identity, is abrupt and ideological. We will see that it is exactly by deepening the commitment to formalism and determination that Badiou finally opened the path to indeterminism and non-self-identity of the entire situation of being. Non-identity is the law of being and not of the subject.

But this path goes through tumultuous times through which praxis yet again gains a radical priority for Badiou. At the end of this era, in the early 1970s, the question of the immediacy of praxis and the role of the subject finds a heightened urgency for Badiou, through which he comes to rethink the relationship between structure and subject. And this will take us to the next part of our work.

NOTES

1. Jacques-Alain Miller, "Suture (Elements of the Logic of the Signifier)," Concept and Form Volume One, edited by Peter Hallward and Knox Peden, 91–101. London, New York: Verso 2012.
2. Miller, "Suture," 93.
3. Alain Badiou *Number and Numbers*, Cambridge: Polity Press 2008, 16–31.
4. Gottlob Frege, *The Foundations of Arithmetic, A logico-mathematical enquiry into the concept of number*, New York: Harper & Brothers 1953.
5. Frege doesn't call this principle 'Hume's Principle' in his own writings. The label was instead introduced by Boolos (George Boolos, "The Consistency of Frege's Foundations of Arithmetic," *On Being and Saying*, J. J. Thomson (ed.) Cambridge: MIT Press 1987, 3–20). Frege does cite Hume when he introduced this principle in Frege-1974 §63, he quotes Hume's *Treatise* (I, iii, 1): "When two numbers are so combined as that one has always an unite answering to every unite of the other, we pronounce them equal." The idea in Hume does bear some resemblance to the principle Frege constructs, and so we shall continue to use Boolos' label for this principle.
6. Frege, *Foundations*, §72.
7. This can be shown using the transitivity of equinumerous extensions.
8. Miller, "Suture," 93.
9. Miller, "Suture," 94.
10. It will be discussed later on in Part II that one of many contributions made by Lacan to the structuralist theory was this discovery that the structural stability is a consequence of the lack and exception. Badiou proposes the name "algebra" for the structural dialecticity of materialism under the metaphorical law of the reflection, as the logic of the thesis of identity, and 'topology' for the historical dialecticity of materialism with the metaphor of the asymptote, as the logic of the thesis of primacy. These two map to what Lacan refers to as causality and remainder.

11. Alain Badiou, "Mark and Lack: On Zero," in *Concept and Form Volume One*, edited by Peter Hallward and Knox Peden, 159–185. London, New York: Verso 2012, 159.

12. Badiou, "Mark and Lack," 159.

13. "The nominal movement, the repetitive compulsion that, in the chain of propositions, unravels our disbelief in the True's common patronym, marks nothing but the lack over which this movement glides without resistance or success" (Badiou, "Mark and Lack," 159).

14. Badiou, "Mark and Lack," 159.

15. "The thesis we are defending here aims only at delineating the impossibility of a logic of the Signifier that would envelop the scientific order and in which the erasure of the epistemological break would be articulated" (Badiou, "Mark and Lack," 160).

16. Badiou, "Mark and Lack," 159.

17. Badiou, "Mark and Lack," 159.

18. Tom Eyers remarks: "Logic is rendered by Badiou here as a self-constituting, self-perpetuating 'machine,' impervious to the vicissitudes of the subject or the signifier" (Tom Eyers, *Post-Rationalism: Psychoanalysis, Epistemology, and Marxism in Post-War France*, London: Bloomsbury Academic 2013, 84). This remark does not match Badiou's project and is more aligned with the constructivst project. Logical machine is not a determinate process. The whole point of the Turing machine is the demonstration of the fact that the determinate algorithmic process may have indeterminate results.

19. Although it is not clear whether this matters to Badiou and the algorithm he presents whether S is finite or infinite, but it does to the algorithmic behavior of the machine. In other words, it is important to know whether the next operation starts when the first operation halts or not. With the current specification, there is no requirement for such dependency in the function of the two operations: the second operation can start as soon as the first operation reaches an output. This in software design is called the trickle-feed.

20. Badiou, "Mark and Lack," 162.

21. Stephen Hawking, *God Created the Integers*, Philadelphia: Running Press 2007.

22. The second incompleteness theorem, an extension of the first, shows that the system cannot demonstrate its own consistency. The second theorem disproves the second problem.

23. One should of course add Russell's paradox to this list.

24. This is an important distinction from the point of view of Badiou's later development. The operation of count-as-one is an operation that belongs to situations not being qua being as inconsistent multiplicity.

25. Badiou, "Mark and Lack," 167.

26. There is a powerful consequence of this method of stratification, which we will also hint in Part III. In set theory there is a clear distinction between construction of sets extensionally (by picking elements from other existing sets) or intensionally (by declaring a formula that defines a set). Russell's paradox shows the latter definition of sets to be inconsistent—which eventually resulted in having an axiom in set theory called the axiom of Separation. One of the consequences of Russell's paradox is that

there are many well-formed formulae for which a set cannot exist. At the surface this may look like that within logic we have the ability to produce things that do not exist, implying that logic may exhaust a greater domain than ontology. But the method of stratification clears this ambiguity. What matters to logic is the result of the last operation: operation of derivation. The seemingly larger domain is the outcome of the first and the second operations. Therefore, for example, an expression such as the self-belonging set, which is the basis for the Russell's paradox, is filtered out (actually as a non-well-formed expression during the formation operation).

27. We are following Badiou's definition of equal sign. A first order logic with equality is usually taking the equal sign as a primitive in the system and has a number of axioms associated with it, of which reflexivity is one of them (it has more axioms than the ones enumerated above). It is worth mentioning that in certain interpretations of first order logic equality may not be a primitive logical symbol. This logic is referred to as first-order logic without equality. If an equality relation is included in the signature, the axioms of equality must now be added to the theories under consideration, instead of being considered rules of logic. For example, there is no primitive = in set theory, that is, the equality of two sets must be defined based on the axioms of the set theory and $\in$ operator, which is its only primitive operator.

28. Badiou terms the functions or mechanisms of concatenation, formation, and derivation M1, M2, and M3 respectively. We shall also follow that convention.

29. We can also say that *Ar.R* means *R* is reflexive.

30. Badiou, "Mark and Lack," 170.

31. Badiou, "Mark and Lack," 171.

32. "Like Lacan's accounts of Gödel's theorem and the semantics of implication, Jacques-Alain Miller's discussions of Frege and Boole are ambiguous in that they combine, simultaneously and indistinctly, what pertains to the effective construction of a logical mechanism with what pertains to the (ideological) discourse through which logicians represent their constructions to themselves" (Badiou, "Mark and Lack," 165).

33. Badiou, "Mark and Lack," 173.

34. Hao Wang (1921–1995) was one of the few confidants of Kurt Gödel. He was also Stephen Cook's PhD thesis supervisor.

35. Badiou, "Mark and Lack," 171.

36. Jacques-Alain Miller, "Action of the Structure," Concept and Form Volume One, edited by Peter Hallward and Knox Peden, 69–84. London, New York: Verso 2012.

37. Badiou, "Mark and Lack," 172.

38. Eyers, *Post-Rationalism*, 87.

39. We shall return to this "liberating" aspect of axiomatic thinking when we review Albert Lautman's thoughts in Part III.

40. Badiou, "Mark and Lack," 173.

41. Badiou, "Mark and Lack," 174.

42. Leonard Lawlor, *Derrida and Husserl, The Basic Problem of Phenomenology*, Bloomington: Indiana University Press 2002, 62–63.

43. Badiou, "Mark and Lack," 175.

44. Badiou clarifies this in the following passage: "But that a crisis in the (ideological) representation of science can induce a (positive) reconfiguration of science itself should not surprise us, given that the material of science is, in the last instance, ideology, and that an 'a priori' science by definition deals only with those aspects of ideology which represent it in the latter: a science continually breaking with its own designation in representational space" (Badiou, "Mark and Lack," 165).

45. Badiou, "Mark and Lack," 175.

46. Badiou, "Mark and Lack," 176.

47. Lawlor, *Derrida and Husserl*.

48. Badiou, "Mark and Lack," 174.

49. Alain Badiou, "Theory from Structure to Subject: An Interview with Alain Badiou," in *Concept and Form Volume Two*, edited by Peter Hallward and Knox Peden, 273–290. London, New York: Verso 2012, 277–278.

50. Badiou, "Theory from Structure to Subject," 278.

Part II

SUBJECT AND REAL

INTRODUCTION

Based on the premises we've established in the previous chapter, our conviction is that the development of Badiou's thought during the 1970s goes through an itinerary that was already opened during the 1960s.

In relation to Miller, who presented a theory of structure in the presence of a subjective element, Badiou contrasted an immanent view of logic, mathematics and science. Miller's method of demonstration was to use logic as the emblem of scientific structuralism and to show how there is a fundamental, repressed element in logic, whose impossible status in structure is exactly what enables the latter's operations. Miller called this logic of interaction, between the structure and its repressed core, the logic of signifier, signifier being what stands for this original convoked and evoked impossibility of non-identity. Miller concludes that number theory must contain a repressed core whose disclosure requires analysis. Repression, however, is not forgetting, but rather something necessary for the production of meaning. Through misrecognition, what is repressed is sutured back to the linear movement of its progression, that is, the meaning it produces—its discourse. Hence it is the task of yet another discourse, a scientific discourse in its own right, to disclose the truth within a structural field such as logic. Miller called the latter the discourse of overdetermination—the discourse whose sole purpose is to un-conceal the truth of a given field, and by the same token, reveal its ideological designations. Miller named two such discourses, that of Marx and Freud, and, finally, a unitary discourse that regulates the transformation of these two with each other, the general theory of discourse, whose core is the logic of signifier.

In response, Badiou offered a different view comprising three crucial points. First, by showing that there is no underlying lack in logic, Badiou refutes the existence of a subject in logic. Logical production proceeds through stratification, not by being subjected to an initial inheritance from the empirical domain. Second, Badiou shows that this capability does not derive from logic being a tautology. In fact, logic has its own proper content, and there is a novelty in logic that by no means can be preordained or known at the outset.[1] Likewise, there is a sense of unpredictability in the Bachelardian and axiomatic concept of sciences, immanent to their proper content. Third, in tandem with their immanent content are the immanent methods of developing each science. It is only through the immanent formalism of their methods that sciences improve—something that may be less obvious in experimental sciences such as physics than in speculative sciences such as mathematics and logic, but many historical anecdotes show how physical theories develop through the pure formalism of their methods, for which experimental evidence only becomes available years after.

In summary, distinguishing between theory and ideology, Badiou proposes an *ontological* approach to science rather than the usual epistemological approach. The mark of this approach is a self-reference operating within the discipline of science, whereby the practice of science itself breaks off with internal ideological designations. What drives a given discipline forward is not the seeking of truth, but fidelity to a formalism that underlies this discipline and constantly divides it into the two of theory and ideology. As such, truth—at least truth as the correlation between sign and object—is not what concerns the scientific discipline.

The debate in this next chapter, however, is different. It is not about science but about Badiou seeking to revive, in the face of an anti-subjective, anti-humanistic scientificism dominating French academia from one end, and the ivory tower stance of *Parti Communiste Français* (PCF) from another, a theory of political inventiveness. What can immanently orient a political process towards a new order? We cover this itinerary not just because it produced *Theory of the Subject*, the work we read closely in this chapter, which has been called Badiou's "most passionate, most uncompromising and most revolutionary book," but because in the thoughts of this era, a certain problematic surfaces, this time within the political thought, that bears a close kinship to infinity (and subjectivity). We will see that political thought brings about a unique configuration of processes, what Badiou designates as subjective, under which the possibility of politics is rendered only if these processes can remain faithful to the emergence of a *new consistency*, and only if the anticipated novelty does not merely repeat the already dominant place of consistency, but continues as the thought of novelty and consistency *together*. We will show in subsequent chapters how the thought of politics as a consistent novelty is a thought whose axiomatic presentation is given by the axiom of Infinity: that is, the thought of politics is anchored in the thought of actual infinity, in the same way, the thought of (axiomatic) science is. The work of this chapter, building toward this thought, is to critique the metaphysical assumptions that underlie the dictum, which states that novelty and consistency are incongruent.

But before we embark on this journey, there is one question we should ask ourselves: Why the apparently "sudden" change of heart for Badiou? Why is he going from "the extreme point of strict formalism"[2] in logic and mathematics to political thought? And more importantly, what does this change entail in terms of the commitments Badiou developed through his formalist era, commitments that are significant to our project? Did Badiou end up abandoning those commitments, broadly conceived as the immanent disciplinary formalism which produces novelty?

Much has been said about how Badiou, after the events of May 1968, changed his primary focus from logic and mathematics to politics, and from theory to practice. It is under this that *Theory of the Subject* is seen as a caesura, a hiatus, in Badiou's career, but as he has said about Gödel, this change "does not express a hiatus so much as a reprise."[3]

May 1968, for Badiou and others of his political generation, brought a new condition to the alliance between students and especially the young workers, offering a transformative moment of hope in the politics of France and the possibility for a new engagement with collective action. The possibility of this alliance posed a question for political activists: What theory can explain how people break with their natural social places? For the students were out of place in a factory, and workers were out of place together with students. This was an intra-political, theoretical, rather than philosophical, question.[4] What was at stake was a theory that could show how ordinary people could leave their places in society and form a collective agency together with a new, common objective. The existence of such a theory within political thinking had no possibility, neither for Althusser nor for structuralist thinkers at the time. For the latter, political identity was always inscribed at the discursive level, and consequently, a collective act within two disparate universes of discourse was a nonstarter. For Althusser, the question of political intervention was itself an ideological inquiry, one that finds its true theoretical expression in the Marxist theory of class struggle as its structuring principle. As a result, the practice of politics must follow the theoretical, that is, the scientific destination. It is completely wrong to ask the question from the point of view of a practical contingent need. And the mere fact that this question is being asked, its spontaneity, so to speak, manifests an ideological disposition. This theoretical standpoint, which did not allow Althusser to see politics as an independent domain for either practice or thought, was criticized by Badiou much later as being the result of Althusser's suturing philosophy to politics.[5]

Badiou did not yet enforce a strong separation of philosophy and politics in those days, but it was quite clear to the activists of the time that it was not a philosophy that they were after. And yet the intense concern of those groups of intellectuals, among whom Badiou had found his political allegiance, regarding the unanswered theoretical questions, especially after repressive measures taken by the Gaullist state against the students and striking workers, contributed to a split between Badiou and his mentor Althusser at the end of the sixties. At this time, Badiou began to gravitate toward what he saw as the more practical activist politics of Maoism, in concert with other intellectuals such as Jean Luc Godard and Sartre.

The Maoism that Badiou believed in during the seventies was a politics oriented towards establishing a closer relationship between Marxist theory and practice and forming a more concrete ground for the clash between real social actors. For Maoism, the theory of political action was not a philosophical theory whose aim was to make conjectures about the historical development of societies, but a theory that was integrated within the political thought and act of the immediate situation. This suggests that, for Badiou, even then, politics was a discipline that had its own interiority independent of philosophy. This disciplinary nature was also present in Maoist political thought. More importantly, this disciplinary immanence is of the same nature as what Badiou, in his formalist years, had attributed to science and logic. In that sense, although the domain of thought was, of course, different, and was now oriented toward militancy rather than erudition, its method of thinking remained the same.

In *Theory of the Subject*, Badiou only names one person toward whom he wishes to express his indebtedness: Paul Sandevince,[6] whose real name we came to know afterward as Sylvain Lazarus. Hundreds of meetings and thousands of thoughts with Sandevince/Lazarus left their indelible mark on *Theory of the Subject*. Of course, we have no idea what exactly went on in those conversations, but a work published by Lazarus years later, *Anthropology of the Name*, shows in an explicit sense a taste of this indebtedness: that perhaps in very intense interlocutions, Badiou became completely convinced that politics is a discipline with its own proper interiority or immanence, with its own theoretical and ideological designations, and with its own formalist and intuitive methods. It would perhaps be beneficial to quote Lazarus, from the first paragraph of his work, in order to illustrate this point:

> The aim of this book is to clarify the subjective from within the subject, or, as I put it, "in interiority," rather than call on objectivist or positivist referents. Politics in this renewed sense will stand here as an exemplification of an approach "in subjectivity." Indeed, my fundamental thesis on politics is that it is of the order of

> thought. This is a thesis concerning the character of politics that is not reducible to any space, other that of politics and the need to think about it in its singularity.[7]

In another sense, Lazarus's work years later seems to suggest a programmatic approach to politics very similar to what "Mark and Lack" had suggested for logic and science, and in retrospect, we can detect that this affinity in method, between science as a discipline and politics as a discipline, is too close to ignore.

But exploring this point further, as interesting as it may sound, is beyond the scope of our immediate attention. What is important for us here can be summarized in a minor and a major point. The minor point is that Maoism, as prominent as it was in the thoughts that drove Badiou during the "red years," from the philosophical standpoint that this work pursues, constitutes an extrinsic and not an intrinsic cause: there are good reasons to think that Badiou, because of his earlier work, was predisposed to the type of politics for which Maoism stands. Moreover, the thought of absolute disciplinary immanence had already predisposed Badiou to the thought of development through the division of one into two—as he showed us how Gödel broke one of the axiomatic systems of thought into two of ideological anticipation of consistency in axiomatic systems (Hilbert program) and the logical truth of their inconsistency. This is a scholarly point that should be developed as a thesis in its own right (and not by the current work). The major point, however, is this: the thought of immanence in politics, regardless of being extrinsically (exposure to Maoism) or intrinsically (prior methods of thought, discussions with Lazarus) grounded, is the reason why *Theory of the Subject* is structured around interlocutions (and their consequences) with two masters: Hegel and Lacan. And this thought is also precisely the reason why Badiou can draw new consequences out of Hegel and Lacan.[8]

The present section is dedicated to the exposition of this major point, and for this, we will develop our work through a syllogism, with one axiom and three theorems in mind:

Definition: Discipline (as defined in "Mark and Lack")
Axiom 1: Something like discipline could exist.
Theorem 1: Science is a discipline (proved in "Mark and Lack")
Theorem 2: Politics is a discipline (proved by Lazarus)
Theorem 3: Any immanent system develops through the division of the one system into two (proved by "Mark and Lack" as well as Lazarus)

Syllogism:

- *Major Premise: Hegel is divisible into two*

- *Minor Premise: Lacan is our Hegel*
- *Conclusion: Lacan is divisible into two*

The following chapters are organized to present each step of the above syllogisms.

NOTES

1. As per the consequences of Gödel's, Turing's, and Cook's theorems.
2. Badiou, "Theory from Structure to Subject," 280.
3. Badiou, "Mark and Lack," 176.
4. Badiou describes this exact encounter very nicely in Alain Badiou, *The Communist Hypothesis*, London, New York: Verso 2010 58-59.
5. Alain Badiou, *Conditions*, London, New York: Continuum 2008, 160.
6. Alain Badiou, *Theory of the Subject*, London, New York: Continuum, 2009, xxxix.
7. Sylvain Lazarus, *Anthropology of the Name*, Kolkata: Seagull Books 2015, 1.
8. In this sense, Hegel, like Bachelard, introduced the method, Lacan, like Althusser, worked out the theory, and Mao, like Gödel, progressed the discipline from its interiority.

Chapter 2

Hegel Is Divisible in Two

What role did Hegel play in the transition between Badiou's critique of Miller's understanding of structuralism in the late 1960s to a theory of politics and history in the 1970s? The answer lies in the theory of interiority. One of the main facets in Badiou's critique was that Miller, like many other philosophers at the time, relied on a theory of knowledge, or epistemology, to fill what was perceived as a gap in theories of structure, to account for some primordial lack that underlies the possibility of any structure. All similar theories attempted to supplement structure with an element that does not belong to it. In the case of Miller, his critique of the ideological designation of science requires an element outside of the practice of science per se—that is, the subject—so that we could successfully haunt ideological tendencies built in the name of science. Science itself is only possible when this element is repressed and, at the same time, sutured, to its discourse. Miller showed how this repression and suturing operate in the field of number theory. Revealing this element supposedly provides a means for identifying ideological discourses.

To this approach, Badiou counter-posed a point of view with two characteristics. First, the ideological designations that matter to science are not built outside of science but inside it. As such, epistemological theories do not matter to the practice of working scientists, nor do philosophical or metaphysical theories about the foundation and consequences of a given science. But the beliefs that working scientists hold about what they do matter to them. So, the ideology that matters is interior to the practice of science. Second, the break with this ideology does not require invoking a theory outside of the science proper in order to tell it what to do in order to break free from the ideologies that are built inside it. What sciences do in order to break away from their ideological designations is simply to produce more science. It is the practice of science itself that does the trick. In that sense, sciences don't lack anything.

There is a lack of lack in science—which is a positive assertion—and thereby, there is no subject of science.

We use the term "discipline" to elicit this sense of interiority and juxtapose it to discourse, the term used frequently by Miller et al., that elicits a sense of exteriority. According to Miller, we need a different discourse, in fact, a theory of discourse, in order to identify ideologies in the sciences. We can formulate Badiou's rejoinder to Miller in the following way: ideological designations are not identified and purged through discursive critique but through disciplinary practice. "Mark and Lack" provided a thorough demonstration of this by focusing on one discipline studied and critiqued by Miller in his 1965 essay, the discipline of logic (or mathematical logic, as used by the number theory, to be more specific). Badiou was able to show, by using the operations that are specific to the field of mathematical logic, operations that he termed stratification, that there is no need for an extra-disciplinary intervention in order for a discipline such as logic to be consistent and complete.

But a problem arises here. Does the doctrine of the pure immanence of scientific disciplines result in avowing the One of a discipline? We know, through our close reading of "Mark and Lack" in the previous chapter, that the thought of One is not what Badiou advocates. Although a discipline stands for one thing, it is not a completely homogeneous entity. But how is it possible for Badiou to express this in his theory? We know that he used the concept of infinite stratification in order to show how a lack in a given stratum could be addressed by a supplement in a different stratum, but what is the exact theoretical underpinning of this stratification: that is, how is it that a discipline can allow for infinite stratification?

The above inquiry, which may point to a structural deficiency in the doctrine of disciplinary interiority, attains more relevance by a particular historical sequence: May 1968, which for Badiou inaugurated a shift of focus from science to politics.

We remarked earlier that Badiou counts himself heavily indebted to Sylvain Lazarus, not just because of their comradery in action but more so in theory. In his 1996 book *Anthropology of the Name*, Lazarus provides a powerful argument for politics as interiority. Lazarus develops an idea of politics that is post-classist and presents politics as an engagement that is not "objective." That means politics does not have an object (e.g., the State or revolution or class); it is (purely) thought, and it is (purely) subjective. He captures the interiority of politics through two statements that are tightly intertwined: *People think*, and *Thought is a relation to the real.* These two statements, in relation to each other, carve out a space for politics as pure immanence, something that is designated to the space of the subject with no object. The statement "people think" designates that politics is of an order of thought, and there is something out there—people—that carries out this thinking. Unlike political science in which the sociologist thinks or in the classical political theory that the party thinks, in Lazarus' model, it is people who think—and in this case, the word "people" has such a minimum determination that, in fact, the statement "people think" effectively means that "there is thinking." So, the practice of politics is inaugurated by the activity of thinking on the side of people, and people always think.

The second statement, "thought is a relation of the real," designates the non-objective aspect of politics. Thought is a relation "of the real" and not "to the real" because a relation to the real makes the real an object for thought, and politics has no object. So, by thought being of the real, Lazarus is emphasizing this aspect of politics that it is of the order of thought ("people think"), as this order exists in the real ("people think" is in the real). It is interesting to also pay attention to the choice of name for the theory that expounds the idea of politics as interiority, which starts with "anthropology" in its title. Lazarus states that the choice for anthropology, as opposed to, say, science, is due to the investigative aspect of anthropology, the investigative aspect that is not preordained by binding to an object, unlike sciences. Sciences are objective because they are bound to something that is not interior to their thinking. Therefore, the choice of anthropology vs. science emphasizes the non-objective nature of the theory that demonstrates politics as interiority. But what about the other part of the title: "of the name"? What does "the name" nominate? Why not "anthropology of politics"? The latter title ("anthropology of politics") suggests that politics is a name about which this theory is performing investigative work. But politics is an unnameable name. Why? Because politics is singular, and something singular and interior is unnameable: something cannot name itself from within itself. Naming politics would require an outside of politics that speaks about politics. Naming also means that politics is co-thinkable; that is, there is a more general field of discourse that uses the name for circulation and conceptual exchange, assignation and polysemy, resulting in de-singularization of the name, both properties that are disavowed by politics as interiority and singularity. This theory speaks to politics from inside the thinking of politics, and within politics, politics itself is unnameable. Hence "of the name" simply designates the topic—politics in its interiority—as an unnameable name.

The classical conception of politics as an objective enterprise is under the thought of totality. The objective field of the classical theory of politics is a heterogeneous field, as heterogeneity is the nature of the objective—as it is grounded on the division of subjective and objective. Hence politics in its classical register is a totality that "proposes to co-think the whole, that is to say, to surmount the heterogeneous character of the objective and the subjective."[1] This results in subjective such becoming unthinkable. Why? Because subjective, in this case, will be only thinkable in relation to objects, not as such, not as a purely subjective process, which in his book he supplements by an axiom, which states the subjective process as such is thought.

If we assume that Badiou subscribed to the interiority of politics in the early 1970s, the aforementioned structural deficiency must have been further compounded. Why? Because not only must we have a theory capable of expressing the interiority of scientific disciplines, but this theory must also be sufficient to account for the interiority of politics.[2]

This is where Hegel, as *the* philosopher of interiority, comes into the picture. If we want a thinker of subjective interiority with no necessary interplay with the other, that is, a thinker of interiority without exteriority, then that thinker would be Hegel. Our proposal is that Badiou's allegiance to Hegel at the time of *Theory of the Subject* is much more than what his Sartrean or Marxist backgrounds would donate: this allegiance is more strategic and structural rather than derivative. Hegel provides the conceptual framework to think and articulate interiority in a way that the Badiou of "Mark and Lack" was not able to: as a single, unified theory spanning heterogeneous disciplines.

At the end of "Mark and Lack," one question remains unanswered: Where is Badiou speaking from when he speaks about disciplines that are not ideological in nature? The implied theory has not yet been spelled out. We know some properties of this theory. First, it is not metaphysics: the notion of interiority should not espouse the concept of a self-evolving One. Second, it is not within the structuralist horizon. So, where is it? The issue of immanence and interiority, which was left open, and the ensuing necessity of constructing a unified theory across science and politics, which is neither metaphysics nor structuralism, leaves open the unique place from which Badiou could arrive at the concept of structure. How could he arrive there? This is the problem for which he turned to Hegel for a resolution.

Badiou needed to be in a very special place with respect to the French structuralism of his time. He needed to be beyond metaphysics but at the same time *prior* to the concept of structure. At the end of "Mark and Lack," he tells us that Miller presupposes too much about the concept of structure. He goes on to assert that only historical materialism can truly say what a suture to a structure is at a given moment.[3] This would suggest a moment that is prior to the concept of structure. Hegel's *Science of Logic* (*Logic*)[4] allows Badiou to grasp such a moment, since by beginning at the category of Something in *Logic*, as we shall see shortly, Badiou is not beginning at the level of the metaphysical Hegel, rather he develops the dialectic of place and force (according to the first division of Hegel), which is present in *Logic* but operates under the cover of Hegel's metaphysics. The interplay between place and force is what gives Badiou a conditioned logic of structure. So, strategically aligning with Hegel will give Badiou a prior footing that allows him to think about the problem of structure. This tool actually allows him to speak about structure and membership without even committing to structuralism per se—to remain at the level of a semi-topological language. Badiou succeeds in localizing this new position prior to the structure by theorizing structure as the interplay between place and force. The structure becomes here the regional thinking of limits of force within the place. This interplay of place and force is such that Badiou does not have to commit to one relation between them. Instead, this relation designates a more general theory than structuralism; structuralism would be one particular relationship between place and force.

As the following will show, using Hegel proved to be more laborious than one could have imagined, and this is due to the effort to avoid Hegel's idealism. A rational theory of interiority in Hegel is intermixed with Hegel's metaphysical enterprise, at the center of which is the priority of the thought of the One. In order to get to the rational kernel of his master, Badiou had to go through a process he refers to as splitting Hegel. We will show this process in three stages:

1. Hegel *can* be split: there is a rational kernel in Hegel's logic—in evidencing the possibility of dividing Hegel, this chapter will show that there are indeed two dialectical matrices at the beginning of Hegel's *Logic*, one metaphysical and the other rational, captured in the dialectics of One and Two.
2. Hegel *must* be split once—in investigating the aforementioned rational kernel, we discover a new grammar required for differentiation: the grammar of place and force.
3. Hegel must be split twice—in investigating the interplay of place and force, we encounter an indeterminate path of circularity captured by the operation of periodization.

Badiou performs the above three operations by closely reading the first 135 pages of Hegel's *Logic*[5] in his commentary to *The Rational Kernel of the Hegelian Dialectic* (*The Rational Kernel*)[6] (categories of Being-Nothing and Becoming, and the beginning of the category of Determinate Being) and *Theory of the Subject* (categories of Something, Limit, Finite and Ought).

This chapter will present Badiou's theory of force, a necessary step toward discussing the subject and Badiou's critique of Lacan, which is the topic of the next chapter.

TWO DIALECTICAL MATRICES: UNCONSCIOUS MATERIALISM OF HEGEL

We should remind ourselves of how fundamental the thought of One is for Hegel. It is the original status of this thought that underlies other fundamental features of Hegel's logic, such as placing the category of quality prior to quantity. This is so because it must be possible for Hegel to make the claim that One is a qualitative feature of Being and not a quantitative feature. As a corollary to the original status of One, we can assert that there is no Other as the radical otherness to One; for radical otherness, since it stands in absolute otherness to One, cannot be derived from One; if there were such radical otherness, then we would have Two (One and Other) as a more original thought. Furthermore, there is a difference between other and Other, in which only the former is derivable from One. Thus, the derivation of quantity from quality in *Logic* is, at the same time, on a smaller scale, the derivation of other (and others) from One. Another important feature of Hegel's logic is that the thought of One dictates the method of its progression. It is not the case that the thought of One and the thought of method, that is dialectic, are two different things which were later merged. In fact, it is completely legitimate to claim that there is no thought of dialectic for Hegel separable from the thought of One, and it is the latter that grounds the advent and the laws of the former. As a result, there seems to be no thought more original than the thought of One in the entire Hegelian logic. Yet there is another consideration that comes before the thought of One in his system: the thought that philosophy, or rather logic, as the science of pure knowing, must be without any presupposition.

> But if no presupposition is to be made and the beginning itself is taken *immediately*, then its only determination is that it is to be the beginning of logic, of thought as such. All that is present is simply the resolve, which can also be regarded arbitrary, that we propose to consider thought as such. Thus the beginning must be an *absolute*, or what is synonymous here, an *abstract* beginning; and so it *may not presuppose anything*, must not be mediated by anything not have a ground; rather it is to be itself the ground of the entire science. Consequently, it must be purely and simply *an* immediacy, or rather merely *immediacy* itself. Just as it cannot possess any determination relatively to anything else, so too it cannot contain within itself any determination, any content; for any such would be a distinguishing and an interrelationship of distinct moments, and consequently a mediation. The beginning there is *pure being*.[7]

Part of the requirement of no presuppositions pertains to the problem of "beginning" (*Anfang*) itself. The concept of beginning brings an additional specification that Hegel would like to do away with. Beginning implies starting with a minimal premise, or nothingness altogether. For Hegel, such a concept of beginning is paradoxically too complex to serve as the beginning of thought. The naive concept of beginning is that of "nothing from which something is to proceed."[8] It thus takes for granted that what is being thought is the beginning of something yet to emerge. At the start of philosophy, however, Hegel does not assume that there will be anything beyond what we start with.

All we may understand there to be is sheer indeterminate being, which may or may not prove later to be or to have been, the beginning of something more. This is perhaps why Hegel's account begins not even with a full sentence, but a sentence fragment: "being, pure being, without any further determination."[9]

So, it is the demand to start with no presupposition, with even the expectation of being the "beginning" of something else extracted, that leads to the thought of pure being as the absolute beginning of logic. But without further qualification in *Logic*, why must the thought of pure being coincide with the thought of One? For Hegel, the thought of the pure being is a singular thought, and therefore it is quite legitimate to think of the pure being as One.

It is beyond the scope of this chapter to show that a significant part, if not the most significant part, of Badiou's philosophy is to break the relation between the beginning of philosophy and the thought of One. This is the ground for the revolutionary idea that the logic of being is not, in fact, the logic that is traditionally developed by, or associated with metaphysics. The logic of being is rather captured by mathematics, hence the formula "ontology = mathematics." These are among the topics that we explore in part III. But for now, we want to emphasize the fact that for Hegel, to begin as a science, logic must start with the thought of One, that is, the pure being as one.[10]

Having established the beginning, the great *aporia* in Hegel's *Logic* is how, from One, can we derive multiple (or simply, two)? How does the One beget many? This is the very place where Badiou's investigation of Hegel begins, both in *Theory of the Subject* and in *Rational Kernel*. Badiou mentions right at the beginning of *Theory of the Subject* that, while nothing can be said of One without already engaging qualitatively, Hegel "is at pains to engender the multiple, the denumerable, insofar as his idealist propensity pushes him always to obtain everything on the basis of a simple term."[11] Inasmuch as deriving One from multiple is difficult (e.g. the jump from quantitative to qualitative), it is far more difficult to derive multiple from One—"to prove the marvels of nature by God is much more complicated since God is necessarily the marvel of marvels."[12]

The short answer is that, for Badiou, Hegel doesn't derive two from One. The beginning of *Logic* involves a "smokescreen" and "sleight of hand," to create the illusion that we have accomplished the derivation of two from One. However, at the end of the first cycle of the progression, when we arrive at Becoming, right before Determinate Being, we have not accomplished this task the way Hegel wants us to believe.

Let's have a look at Badiou's comment in *The Rational Kernel*, which involves Badiou's remarks on the very first categories of *Logic*, Being, Nothing and Becoming. As we know, *Logic* starts with the concept of the indeterminate or pure being, the category of Being. This concept is so indeterminate that it will vanish into the concept of pure nothingness, into the category of Nothing, and from there, we arrive at the category of Becoming. But as Badiou remarks, Becoming can only be legitimately derived if Being and Nothing keep vanishing into one other, and this could only happen if they are qualitatively the same. In that case, Becoming stands for an iterative process that elicits the sameness, and not the difference, of Being and Nothing. This process at the beginning of *Logic* does not represent a dialectical progression. Why? Because there is nothing new, no qualitative other, no new truth that takes place as its result. We are basically positing a concept, which due to its minimal determination, is effectively the same as not positing anything at all. Put differently, Becoming is no more determined (or in-determined) than Being.

This is what Badiou rightly calls the problem of "commencement" in *Logic*.[13] In the quote from the beginning of *Logic*, Hegel's condition for beginning is an absolute/abstract/pure beginning with no determination and no mediation. The thought, in this case, is equivalent to having no thought and having no thought is equivalent to having this thought. So, the beginning posits a thought that is effectively equivalent to positing nothing. But at the same time, nothing is equivalent to positing the abstract thought of the pure being. This is the problem of idealism—the problem of thinking pure being according to the metaphysical category of One. If Hegel were less rigorous and had started with a particular determinate content (substance or monad, for example), rather than an objective determination (thought, intuition, sensation ego, subjectivity, etc.), then the problem of commencement would be either less severe or non-existing—as the case had been for philosophers before him (Spinoza, Leibniz, or Kant for example).

What is this problem of commencement that Hegel is facing? It is the derivation of two from One: thinking the abstract beginning under the category of metaphysical One does not provide any leverage for thought to derive multiplicity from the purity of One. Nothing is not an other to Being, and Becoming is a name referring to their sameness, so we have three marks for the same pure, abstract principle. In other words, it is the sameness of the concepts of Being and Nothing that makes this process work—and if the dialectical process is to engender an other out of what has been posited, based on the contradiction within it, this process proves to be non-dialectical: there is no difference, no new determination, no two.

Now, according to Badiou, Hegel was aware of this problem of deriving two from One and stressed that, although Becoming is derived from the sameness of Being and Nothing, we should still consider them different. Here is what Hegel says:

> *Pure being* and *pure nothing* are, therefore, the same. What is the truth is neither being nor nothing, but that being—does not pass over but has passed over—into

> nothing, and nothing into being. But it is equally true that they are not undistinguished from each other, that, on the contrary, they are not the same, that they are absolutely distinct, and yet that they are unseparated and inseparable and that each immediately *vanishes in its opposite*. Their truth is, therefore, this movement of the immediate vanishing of the one in the other: *becoming*, a movement in which both are distinguished, but by a difference which has equally immediately resolved itself.[14]

Despite being the same, Hegel enjoins us to think that Nothing is the other to Being, and vice versa. Being and Nothing are utterly different from one another but collapse logically into one another because of the indeterminate immediacy of their difference. The fact that Being and Nothing are indistinguishable in their immediate difference is evident in their immediate disappearance into one another. On the other hand, the fact that they are immediately different in their in-distinguishability is evident in the immediate disappearance of each into the other.[15] As such, there seems to be a double relationship between Being and Nothing: they are the same, so positing one is equal to positing the other, like two marks pointing to the same content; but at the same time, positing one causes its own disappearance and the appearance of the other. Becoming is the category that watches over this double relation. If positing Being did not cause the disappearance of Being into Nothing and vice versa, that is, if there were no difference between Being and Nothing, the way the iterative conception of Being and Nothing suggests, then there is logically no room for a separate category of Becoming—Becoming would merely be a mark of the iterative alternation of Being and Nothing.

But, the subtle point that Badiou is evoking in his very concise remarks seems to be this: if we have to think about the difference between Being and Nothing, which he agrees with Hegel we should, this difference cannot be produced out of the category of One: One only begets One—as Hegel himself showed, by going from One of Being to One of Nothing and vice versa, naming this alternation the One of Becoming. In the alternation of Being and Nothing, there is no other. If there is an other, it is under a different category than One, not One itself. Pure being is different from pure nothing, but that means that they are two things rather than one. The category of One cannot circumscribe both. Nothing cannot be derived as the other of Being through the arguments that Hegel provided in support of their sameness.

So where is the second relationship coming from? Badiou suggests that there is no place for the second determination/relation other than simply positing/forcing it. Referring to the passage we quoted from *Logic*, Badiou emphasizes how Hegel forces the second relation in which Being and Nothing stand as other to each other: Being and Nothing indeed stand in opposition to each other, but the point is that this opposition cannot be derived in continuity from Being to Nothing. This opposition is assumed, not derived.

What is derived is rather their sameness. Hegel derives Nothing on the basis of its sameness with Being and subsequently "reminds" us of their difference. This intervention is what Badiou refers to as Hegel's forcing, whose consequence is a secondary category of Becoming, one in which both Being and Nothing are simultaneously present. Therefore, continuity (One) is derived, but simultaneity (two) is forced. From Badiou's perspective, this marks the genesis of the simultaneity of Being and Nothing in the entire Hegelian logic.[16]

That Hegel was aware of this is evidenced both immediately prior to the start of the first chapter and also in part C ("Becoming"), section 2 ("Moments of Becoming"), as well as section 3 ("Sublation of Becoming") of *Logic*. In all three sections, Hegel expresses the double determination of Becoming: Becoming of succession (of Being and Nothing iteratively) and Becoming of contradiction (of Being and Nothing simultaneously). In moments of Becoming, the category of Becoming will watch over two moments of ceasing-to-be (Being) and coming-to-be (Nothing) simultaneously, not just as vanishing terms but as moments of Becoming—the language of moments, which implies stable determination, begins to take over the language of vanishing, succession, and continuity. As Badiou mentions, this will mark the beginning of categories comprising contradictory moments, and the eruption of these contradictions is what dominates the rest of *Logic*.

The point is this: Hegel starts with pure being clearly as a variation of the One, and yet, by the time we arrive at the sublation of Becoming, we have two stable and contradictory moments carried forward for the rest of his logic. But two was never derived from One; Hegel posited it in such a way that he can claim them under One again, as a double determination: continuity (One) and simultaneity (Two).

So, when Badiou says, "Hegel in any case is at pains to engender the multiple, the denumerable, insofar as his idealist propensity pushes him always to obtain everything on the basis of a simple term," he must have in mind the foregoing attempt by Hegel to somehow include the determination of two under One, and when Badiou refers to this attempt as a "smokescreen," he is referring precisely to the moment when that the contrariness of Being and Nothing is posited and brought under the One.

This should give us some hint as to why *Theory of the Subject* does not start with the categories of Being-Nothing-Becoming, nor even Determinate Being, but with the category of Something. This helped Badiou to avoid the issue of "commencement" and postpone it to some other time.[17]

Badiou elicits self-relation in the category of Something in his analysis of Hegel. He will later name this the operator of count-as-one. We can map the category of Something in Hegel to what Badiou calls, in his subtractive ontology, a situation. On the basis of self-relation, The category of

Something or a situation does not require anything external. We will return to this point when we review Badiou's critique of Hegel's dialectic in the subsequent sections.

Hence, logic starts with a notion of the immediate, indeterminate being, but speculation into this concept will lead to the realization that the truth of the concept of immediate being is immediate nothing, and vice-versa. In this way, logic arrives at the notion of Becoming. Further speculation discovers the truth of Becoming to be the Determinate Being, which is so because, in Becoming, we observe the coexistence of the two. The speculative progress from one logical concept to another is based on discovering the truth of each concept. It is not due to an intervention of any kind that we discover the truth of Being to be, in fact, determinacy, the truth must already be embedded in the concept of pure, immediate being. By the time we reach the category of Something, we notice that the opposition of terms performs all the heavy lifting at the beginning of *Logic* (and, for that matter, throughout). From Being to Nothing to being and non-being, to Reality and Negation, and, finally, in Something and Other, the positing of each term also posits its opposite, and it is never the case that a term stands alone: "It is Two that gives its concept to the One and not the other way around . . . there are all sorts of contortions on Hegel's part that serve to mask this recognition."[18]

This means that Hegel's *Logic* is not one but, in fact, two dialectical matrixes:

- *Dialectic of One:* this is the process of One time, two
- *Dialectic of Two*: this is the process of two times one

Both dialectics are present in *Logic*, with the Dialectic of Two as the main motor of progression, and the Dialectic of One as what maintains the impression of the overall system of thought—the ideal of Spirit. Both dialectics have the same structure: that which is posited and the operation that ensues. In the Dialectic of One, One is posited as the "absolute" or what is synonymous, as "an abstract beginning."[19] With One given as such, dialectic will perform the operation of doubling (two) of the same One, it repeats (doubles) One: one is the qualitative and two is quantitative.

In the Dialectic of Two:

> The analysis of the beginning would thus yield the notion of the unity of being and nothing—or, in a more reflected form, the unity of differentiatedness and non- differentiatedness, or the identity of identity and non-identity. This concept could be regarded as the first, purest, that is, most abstract definition of the absolute—as it would in fact be if we were at all concerned with the form of definitions and with the name of the absolute. In this sense, that abstract concept

would be the first definition of this absolute and all further determinations and developments only more specific and richer definitions of it.[20]

The fact that in this dialectic, what is posited is Two (being and nothing, identity and non-identity, etc.), and the operation is to make one out of the two (or to make unity out of the opposites) is also evident in *Logic*. In this way, one is derived, the outcome of an operation; Two is what there is—as the most abstract definition of the beginning. In this case, two are qualitative and one quantitative since it is the result of an operation that counts the two under one.

It is important to note the difference between the posited One and the derived one of the two dialectics. The posited One is substantial, whereas the derived one is structural: the former is a substance such as God, matter or spirit, whereas the former is a structure with parts such as the universe or whole. The difference between these two will prove to be of utmost importance to the overall strategy that is pursued in *Theory of the Subject*. We will see that later in this section.

The first dialectic presents the logic of repetition, "condemned to the sterile operation of a coordinated iteration."[21] It attempts to derive two from one, and under the dominance of the metaphysical category of One, it thinks of progression as repetition rather than novelty and thereby represses the innovative force of dialectic, whose sense of inventiveness is procured from two (contradiction). It presents the progression as succession. The second dialectic presents the logic of contradiction, which stands for "an impossible passage from the successive (being-nothing) to the simultaneous (being and nothing)."[22] The progression, according to the former, is iteration or repetition, and according to the latter, is the relation of contraries—annihilation (Being and Nothing), coexistence (being and non-being) and mutual determination (asymmetrical: Reality and Negation, symmetrical: Something and Other). In the Dialectic of One, one is posited, and two is derived as the result of the operator of the succession. In the Dialectic of Two, two is posited, and one is derived as the result of the operation of unity of contraries.

So, what is Hegel's position with respect to these dialectics? Hegel's text is amazing in this regard. It interlaces Being-Nothing and Being and Nothing at the beginning of *Logic* quite often. But by the time it reaches the category of Becoming, there is another contradiction that is evident between the aforementioned dialectics. Becoming is not only the sublation of Being and Nothing under the law of the one (Determinate Being), it is also the sublation of iteration and contradiction (Dialect of One and Two, respectively) under the law of the one (iteration, Dialectic of One). The sublation, therefore, sublates not only two under one but also contradiction under iteration. Why? Because, as we've claimed, for Hegel, the category of One is the fundamental category of thought. At the beginning of *Logic*, Hegel emphasizes that the beginning of logic must have objectivity:

> The *principle* of a philosophy does, of course, also express a beginning, but not so much a subjective as an *objective* one, the beginning of *everything*.[23]

Despite the aforementioned "objectivity," Hegel makes the silent subjective choice to think of the beginning under the category of One (Being), to suture the beginning to the category of One, a choice that, under the category of Becoming, is manifested as the sublation, or (as Frank Ruda has offered[24]) repression, of contradiction under iteration. But "*Hegel's greatness for Badiou is to constantly point to that which he represses*."[25] Hegel, as we've seen, keeps returning to a contradiction in one of its forms in every category of *Logic*. The constant return of what is repressed manifests, or resembles, the unconscious operation, which leads us to conclude, as Ruda did, "For the early Badiou, *Hegel is an unconscious materialist as his thought embodies the contradiction between contradiction and the perpetual defense against it*."[26]

What does Badiou's choice of Something rather than Being as the starting category imply for the interiority of a discipline in general and the discipline of politics in particular? We think this move allows Badiou to preserve the dialectic of interiority (as per the theory of immanent contradictions) while decoupling it from the One of being (as per the problem of the indeterminate as the immediate). In *Theory of the Subject*, despite lacking a theory of beginning, Badiou nevertheless has the correct intuition to avoid the metaphysical stance that plagued the dominant theories of beginning. It is only later, in *Being and Event*, that Badiou is able to work out a non-metaphysical theory of beginning, which is the focus of part III of this work.[27]

IDENTITY AND PLACE: SPLITTING HEGEL THE FIRST TIME

The two dialectical matrixes presented above inform our understanding of Badiou's division of dialectic into alienation and scission at the beginning of *Theory of the Subject*. The upshot is this: for Hegel, there is only one quality: pure being—Nothing is not another qualitative state opposing pure being, that is, lack of all quality altogether per se, which is a qualitative difference. In other words, we are simply dealing with one qualitative category (or simply One), and there is no qualitative differentiation possible between various categories of logic. So, what can give us the license to think about the differentiation between Something and Other, or even Being and Nothing for that matter, if they are not qualitatively different? This question would be easy to answer had we posited two: that is, two qualities—Being and Nothing as two differentiated qualities, for example. Then all the differentiations of subsequent categories would borrow from this initial difference. But that is not Hegel's answer in *Logic*. The question Badiou poses is based on the Dialectic of One, involving succession, iteration, and repetition: how is positing Something and Other even possible—or Being and Nothing, for that matter? That is, if there is no qualitative difference, then what does it truly mean to say Being and Nothing are different, even to the minimal degree according to which they vanish into each other? And although this question is valid even for Being-Nothing under the Dialectic of One, it has a much higher relevance when it comes to Something and Other, for unlike Being and Nothing, they coexist and mutually determine each other. So once more, the question is, at the dawn of qualitative, or in the situation that "alterity has here no qualitative support,"[28] how is it legitimate to speak of Something and Other? What is the statement of their difference? Or in other words, if what is posited is One, how does the succession function that repeats the same One, repeat it as recognizable and differentiable two (i.e., Something and Other)?

It is important to emphasize again that it is only by looking at Badiou's foregoing critique of Becoming that we may fully appreciate his strategy to start from Something. Badiou recognizes the issues pertaining to the beginning in *Logic*, and by starting from Something and Other, he could avoid the problems of the beginning, and the metaphysical commitments entangled therein.

Badiou's answer is that it is impossible and, therefore, illegitimate! Hegel only makes it look possible because "there are all sorts of contortions on Hegel's part that serve to mask this recognition,"[29] allowing him to maintain the appearance of a simple succession.

The minimum requirement for granting repetition is that the repeated thing must appear in a different "place." In that case, "*This* only differs from *that* by the statement of the difference, by the literal placement. One could name this infamous stasis of the contradiction the indexical stasis. There is A, and there is A_p."[30]

This approach recognizes two things: an identity (*A*) and a space of placement, or simply a place (*P*). An identity has the power to posit itself, at a distance from itself, and a place has the capacity to accept the placement. Here is where we finally encounter the two that is qualitatively different: identity and place, and this, according to Badiou, is the minimum division that must be admitted if we want to speak about something. So, the true contrary to Being, to any posited identity in fact, is not Nothing, but the place of positing *P*. Therefore, when Hegel posits Being, the hidden assumption for this positing is that there is somewhere that this identity has posited—and this is not Nothing, which as the repeat of Being is the different thought of the same faint mark of the Being as the placed identity that is almost equal to not positing anything. Therefore, nothing finds two senses in *Logic*: the sense attributed to Nothing, the repeat of Being, and the sense according to which there is pure nothingness—lack of any identity, pure place in contrast to pure identity. Nothing is the thought of a mark so faint that it is analogous to not positing anything at all, whereas the nothing of place is of a different sort—it is a qualitative difference that serves as the ground of any marking. Place constitutes the determination that an identity receives. Every identity is a placed identity, and as such, every identity is a determinate something. In this sense, every posited identity is conflicted and split: according to its pure self and the place of its identity, the split between *A* and *A* according to its place in *P*, the existent and its place. Badiou presents this in the following definition:

$A = (AA_p)$

The above definition means that everything is split and is made of an internal contradiction or scission of what it is and how it is determined by where it is placed, in *P*. Additionally, unlike what Hegel presents, none of the contradictions within the progression of logic really matters to the dialectical process at the end. They all serve as a diversion that prevents us from seeing the true contradiction between identity and place—the contradiction that indeed does all the dialectical work, and what is truly behind all the change. We see that clearly in the opposition of Something and Other. Contradiction is not the relation that the terms of logic have with each other (being and non-being, Quality and Negation, Something and Other, etc.); contradiction is in the relation of term and place. Hegel is, therefore, right to say that each of these terms is truly the repetition of its predecessor. Hegel is right to say that each term emerges from the previous in a chain of succession. Hegel is right because the true difference is not coming from the thinking of identity alone (as Hegel postulates) but from the kind of impact that the space of placing has on the identities, that is, the categories of logic.[31]

So, when we say that everything is split (AA_p) we are not yet speaking of the determination of a particular term or thing. Instead, we are claiming that every concept comprises something that is, *A*, and the fact that it is posited A_p. This is so because every thought contains two things: an identity and a space of placement of that identity. The definition (AA_p) created by Badiou is, in fact, a shorthand for the primordial contradiction of identity and space of placement. Hence, all the beginning categories of *Logic* that we have covered should be rethought according to this schema, and when we do that, we find out that every category is a unity over a split identity. In particular, this lets us read anew the moment of determination, as in Determinate Being or Something whose singularity of its existence, is not, of course, due to *A*, the generic term closed in on itself, indifferent to any dialectic. It is rather A_p, that is, A according to the effect of the place into which it is inscribed.

Based on the definition of Something as (AA_p), Badiou further defines two operators: the operator of determination $A_p(AA_p)$ and the operator of limit $A(AA_p)$, representing the operators of unity that make One from two. Accordingly, every determination or limit is a unity of contraries and is the result of an operator of unity. This harkens back to the second dialectical matrix we formerly articulated:

- *Dialectic of Two:* which is the process of two times one

We have posited two (AA_p) and derived one through the operators of unity $A_p(AA_p)$ and $A(AA_p)$. Having thus defined the operator of unity in formulaic form, Badiou uses this form as an apparatus to derive other dialectical concepts (see figure 2.1). The first axiom is that the formulae above are distributive. That is, while the overall formula stands for the unity of contraries, the operators of A and A_p can be applied to both terms of the scission A and A_p.

$A_p(A)$ would stand for determination proper, that is, the effect of P on A as the result of positing A in the space of P. $A_p(A_p) = P$ stands for relapse into the general space: nothing will have taken place but the place—the principle of structuralism in all forms.[32]

By changing the operator from A_p to A, we derive new formula: $A(AA_p)$. This formula is called the determination of the determination by which force reapplies itself to that from which it emerges in a conflictual way. Similarly, this operator is also distributive over its terms: thus $A(A)$, the simple reaffirmation of the identity, against (but outside of) determination, parallel to the replacement into P, and $A(A_p)$, which is the operator of the limit (see figure 2.2). While determination is the effect of the place P on the identity of A, that is, how A is changed by belonging to P, limit determines to what extent A remains the same despite belonging to P.[33]

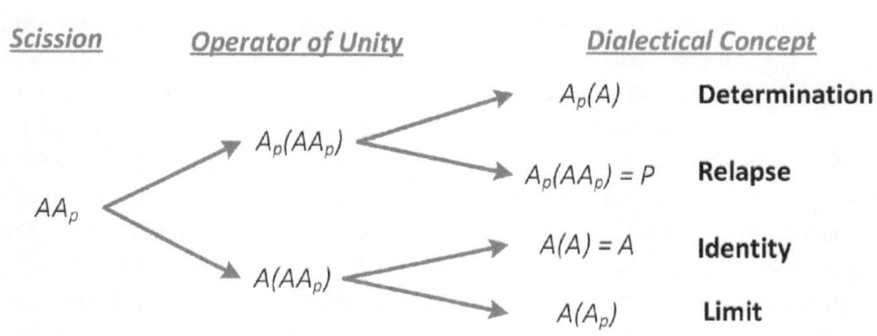

Figure 2.1 Scission and Operators of Unity. *Source:* Author created.

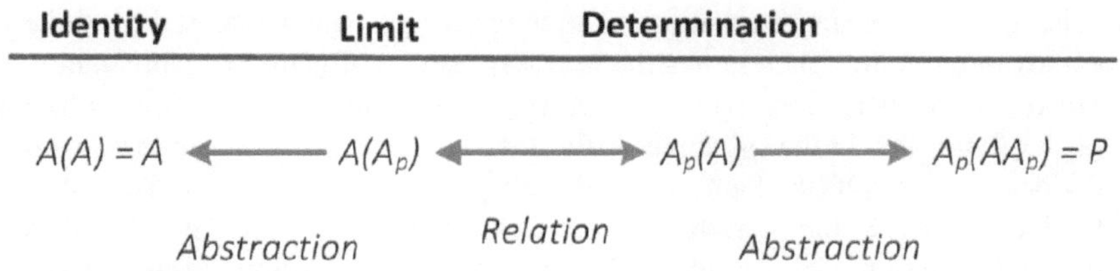

Figure 2.2 Operators of Unity and Deduction. *Source:* Author created.

This sequence manifests that positing a scission as the beginning of logic (Badiou only commits to determination and not the prior categories, and he thus starts with Something) and deriving unity through an operation may indeed manifest the unconscious function in the materialist Hegel. The scission is not a scission of terms, but a scission of identity and place in each term. Each category of logic is indeed a scission comprising an identity and the space of placement. The unity of each category is a derivative of two functions that determine to what degree the identity has limited the effects of the place, and to what degree the place has determined the identity of that category. Although Badiou only shows this for the category of Something, showing how thinking Something and Other in *Logic* would only make sense if we adopted the Dialectic of Two under the strict assumption of scission and derived unity, we believe that a similar analysis may be performed for all other categories of *Logic*. But this task is beyond the scope of the current writing.[34]

CIRCULARITY AND PERIODIZATION: SPLITTING HEGEL THE SECOND TIME

What the previous diagram elicits is that neither determination nor limit allows the dialectical relation to exceeding the place or the identity. In a sense, both place and identity are boundaries according to which any determinate something is placed: A and P sketch the boundaries of $A(A_p)$ and $A_p(A)$ so that they always return to (AA_p). We have already seen an example of complete determination in the earlier chapter of this work: structuralism is the theory that is based on the complete determination of any member that belongs to the structure. In this case, the structure is the place, and a member is the identity placed in the structure. Therefore, a complete determination means that an identity placed within the structure will reinforce the structure itself—$A_p(A_p) = P$. For the opposite situation, limit, Badiou provides an ingenious example in *Theory of the Subject*: Christianity, especially how Hegel interprets it. Take P as the place of finite and A as God or the infinite, the radical out-of-place. How could the relation of the infinite and finite have any meaning, let alone be dialectical?

> What gives it meaning is its historicization in scission, which makes the infinite ex-sist in the finite. Therein lies the necessary stroke of genius of Christianity. For this to happen, God (A) is indexed (A_p) as specific outplace of the splace of the finite: this is the principle of the Incarnation. God becomes man. God divides into himself (the Father) and himself-placed-in-the-finite (the Son). A is the Father, and A_p the Son, that historic Son by whom God ex-sists. God thus occurs as scission of the outplace, $A = AA_p$, God = Father/Son, a scission that the Council of Nicea, the first of the great modern politico-ideological conferences in history, will designate as sole existence—as unity of opposites—in the well-known dialectical axiom: "The Son is consubstantial with the Father."[35]

This example is very illuminating for Badiou. In his view, it displays the schema of the dialectic movement that Hegel is setting up in *Logic*. What is so crucial about this schema is the fact that it is circular: at the end of it all, we obtain only the pure scission of the Father and the Son as the integral concept of the redemptive absolute. $A(A_p)$ ultimately does not exceed $A = (AA_p)$: it only leads back to it. An identity, the Father, is placed in the world as the figure of the Son, who is resurrected and ascended to the Father. So, at the pinnacle of this drama, we get the return to the original scission: (AA_p). This return to the origin as a sort of redemption of the origin, in Badiou's eyes, manifests the overall schema of Hegel's strategy: the dead Son reintegrated into the divisible immanence of the Father completes the concept of the Christian God, which is the holiness of the Spirit.[36]

The circularity of this schema ingeniously detaches the notion of infinity from the Absolute and posits it in a way that infinity manifests itself as the truth of finitude. Put simply, for anything finite to exist—that is, to be something—it must also be infinite. But for this notion of infinity to work in the way that Hegel designed it, infinity must manifest itself as a circular schema that, in fact, starts from the category of Limit.[37]

According to Hegel, for a determined being to be, it must also cease to be; that is, it must have a limit, and this is exactly the definition of the finite being: finite being, for Hegel, is what comes to an end through being what it is. Although the logical category of Limit in *Logic* is part of being finite, and Limit logically develops into finitude, Limit and finitude are not the same. A finite being is not the same as having a limit. In being limited, Something is related to the Other, which is the limitation (*schranke*) to the thing. Limit is a relational category, whereas Finite is a self-relating category. A finite thing comes to an end not because of what limits it, but because of the way it logically is: the finitude of a thing is due to the thing itself and not something else that stands outside of it as its limitation.

Here, Hegel introduces the category of Ought, manifesting the truth that, although a finite thing ceases to be in order to be finite, there is determinacy in the being of the finite thing that points to what it ought to be, although it cannot reach this form because of its finitude. Something enjoys a proper determination by logically gaining a limit. Unlike when Something, as a determinate self-relating being, has its determinant quality in itself (as self-relating reality), in a proper determination, the thing finds its determination primarily in its limit and only secondarily in the positive quality that it is. In this case, Finite, that which ends on its own account, is primarily a negativity and only secondarily a positivity. This falling short logically gives rise to the notion that Hegel refers to as Ought: a finite thing is only secondarily what it is.

A finite thing ceases to exist on its own, and not because something else ends it. In positing itself in what it is, a finite thing will posit its own non-being.[38] The proper determination of a finite thing is in its limit and what lies beyond that limit, the negation of the something that the finite thing is. In other words, a finite thing is drawn to its non-being in order to posit its own being. The definition of determination is when a thing, aside from having a positive and self-relating quality that defines its determinacy, is delimited by what is "outside" of it and negates it. A determined thing is a thing that is primarily defined by the delimitation of the negative force that is its non-being. Therefore, when a finite thing posits its own being, it also logically posits its non-being, which is why a finite thing ceases to exist. So, if ceasing to be is how the being of a finite thing is, what is it that actually ceases to be? Hegel reminds us that what ceases to exist is the positive quality intrinsic to the thing (counter to negation). Finitude is indeed finitude of this intrinsic quality, of it falling short of what it should be. So, the same thing, the delimitation that determines the intrinsic quality of a thing, at the same time, causes this positive quality to fall short of what it should be. The thing, as Hegel says, bears within itself an "ought-to-be," which marks both of these traits: the limit and the positive quality.[39]

In chapter 8 we will return to the exact way that Hegel derives the notion of Affirmative Infinity from the category of Finite in *Logic*. It is important to emphasize that the notion of Infinity at stake for Hegel is not the infinity of an absolute, the infinity of the One, but the infinity that requires the mediation of finitude for its realization, and as such, always bears the mark of finitude (it is the infinity of a determinate being, which is not infinite itself; it is not the Absolute Being). The Affirmative Infinity is thus more than just manifesting the infinite, it also manifests finitude: it is exactly due to the finitude of the Determinate Being that Hegel is able to derive Affirmative Infinity.[40]

The circular schema is thus intertwined with a logical procedure through which the Finite realizes its infinite truth: it ascends to the level of infinite and "will sit on the right side of the Father." Two things are wrong with this circular schema. First is the very fact that the circular schema re-posits the origin: we return to the same scission that we started with—Something is a placed Identity, and the procedures provisioned by the circular schema loop back the whole process to the same Something. This is the problem of new. No new thing takes place as a result of this schema. Yet, there is an illusion of Affirmative infinite that has been maintained—this is the self-sufficiency of the One.[41] Second, in relation to the aforementioned issue, this schema is incapable of producing multiple and thereby incapable of rendering any true history. Time turns out to be the unfolding of the same. The sense of infinite in the circular schema neither presupposes nor engenders, the being of multiples. As a result, history is not the history of coming and ceasing to be of multiplicities—the sense that Badiou assigns to true history—but a compromised concept modeled after the Christian God.[42]

The above concerns are better understood as two strands in a double torsion. On one side, we have the finitude of infinity—infinity is "finitized" by the schema of circularity, such that infinity is the infinite return of the same (finite), and thereby ironically, the thought of infinity comes as the foreclosure of the thought of the new. On the other side, history is "infinitized" to manifest the unfolding of the infinite being, and thereby the concept of history requires the foreclosure of the thought of the (finite) multiplicities.

Here, Badiou proposes to split Hegel a second time. To this double torsion and the circular schema that generates it, Badiou posits the schema of periodization. It is important to pay attention to the manner in which Badiou anchors the notion of periodization—which yet again attests to Badiou's fidelity to Hegel as his master: being a true Hegelian, Badiou is forced to take a road that Hegel did not take himself. We've seen how, in the case of Something and Other, complete fidelity to the Hegelian dialectic forces us to posit two against one—even though the idea of the two was present in Hegel (unconsciously) all along. In the current case, Badiou will also go back to Hegel in order to split him. By showing that Hegel, on the point of two different schemas of looping back, is, in fact, "standing on both edges of the knife—two dialectical matrices, as always."[43] It is interesting that—to the credit of his master—Badiou quotes Hegel three times, from three different fragments, to show that, despite the appearances—that is, despite the attempts to present the functioning of the circular schema as a smooth and flat function—Hegel was well aware of the discontinuity near the end of the loop, just before the completion of the circle and just prior to the return of the origin, that there is a point of singularity, a point of a qualitative shift, a jump from theory to practice and from conceptualization to effectuation. Hegel was exactly aware that the two ends won't meet, and that, in the end, the return home requires an intervention—an act—that alters the automatic and preordained succession of Dialectic of One.[44]

Through this intervention and qualitative shift from theory to practice, we can amend the previous conclusion that what comes at the loop-back point is not what was but something new and truly unforeseeable. This constitutes a passage from one stage of development to another—the schema under which the dialectic operates does not manifest itself as a giant circle, but a spiral, one in which every stage of development never "meets" the point of origin.

There is, thus, no perfect circle. There is no return to the origin, and with this breakage of the perfect circle, there is no infinite as the truth of being finite. The powerful consequence for us resulting from the affirmation of periodization against circularity is the confirmation of what Christianity knew all along: infinity is the radical otherness of finite and not its conceptual succession; there is a radical break that divides the two. Infinite comes to the finite arrangement of a historical situation via a radical break, a rupture, a subjective act, or, as Badiou borrows from Lacan, his other master, "the real." Here again, Badiou is showing to be more of what Hegel passionately aspires to: Badiou is more Christian.[45] This assertion leads us to one of the main theses of this work, which we will elaborate on in the coming chapters: the infinity of history or the world is not the infinity that can be conceptualized or theorized, because it is a matter of practice rather than theory. No apparatus can capture the exact infinitude of what there is in conceptual forms.[46]

Therefore, in short, our theoretical gain is this: once interiority has been decoupled from the One (as demonstrated in the first two sections), the problem becomes if interiority can go "toward" the indeterminate (periodical) or only "return" to it (circular). The foregoing shows that Hegel has the means to think the first but must privilege the second due to his commitment to the grammar of the "indeterminate immediate" in his dialectical logic.

TOWARD A THEORY OF FORCE

Then again, the same question arises: How can this qualitative shift "in the vicinity of completion"[47] be theorized? How is it that the real passes beyond? How is it that it periodizes rather than running in circles?

> We need a theory of the pass of the real, in the breach opened up by formalization. Here, the real is no longer only what can be lacking from its place, but what *passes through by force*.[48]

This is the venture Badiou promises us, and this is what his effort in *Theory of the Subject* is all about. From this point of the impasse of formalization, he ventures to the point that formalization is the "im-passe" of the real—the point that real passes through by force. The way we understand this point of venture is this: ideology (and metaphysics at its core) wavers at the point of surpassing the One, and even if it concedes to two at the beginning as a premise, it finally precipitates toward the conservation of one of the two. Hence, we say that we will need a split of a split; we need a new dialectic:

- *Dialectic of Four:* this is the process of two times two

Our claim is that *Theory of the Subject* is the exposition of the Dialectic of Four, and it is the path of this exposition that will finally lead Badiou to his epoch-making breakthrough in *Being and Event*. Both materialism and idealism vacillate between the Dialectic of One and the Dialectic of Two, but they arrive at the same conclusion: one. Dialectic of Four recognizes this and provides a theory for this vacillation, as well as a theory to surpass the thought of one. Dialectic of Four is the law for how the task of thinking in science and politics, from within each discipline, surmounts the One of discipline—what we have previously termed the disciplinary engagement through which science produces more science and politics, by the same token, shall produce more politics.[49] Table 2.1 summarizes these three dialectics.

And yet again, in order to find the constituent elements of this theory, we don't have to look too far away from *Logic*: "Everything is here."[50] We should first consider that, in relation to two terms which Badiou calls contradiction, there exist the following three moments:

1. Difference or a scission: the two terms have a difference of some sort.
2. Correlation: the terms are related to each other and not disjoint.
3. Position: they have different statuses in relation to each other, statuses that may be invariant or reversible.

Next, each of the above moments has weak and strong modalities, and based on these modalities, two categories of contradictions:

1. Structural contradiction: weak difference (difference of placement), weak correlation (scission), and weak position (invariant).

Table 2.1 The Three Dialectics

Dialectic	*Principle*	*Process*
One	It posits One	It concludes with One (One of the substance)
Two	It posits two	It concludes with one (unity of the two in one of the structure)
Four	It posits two	It concludes with two (weak and strong unities)

Source: Author created.

2. Historical contradiction: strong difference (qualitative heterogeneity), strong correlation (struggle), and strong position (reversible asymmetry)

The point is this: Dialectic of Two concludes with a unity of two, and using Hegel's *Logic,* Badiou posits two ways of understanding this unity: under the schemas of Whole-Part and Force-Force. To explain this, Badiou refers to a passage from the second part of *Logic*, in discussing the category of Essence and Appearance. The discussion of Whole-Part is the first moment in Appearance, and Force-Expression and Inner-Outer as the second and third moments of Appearance. All moments in Appearance involve a relationship between two sides, and through this relationship, they make one thing. Thus, as the name portrays, the appearance of something as one thing has different moments:

a. What appears in different ways (Part) is one thing (Whole)
b. What appears as having different expressions is one force
c. What appears outside is related to what is inside

The relation operative in Whole-Part is a mechanical relationship where the Whole and Part are not essentially related to each other: Part is self-subsistent on its own, and ambivalent to whether it is part of a whole or not.[51]

Badiou's claim is this: theory of contradiction under the schema of Whole-Part has a problem, according to Hegel. The problem is that the unity of terms will eventually collapse into one; it will re-posit one. This is the metaphysical collapse of the Dialectic of Two, whose procedure is to derive unity from the two. This logical interaction, in the theory that involves place *P* and identity *A* resulting in the splace and outplace (with splace as the dominant term), culminates in the vanishing of outplace in splace.[52]

This logic, the vanishing of outplace into the splace, is the foundational logic of all forms of structuralism. Understanding structuralism under the logic of Whole-Part explains why ultimately, structuralism calls for a strict determination of terms under the law of the structure, such that ideally, each term repeats the structure as a whole. Hence, Badiou identifies the weak modality of contradiction as Structural Dialectic (SD). SD, as the determination governing the weak modality in the contradiction of terms, is therefore not the determination that can provide a theory of periodization, because SD is the theory under which the dominant term keeps returning and repeating. Therefore, although SD describes how thought is under the dominance of one, it does not allow us to theorize anything that goes beyond one.

Whole-Part is not the only schema in *Logic* under which the problematic of relation of terms is involved. After Whole-Part, Hegel introduces the second moment of Appearance as Force-Expression, and as is usual with the progression in *Logic,* the second moment, Force-Expression, should emerge out of the first moment. The passage in Logic, which is right before the "strongest passage" Badiou quotes,[53] tells us two important things:[54]

- The *Conditioning* aspect: the Force-Force relation involves a mutual impulse
- The *Determining* aspect: this mutual impulse grafts each Force as (part of) the determination of the other

The first point states the conditioning aspect of the Force-Force dyad: that, at some level, the dyad comprises terms that are both active in relation to one another. The second point states the determining aspect of the Force-Force dyad: that the externality of each Force is an internal moment of the other.

This inwardness of Force finally breaks its immediate determination, transforming the negation outside of Force to one internal, so that this determination becomes solely based on mediated inwardness. This is a familiar movement in *Logic*, as we have seen in the category of Finite, and reveals the truth of Force to be Affirmative Infinity. This logic of force is important to our problem as it does not result in a vanishing of one of the forces in the relation to two contrary forces. The Force-Force contradiction, far from resulting in the annihilation of one by the other, shows the very existence of Force to be determined by the other; the condition of possibility of one force is its other force.[55] Furthermore, there is an inherent presence of one force within the other force. Each force contains the determination of its other. This double feature of force, mutual conditioning and determination, provides a level of heterogeneity that is absent from the structural relation of Whole and Part. Understanding the correlation of terms according to the schema of Force-Force gives us the minimum condition according to which the result of the correlation is not one of the terms: the correlation is properly schematized as the possibility of the mutual existence of two forces.

Historical Dialectic (HD) is the name Badiou gives to the correlation under the Hegelian schema of Force-Force: HD schematizes the correlation of terms when the modality of terms is strong, that is when there is no vanishing of one term, thus provincially qualifying HD as the foundation upon which periodization may be theorized.

Before we extend this line of thought, one more clarification must be made. If SD is based on Whole-Part and HD on Force-Force, and if, according to *Logic* Force-Force is the truth of Whole-Part, can't we also say, along with Hegel, that HD is the truth of SD? Badiou's answer is no. SD and HD are two distinct ways to schematize the correlation of terms based on the modality of contradiction. Hence, Badiou takes a stance here that separates him from Hegel. Badiou argues that Force-Force is not a derivation of Whole-Part. He thinks Hegel has gone too far in deriving force from the relation of whole and part.[56]

Force schematizes a sort of relation between two terms which entails a qualitative binding between them, such as bourgeoisie and proletariat classes, for example—each force is grafted onto the other, and in their relationship, they condition and determine each other. The relation of the whole to its parts is the relationship of belonging. There is no qualitative correlation between them. The crucial point here is this: for Badiou, unlike Whole-Part and Force-Force in *Logic*, SD and HD each exist and interacts with the other simultaneously, without one replacing the function and potency of the other. Force influences the logic of splace-outplace, while at the same time, the relationship of splace-outplace finds distinct constructions according to the inclusion of the parts into a whole. This is an important qualification. For Badiou, one relation cannot exist without the other. Another important qualification that separates schemas of SD and HD is that the former preserves unity through one of the terms—it is a numerical unity of one, whereas the latter maintains unity through a properly dialectical procedure whereby the terms coexist and interact with each other—it is a unity of numerical two. Finally, it is important because the Dialectic of Four concludes with two: two of the method and two of the premise.

As we shall see later, in this break from Hegel, Badiou follows Lacan, his other master. The strong modality is the impasse of the weak schematization: an impasse, that, in Badiou's account, Hegel, in "a striking anticipation of Lacan" had seized in "the pulsation, the vacillation, and the alternating eclipse"[57] of Whole-Part.[58]

The departure from the "error of Hegel's truth" by preservation of both schemas allows Badiou to expand the operations within the Dialectic of Four in important ways. In the same way that HD is anchored in SD, force is anchored in splace. In that sense, the terms splace and outplace have a corresponding force of their own. Thus, while splace and outplace at the level of SD have a particular (weak) correlation and interaction, the same terms also correlate and interact (strongly) with each other as forces. In that sense, there is a force that "rules over the unity of splace (force in the position of the State, or of the symbolic)" and another that "reveals the outplace (force in the position of the revolution, or of the real)."[59] This expansion gives us a new way to schematize the determination of an identity *A* in a place *P*. We've said that an identity finds its determination when it gets an index from the place *P*, an index whereby *A* is placed (an outplace) within the space provided by *P* (the splace). We can now also state that what is being placed is a force (force of the identity of an outplace or an essence according to the *Logic*) against another force (force of the place of a splace, or another essence). Understood as force, outplace as affirmative action (expansion) is "placed within the overarching structural aspect of the other force"[60]—that is, the force uniting the splace. This means that Determined Being in *Logic* can be defined through two schemas of determination: splace-outplace and force.

Now, this placement of force in the splace makes the force impure because it is always placed. Something of the quality of force becomes homogenized with the splace.[61]

But the key here is this: force is not reactive, but rather a quality not completely defined by the determination it receives. The impurity becomes part of its determination, but as we have seen, the more this determination persists, the more the force enacts its own affirmative character. The unfolding of the affirmative character of force is the process by which force "concentrates" its qualitative identity. Following Hegel, this affirmative procedure must have internal and external modalities, but nonetheless, it purifies the qualities in opposition to the "impurity" that the force receives from its determinative placement.[62]

This is a familiar movement in our study. In the previous chapter, we encountered the concept of formalism as the method through which sciences advance according to the internality of their disciplines. Now, the schema of force and the process of concentration by which a force purifies its affirmative quality give us a logical schema which clarifies the function of formalism. Formalism constitutes the process by which a science achieves a concentration of scientific force in opposition to the splace of the body of the science, what the science thinks about itself. We have explicated how formalism allows science to break away from its ideological designations by creating more science. This is the destiny of the purification of force: "The purification of force amounts to the concentration of its newness."[63] It is only by committing to a formalism that science tears itself apart from intuitions that hold sway over it. Concentration and newness are formal and final causes of the same procedure: the purification of force. The same procedure applies to politics. Badiou brings numerous examples in *Theory of the Subject* to show how an identical logic is applicable to politics: the outplace of a political force purifies itself in the very struggle that attempts to subjugate that force into submission to the dominant social forces. The logic of purification is the same: the force concentrates through this procedure, and concentration brings about newness.

We can appreciate now how this schematization helps us to analyze new conditions. For example, the same process of purification, concentration, and newness applies to the force that unites the splace: it is not the case, for example, that politics is only about the purification of the oppressed class. It is also about the purification of the dominant class.[64] This schema allows us to formalize what still escapes progressive forces: the realization that it is not only the progressive forces that make a political movement.[65]

So, what this schematization (that we have called the Dialectic of Four) has been able to achieve is this: (a) The terms of the contradiction are doubly determined: as to their place (splace/outplace) and as to their force. (b) Force is doubly determined: objective and subjective.[66] Two terms are entangled in four procedures—thus the Dialectic of Four. And finally, this newfound dialectic relates to the question of periodization:

> • that splace be caught up in the destructive flagrancy of the outplace;
> • that the subjective aspect of force attains a threshold in its qualitative concentration.
>
> This is the double precondition for the advent of a subject-process.[67]

Our take on these rather cryptic points is this: the process of concentration of subjective force brings force to a threshold at which action is what shows itself at the vicinity of completion. The dimension at which the force wretches itself away in what Hegel calls the infinity of force. To speak of force in its infinity is to speak of the action as correlation; it is to name the primacy of practice.[68] Badiou calls this the precondition of the subjective process, that which demonstrates how going beyond the circle is finally possible.

The Dialectic of Four embodies two dialectical matrixes: SD and HD. But as much as these two dialectics are manifested through logical schematization, the encounter with these dialectics is not primarily a matter of theory but political practice. It is the immanence of politics, as the practice of thought (in the sense conceptualized by Lazarus), through which these dialectical matrixes have been originally encountered and experienced. Understanding "immanence" in this context is important here. It is precisely analogous to the immanence of mathematical practice we have seen in the movement from Cantor to Gödel to Cohen. The theorization of this immanence, by Bachelard and later by Badiou, is always secondary to the practice itself. This is no different when it comes to the immanence of politics. Badiou, following Mao, speaks of two contradictions: fundamental and principal. Mao, by devising a theory of contradiction and dividing it into base and motor, was not creating a theory of discourse of politics; he was practicing politics. It is precisely within this practice that his thought of contradictions of base and motor should be located. In this sense, Mao played the same role in political discipline as Gödel did in the discipline of mathematical logic. The division of contradiction is not an answer to theoretical questions outside of the practice of politics; it is the practice itself. This is also evidently true regarding the interplay of the two contradictions with each other. Again, as much as we have pursued this thought in its logical form, the interaction itself is a matter of practice of politics.

You find this point clearly when Badiou speaks of the knotting of the two contradictions.[69] The emphasis Badiou places on the knotting of the working class indicates that what is at stake is practice. But this is so especially in light of the reading of Hegel that we've presented: after the decoupling of interiority and the One, by letting go of the indeterminate as immediate, and after showing that this allows us to explore the indeterminate as the orientation of a sequence (rather than as a circular return), we have now gained from our new reading of Hegel the understanding of "a" grammar—that of forces—which is simultaneously dialectical (preserves within the interior the place for heterogeneity and contradiction) and opens to the disciplined inscription of a different consistency. So, we do get the elements we need—except that we must unearth this historical dialectic from the interstices of Hegel, where it does not yet have a proper theoretical form.

This observation is what takes us to the next step in this inquiry: the theory that explains this knotting of the two contradictions—in our case, the two dialectical matrixes: SD and HD.

CONCLUSION

We set out to show that Hegel is divisible in two. We showed this by dividing Hegel into three steps:

- Hegel's premise is not only (the) One but also two: place and identity.
- Hegel's schema is not only circularity but also periodization.
- Hegel's contradiction is not only structural but also historical.

The question that we now turn to is one raised primarily by political practice: knowing that these matrixes can be knotted together, how could this knotting be explained?[70]

This is the point at which we turn to Lacan, who explicitly theorized these two sequences. Whereas in Hegel, both structural/algebraic and historical/topological sequences do exist, they are unconscious of one another. The early and late Lacan, however, each explicitly theorizes these sequences in distinct theoretical "paradigms."[71] Lacan becomes our Hegel because, unlike Hegel, Lacan does have a theory of SD and HD; he proposes an explicit topological theory of the subject in contrast to an earlier algebraic theory that he had already developed before. From the unconscious One of Hegel, where the duality of SD and HD are present silently, we trace the steps toward Lacanian thought in which this duality shapes two different sequences in theory: early to mid-Lacan and late Lacan.

By turning to Lacan, Badiou, of course, never turns away from Hegel. As we have seen, the original source of this inquiry will always remain in Hegel.

However, Hegel remains an unconscious historical dialectician. So, while the potency and immediacy of the question, in theory, is coming from the practice of politics, the theory itself will have to take its route through the practice of the clinic—and through the master of this practice, Jacques Lacan.

NOTES

1. Lazarus, *Anthropology*, xxi.
2. Badiou alluded to this later when he referred to the difference between his procedure of truth and Lazarus's concept of "historical modes of politics." He then adds: "I explained in Logics of Worlds why my philosophical enterprise is nevertheless compatible with Lazarus's, which puts forward a thought of politics elaborated from the standpoint of politics itself" (Badiou, *The Communist Hypothesis*, 231–232). What remained consistent in Badiou's project in the interpretation offered herein is the idea of immanence or interiority, whose sense in politics was very much the focus of Lazarus's project as well.
3. Badiou, "Mark and Lack," 172.
4. George Wilhelm Friedrich Hegel, *Science of Logic*, Amhurst: Humanity Books 1969.
5. This includes the chapter of "Determinate Being" in the *Logic* up until the category of Infinity, which Badiou takes up in *Being and Event* (Alain Badiou, *Being and Event*, London, New York: Continuum 2007).
6. Alain Badiou, *The Rational Kernel of the Hegelian Dialectic*, Melbourne: re .press 2011, 51–57.
7. Hegel, *Logic*, 70.
8. Hegel, *Logic*, 73.
9. Hegel, *Logic*, 82.
10. Without getting into too much detail, we can say that the mature Badiou (Badiou of *Being and Event* and after) agrees with Hegel's project in *Logic*: to find the beginning. Badiou also agrees with Hegel that the proper beginning for philosophy is some notion of pure being. What differentiates Badiou and Hegel is the latter's commitment to the category of the One: the de facto suturing of the thought of pure being to the One. This slip is not particular to Hegel and has its pedigree in the entire philosophical history since Parmenides. Hegel, as the last great metaphysician, belongs to the long line of philosophers who believed that pure being is only thinkable as a metaphysical category, and it is by adopting this thought under the framework of metaphysics that it becomes equivocal to the thought of the One. The genius of Hegel, however, was to take this equivocation of pure being and the One to its final conclusion—that the thought of the One alone must produce its own laws and its own progression—that there is no law of thought, no subject, outside of the One that could write the history of the One according to laws of logic (i.e., Aristotelian logic). Logic and being are the same, and therefore the categories of logic themselves must be rewritten by the thought of the One. The One is not only the substance but also the subject. This is why when Badiou proposes mathematics as the alternative

paradigm to metaphysics to think of the pure being, he not only changes the underlying field of ontology from metaphysics to mathematics, he not only de-sutures pure being and the One, he also abandons the need for dialectics, as the inner law of how the thought of the One emerges as different truths. This is why Badiou's ontology (that is, mathematics) is not dialectical. It is also important to briefly mention three more thoughts. Despite this difference, however significant, Badiou follows the same project as Hegel does in *Logic*. Their logic and ontology produce different contents, but they start from the same premises and are dominated by the same concerns: it is the same project, with the same point of entry and the same inquiries, but one takes this project forward under the framework of metaphysics, the category of the One and the procedure of dialectic and the other forwards the project under the framework of axiomatic set theory, the category of multiple and the procedure of generic. So, when Badiou says that he finally abandoned Hegel (Badiou, *The Communist Hypothesis*, 237–238), one can see that the truth, in fact, is that he abandoned Hegel in order to preserve the Hegelian project of the beginning—or to save Hegel one must abandon his metaphysics and dialectics as the grounds for ontology. Perhaps that is why Badiou expressly tries to measure himself against Hegel and *Logic*. Secondly, Badiou's work in the two volumes of *Being and Event* (*Being and Event* and *Logics of Worlds*) may be more than just a parallel to the *Logic*, a detailed analysis may be able to show a more intricate relationship between the two—something that requires a different work altogether. Thirdly, as Badiou has mentioned (Alain Badiou, *Manifesto for Philosophy*, Albany: State University of New York Press 1999), the untethering of pure being and metaphysics and tethering it to mathematics is an epochal decision—a decision that in our times will allow the resurrection of ontology from its rubbles after deconstruction and the linguistic turn. As formerly mentioned, these points will be among the topics that we will take on in part III.

11. Badiou, *Theory of the Subject*, 4.
12. Badiou, *Theory of the Subject*, 5.
13. Badiou, *The Rational Kernel*, 51.
14. Hegel, *Logic*, 82–83.
15. Stephen Houlgate, *The Opening of Hegel's Logic, From Being to Infinity*, West Lafayette: Purdue University 2006Houlgate-2006, 263–283.
16. Badiou, *The Rational Kernel*, 53–54.
17. It is important to mention that even here, although "silently," we are confronting the axiomatic as the form of thought. It is interesting that this remains unthought in *Theory of the Subject*—as it does in "Mark and Lack," where Badiou was correct in criticizing his interlocutor, but did not develop the standpoint from which this critique was made. Here we get a decision to start with Something, not pure being, but no theory of why this is so: this would have to be a "meta-ontological" theory, which Badiou only develops in *Being and Event*. We will return to this topic in part III, as the problem of the beginning and the thought of the axiom will become "half the battle" regarding the axiom of Infinity. As we have already mentioned, there has been a theoretical expansion on Badiou's part in *Theory of the Subject*, where he begins to rationally account for what was previously presented solely in critical form, and this is what rationalizes the passage we are trying to build from part I to part II.

Similarly, we can anticipate there will be a rationalization in moving from part II to part III concerning the decision of beginning.

18. Badiou, *Theory of the Subject*, 5.
19. Hegel, *Logic*, 70.
20. Hegel, *Logic*, 74.
21. Badiou, *The Rational Kernel*, 54.
22. Badiou, *The Rational Kernel*, 54.
23. Hegel, *Logic*, 67.
24. Frank Ruda, *For Badiou: Idealism without Idealism*, Evanston: Northwestern University Press 2015, 135–140.
25. Ruda, *For Badiou*, 140.
26. Ruda, *For Badiou*, 140.
27. We will discuss a theory of the beginning that is de-sutured from the thoughts of the One and the ground in part III of this work (see chapter 6, "Axiom," and Conclusion, "Thinking Discipline").
28. Badiou, *Theory of the Subject*, 6.
29. Badiou, *Theory of the Subject*, 5.
30. Badiou, *Theory of the Subject*, 6.
31. Badiou, *Theory of the Subject*, 7.
32. Badiou, *Theory of the Subject*, 11.
33. A simpler way to articulate it would be *Ap*(*A*) determines how much of *P* is in *Ap*, and *A*(*Ap*) determines how much of *A* is in *Ap*.
34. It is important to also consider this in light of the crucial intervention made by Althusser regarding Hegel's ideological representation of dialectic, which also relates to the way Badiou treats the Beginning of *Logic*. According to (Badiou's reading of) Althusser in the second part of "On the materialist dialectics" (Louis Althusser, *For Marx*, London, New York: Verso 2005. 161–218), Hegel is wrong to think that the appropriate "grammar" for dialectics is that of the inconsistency of the immediate in dialectical logic. For Badiou (following Althusser's critique), the connection between the indeterminate immediate and determinate mediation does not exist for a materialist: rather, materialist dialectics posits pure mediation at first, asserting that, to begin with, a mediation of nothing (void) is the only way to dialectically attach oneself to the indeterminate immediate (pure multiple). We think that we can understand this inversion of the dialectical relation in terms of the mediation that the place, as the pure void, has in the determination of the identity, as the indeterminate immediacy.
35. Badiou, *Theory of the Subject*, 15.
36. Badiou, *Theory of the Subject*, 17–19.
37. This thought is elaborated in a much fuller context in *Saint Paul* (Alain Badiou, *SAINT PAUL: The Foundation of Universalism*, Stanford: Stanford University Press 2003). From the standpoint developed therein, we could clearly state that the problem with the circular view departing from Being rather than Something is that nothing needs to happen after the crucifixion: Christ is God simply because he has been revealed to be the mediated form of what was indeterminate in the immediate presentation of God as Being. For Paul, on the other hand, this is not so: he began with the situation (the Something, which was his own being, placed in Judea,) and

that is why the indeterminate referent of the declaration "Christ has resurrected" was not immediately given, but had to be constructed as a community of believers—so something (indeterminate) still had to take place for the crucifixion to become what it "will have been." So this connects us back to the initial problem: there is no need—or better: no space—for a "new interior" if the indeterminate is given in the form of the immediate; the only indetermination by which we can orient ourselves is a "past" one (from the standpoint of the intermediate), so an immanent discipline of the new is only possible—or necessary—if we place the indeterminate as the product or future of the passion, rather than in the position of God the Father. That would be the Pauline solution, which is conceptualized under the operation of periodization that we shall see shortly.

38. This is like the "death drive" of the finitude.

39. Hegel, *Logic*, 116–137.

40. Hegel makes powerful remarks that reveal his strong motivations toward the subjective nature of these logical categories (see Hegel's remarks regarding the category of Ought and the fact that while Ought is the transcending of Limitation, it is only finitely so (Hegel, *Logic*, 35)). This is what Žižek believes to be the core of the Hegelian subject: the Hegelian subject is the cut that exposes the incompleteness and failure of Determinate Being (its truth as Ought—a pointer to an inherent deficiency or having fallen short); it is the relation of the negation of the Determinate Being to itself.

41. "If one opts for circularity one basically claims that what in a dialectical move is attained is nothing but that which will have been there from the very immediate beginning (the split between the finite and the infinite, the constitutive scission)" (Ruda, *For Badiou*, 143).

42. Badiou, *Theory of the Subject*, 19.

43. Badiou, *Theory of the Subject*, 19.

44. Badiou, *Theory of the Subject*, 18.

45. In line with the aforementioned reference to Paul, we could alternatively say that Hegel is "less" Christian in that his view culminates in considering Christians unnecessary to the truth of Christianity, while for Badiou, the infinity of God would still depend on what happens after the resurrection—in the Pauline discipline of the community.

46. "Like the Hegelian absolute, a world is the unfolding of its own infinity. But, unlike that Absolute, the world cannot internally construct the measure or the concept of the infinite that it is" (Badiou, *Logics of Worlds*, 309).

47. Badiou, *Theory of the Subject*, 19.

48. Badiou, *Theory of the Subject*, 23.

49. As we shall see, Badiou calls this the process of concentration.

50. Badiou, *Theory of the Subject*, 32.

51. Hegel, *Logic*, 518.

52. Badiou defines splace as a place that is used for the placement of various identities. So splace is a place in which identities are placed. Outplace are identities that are placed, the identities that have received the determination of a splace. According to this definition, *A* is the identity, *P* is the place, and *Ap* would be the general schema

of outplace. When we speak of, outplace and splace, we are not speaking about the general schema of *A/P*. We're speaking about a particular placement of a particular identity.

53. Badiou, *Theory of the Subject*, 32.
54. Hegel, *Logic*, 522–523.
55. This constitutes Hegel's reading of Newton's second law.
56. Badiou, *Theory of the Subject*, 34.
57. Badiou, *Theory of the Subject*, 33.
58. Badiou may be referring to a passage in *Seminar XI*, where Lacan uses the term "temporal pulsation" to refer to the way the unconscious manifests itself in a clinical setting: "unconscious is what closes up again, as soon as it has opened, in accordance with a temporal pulsation" (Jacques Lacan, *The Four Fundamental Concepts of Psychoanalysis, The Seminar of Jacques Lacan Book XI*, New York: W. W. Norton & Company, Inc. 199, 143).
59. Badiou, *Theory of the Subject*, 37. The correlation of the two schematizations together is of utmost importance to Badiou. He remarks that forgetting this correlation will result to either rightist (on the side of weak schematization) leftist (on the side of strong schematization) tendencies.
60. Badiou, *Theory of the Subject*, 37.
61. Badiou, *Theory of the Subject*, 38.
62. Badiou calls this process of qualitative concentration of force the subjective process and, correlatively, the process whereby force is placed and thus becomes impure the objective process. See Badiou, *Theory of the Subject*, 41.
63. Badiou, *Theory of the Subject*, 39.
64. Badiou, *Theory of the Subject*, 42.
65. This realization finds its own formalism: Badiou calls a force subjective if the force designates an outplace, and conversely objective if the force designates the splace. This schema also relates to the subjective schemas in *Logic of Worlds*.
66. Badiou, *Theory of the Subject*, 44.
67. Badiou, *Theory of the Subject*, 45.
68. Badiou, *Theory of the Subject*, 35.
69. Badiou, *Theory of the Subject*, 26.
70. The term "knot" here refers to Badiou's usage in the latter quotation from *Theory of the Subject* and is unrelated to Lacan's Borromean knot. For reasons that will become clear in the next section, the knot that Badiou is after was not developed by Lacan, although the road to it will go through Lacan's thought.
71. To borrow a term from Miller, who, as we shall see shortly, reconstructed the movement of Lacan's thought in six paradigms. We shall use his reconstruction to elicit the two paradigms that are important to our study.

Chapter 3

Lacan Is Our Hegel

For Badiou, the practice of politics struggles to formulate a revolutionary political theory that is not only capable of capturing the complexities of the present situation but also capable of providing new pathways for revolutionary forces. From this perspective, revolutionary politics is in a crisis of theory, and this crisis invites serious political thinkers and practitioners to rethink the premises of political activity. This task to Badiou is nothing short of going back to the basic categories underlying this thought: dialectic, contradiction, history, and subject. What Badiou discovers in this path is a rather staggering result that helps to shed light on reasons why revolutionary politics—after two world wars, the rise of global capitalism, and the monster that was the Soviet party in power—is not equipped to think and absorb the depth and scale of what has transpired. The reasons boil down to this: Marxism, as the main theory underlying revolutionary politics in the capitalist era, inherits an underdeveloped dialectical theory that pertains to two dialectical matrixes that, while originally present in Hegel, never came to full articulation and, due to this lack of theoretical depth, it has become impossible to either understand the agility and resilience of capitalism to recover from the rubbles it created by its wars, as well as the defeat it suffered by socialist revolutions in Europe and Asia or the transformation of the revolutionary political party of Lenin to its complete opposite.

Coming to realize this impasse and empowered by strategies and tactics of political practice in China during the cultural revolution, as well as the events in May 1968 in France, Badiou sees both the need as well as the possibility of laying the ground for a new theory that can explain what occurs prior, during and after the revolution as well as agents and masses who participate in those sequences. No doubt, this possibility is indebted to the development of structuralism in France, some of which we engaged with in the previous chapters.

But the other mainspring of thought was the discovery of the unconscious—and the consequences that this discovery had with regard to the notion of the split subject. To Badiou, Lacan was that theoretical force that merged these three mainsprings: dialectic, structuralism, and psychoanalysis, into a stream of thought that is perfectly suited to nourish the revolutionary theory. This unlikely blend of disparate sources of thought, according to Badiou, enabled Lacan to develop the two dialectical matrixes as pitted against each other, at whose crossroads the subject decides its fate. Therefore, Badiou's turn to Lacan is no less vital than Marx's turn to Hegel, as both Hegel and Lacan are the master dialecticians that have developed the thoughts from which the revolutionary political theory should emerge.[1]

The unpacking of the above thesis, which is the main purpose of this section, should provide a precise understanding of each of the three distinct concepts in the phrase "idealist dialectic of our time." That is, the sense in which Lacan is our Hegel is dependent on these three exact notions: dialectic, "our time," and idealism. In other words, we have to prove the following three theses:

- Lacan is dialectician.[2]
- Lacan is of our time.
- Lacan is idealist.

In chapter 1, we have provided a precise sense in which Lacan should be construed as a dialectician. By going through the main manifestos of discourses of overdetermination, we uncovered how the general logic upon which these discourses are founded is built around the dialectic of subject and structure and the fact that one cannot exist without the other. It, therefore, comes as no surprise that both instances of this discourse, Marxism and psychoanalysis, are heavily rooted in the dialectical thought in which the relation of their opposing terms do all the theoretical heavy lifting. As such, our discussion in chapter 1 provides sufficient evidence for positing the first thesis in the above list. So, our main concern in this chapter will be to prove the other two theses: that Lacan is of our time and that Lacan is an idealist. This task would require us to define the meaning of these two terms: "of our time" and "idealist."

When Badiou refers to Lacan as our contemporary, as a thinker of our time, he is referring to the sense that Lacan is a thinker of SD as well as HD. This chapter will establish the fact that what we saw in the previous section as two dialectical matrixes in Hegel becomes full-fledged dialectical sequences in Lacan: sequences in which a particular paradigm of thought dominates Lacan's teachings. We will further show that these paradigms of thought, while fully developed in their theoretical edifice by Lacan, do not finally form a knot and thereby will eventually fall short of being able to account for the *aporia* at the center of the revolutionary political project—and this is precisely the sense in which Badiou thinks that Lacan remains an idealist.

LACAN IS OUR CONTEMPORARY

For Badiou, Lacan is contemporary because he is effectively the Hegel of our time, against which we can pose the dilemma of the immanence of practice, as has been developed in two distinct disciplines: science and politics.[3]

Lacan, in September 1953, in his famous Rome Discourse, three months after his resignation (in June 1953) as the president of *Société Psychanalytique de Paris* (*SPP*)[4] outlines a reconstruction of the field of psychoanalysis by going through the structural coordinates that the Hegelian system has identified, that is, Hegel's insistence on the fundamental identity of the particular and the universal, an insistence that reveals the extent of Hegel's genius.[5]

This is precisely the point that makes Lacan our contemporary: Lacan's "dialectical modernity" takes up the task of rethinking the relation of particular and universal, the task that, as we have shown, is at the heart of our inquiry. Because of his dedication to the clinic, Lacan managed to revolutionize the way we think about the relationship between particular and universal, and a close examination of his work shows that he remains Hegelian even when he drops references to Hegel.

The clinical reference is of interest here since discipline—much like the clinic—is a wager that something which is the "case" can be more than the particular instance of an already established universal. A discipline implies the possibility that a particular case immanently exceeds its status as a particular, just as the singularity of a clinical case is confirmed, as it challenges the mere application of the universal categories of psychology. This is how Lacan is our Hegel—Lacan brings the torsion embedded in the relation of structure and element to the post-Freud era.

Lacan's thought was a unique blend that had influences from many schools of thought in France. In some ways, Lacan's own intellectual development is a compendium of the intellectual history of the French intelligentsia, starting from phenomenology/existentialism to psychoanalysis to structuralism, while at the same time, in his own words, always looking back at the structuring moments of the dialectical system of Hegel. To better understand how Lacan's thought re-grounds politics, we chose to provide a brief synopsis of its development using some of Miller's texts and by breaking it into three major phases: early, middle, and late Lacan. Early Lacan is the Lacan between 1932, the date of his doctoral thesis, to 1953, the date of delivering Rome Discourse, the middle Lacan starts after 1953 to roughly around his

Seminar XVI in 1968, and we mark the late Lacan as beginning at the time he delivered *Seminar XVI* in 1968.

Early Lacan

In the twenty years between his doctoral thesis and his pronouncement in Rome regarding the future of psychoanalysis, Lacan managed to find a point of convergence between phenomenology and structuralism. From the very beginning of his work in psychiatry, he was oriented toward Husserl's phenomenology, as Karl Jaspers had incorporated it into psychiatry. Structuralism was also a strong orientation in France around that time. Lacan, as early as 1943, started to blend phenomenology and structuralism—his Rome Discourse in 1953 is a combination of the two: Lacan's theory of speech was existentialist and phenomenological, while his theory of language was structuralist.

Lacan was influenced by Karl Jaspers' work on phenomenological psychiatry and used Jaspers' work extensively in his thesis of 1932. Jaspers' model fits a typical clinical presentation of psychosis: once one comes to understand a patient, a part of the content of the patient's psychotic beliefs appears as understandable as any person's way of thinking, despite having other aspects remaining as incomprehensible but yet still explicable.

It is hard to overemphasize the importance of the path that Lacan made from psychosis to psychoanalysis. The experience of studying psychosis educated Lacan about the extremely singular nature of psychosis. One psychotic is different from any other, and that's why the dominant idea of psychiatry was that there is no regularity in psychosis: if there is any inter-regularity, it should be at the physical and biological level, not at the level of thinking processes of each patient. So, psychiatry brought Lacan to the singular, and phenomenology backed up this perspective. But Freud brought him back to the universal as well because the Freudian theory of psychosis hinted at the idea that there is something universal in the processes and the overall logic of psychosis. Lacan's entry into psychoanalytic thinking was, therefore, very much the problem of the connection between universality and singularity. This problem remained central in Lacan's thinking and teaching for the rest of his career.

After his doctorate, Lacan, totally fascinated by psychoanalysis, decided to become an analyst, and he started his own analysis while at the same time pouring over Freud's texts. His experience of analysis combined with his phenomenological orientation led to a unique understanding of the experience of analysis—which he described in a paper he wrote in 1936, "Beyond the 'Reality Principle,'" which outlines, in a phenomenological language, the paramount importance of language in the experience of analysis.[6] In that experience, we are not dealing with a description of a symptom, but a narrative, a discourse, and a speech uttered by the patient. This view is completely phenomenologist as it pertains to the world that is of the patient alone and thereby completely subjective. It is important that any traces of objectivity be bracketed from this experience. The continuity between the 1932 thesis and the 1936 paper was the experience of analysis. During his analysis, Lacan experienced the Jaspersian subjective meaning first-hand and not as an abstract philosophical concept. The *eureka* moment for Lacan happened when he realized that subjective meaning is to be pursued in the subject's speech and not in an abstract internal world of affections as Jaspers and other phenomenologists had thought. From this point of discovery, Lacan's inquiry into language started, which would lead him to Saussure's linguistic theory.

Thus, for Lacan, speech was the main datum for the analytical experience, and this was consistent with the phenomenological view. This brings out the question: What does Lacan mean by language? From 1936 to 1953, there is a progressive enrichment of the concept of language in Lacan's work. But his breakthrough came when he encountered structural linguistics. "In 1936, Lacan considered language to amount to signs."[7] Lacan's take on language was very important. Sign in language signifies something, but Lacan understood sign primarily as signifying *someone*: that is, prior to signifying something, a sign signifies someone for the person who uses the sign. This means that, by using the sign, the person, in fact, is communicating with another. That makes communication or interlocution, the primary function of language.[8] The analysand's speech is something that happens between the analyst and analysand, and the Other here is language itself which, precisely due to its exteriority, remains "other" to the speaker. The job of the analyst is to allow the analysand to speak and to prevent this otherness of language (that which in the language constraints the speech; its determining power) from being sutured in the guise of some supposition about the analyst: something is impossible to say due to the analysand's relation to language, but it is not said due to what the analyst might think, for example. Therefore, the whole point is for the analyst not to allow herself to be taken for the Other of the speaker but rather to get out of the way so that this otherness can become legible.[9]

Lacan's understanding of the role and importance of language deepened during the prewar years, and through this development, he came to realize that many of the important facets of Freud's edifice should be reconstructed according to the new orientation that had opened up to him via structured linguistics. By 1936, Lacan was dividing Freud's theory of language from his theory of libido, and how to formulate the theory of drives based on the theory of language becomes a fundamental problem—a division that still shows itself in the theory of jouissance and signifier in the late Lacan.[10] In "Beyond the Reality Principle," Lacan not only acknowledges the dualism in the Freudian theory, but he also asserts the significance of the concept of libido both in its theoretical and clinical functions.[11]

However, if we take up this position that speaking is primarily to someone and only secondarily about something, we should consider the social character of the language more important than its descriptive character, its capacity to convey information about affairs. This position will also create a difficult encounter with Freud's theoretical reliance on biology. In addition, Lacan sets out to show that drives are completely encoded in language. Thus they are structured as a language. For example, Lacan calls the notion of libido no longer meta-psychological and is inclined to formulate a linguistic basis for drives.[12]

Beginning with his 1936 text, Lacan plotted a course for reversing the relationship between biology and culture: in explaining human behavior, culture is a more primordial power than biology, and this is the distinct difference between humans and the rest of nature. After World War II in 1948, Lacan finally published the second part of the 1936 text, "Aggressiveness in Psychoanalysis."[13] This text focuses on another Freudian primordial instinct, the death instinct, which in the psychoanalysis of the time was gaining an acceptable status due to the events of the war. This theory states that there is not only *libido* but also another primary instinct called the *death drive* responsible for psychical formations. In prior years, Lacan had developed his theory on the mirror stage, and he famously related aggressiveness in human behavior to this stage. In other words, aggression in Lacan's view is always grounded in narcissism.[14] In "Aggressiveness," Lacan builds upon his theory of the mirror stage and the imaginary register by introducing the phenomenological language of intersubjectivity for conceptualizing the analytical experience and the fact that what is essential in communication is meaning as opposed to reference. It is here that Lacan starts to introduce the notion of the subject in psychoanalysis by introducing a bidirectional relation between subject and meaning.[15]

This shows that the notion of the subject appeared in Lacan's theories even before his assertion that the unconscious is structured like a language. The phenomenology of consciousness oriented this concept of the subject, but this orientation, inaugurated by Husserl, did not have the concept of the unconscious. Hence, the theoretical dilemma became how to retrofit the discovery of the unconscious by Freud into a phenomenological theory of the subject, an increasingly difficult endeavor since the phenomenology of Husserl, Sartre, and Merleau-Ponty had become essentially anti-objectivist: consciousness is not an object and cannot be described by the same categories used to describe objects of the world, like substance or force—there is no positive category with which to describe consciousness. Furthermore, every instance of consciousness is associated with a body, and since the body is always localized, every instance of consciousness is also localized. Furthermore, a given consciousness is not something that you come to encounter once and for all; rather, consciousness is a stream, something in movement and evolution. In psychoanalysis, the subject is somewhat parallel to such a localized consciousness, with one major caveat: the subject owes itself to the unconscious rather than the conscious part of the psyche.

The first difficulty with this definition is how to "bootstrap" the subject: How does the subject come to be? The first part of Rome Discourse, on which the first two seminars are grounded, starts with addressing exactly this question. Moreover, the concept of the subject as "nothing that becomes" gives a historical aspect to the subject. The concept of nothingness promotes a specific concept of history, that which is constructed and infused with meaning: the history of the sessions is the reconstruction of the subject.[16]

Furthermore, the concept of the subject as nothingness distinguishes the subject from the ego. Sartre sees the reflecting consciousness as no-thing, as a pure act that can never be grasped on its own, whereas ego, or "I," is an artifact of the non-reflective consciousness. When consciousness uses "I," it is always oriented towards the projects of the world, towards objects extraneous to the consciousness itself. As such, "I," or ego, is always already objectified; marred by the projects that take the consciousness away from itself and into the world that it intends, the ego is captured as an opaque object—a correlate of the objects intended by consciousness with which, borrowing a Kantian phrase, it must be able to accompany as just another object of experience.[17] This makes it clear that the subject must be distinguished from the ego, which is what Lacan's theory of the mirror stage attempted to accomplish. Both the subject and ego have their correlates: the subject's correlate is the place of meaning, the place where all meanings find their affirmation, which in the foregoing, we alluded to as being the role of the listener in the analytic setting. "Other" is the term Lacan gives to this correlate of the subject. Lacan substitutes the letter *S* for the subject and the letter *A* for the Other, hence the correlation between *S* and *A*. The ego also has a correlate: the "alter-ego." Lacan substitutes the letter *a* for the ego and the letter *a'* for its correlate, and hence the correlation of *a* and *a'*.

The idea here is to suggest that the ego and the Other are homogeneous—two instances of the same thing (*a*)—which implies that the alter-ego is not totally other to the subject, just like the ego. It is a move that deflates the otherness of the alter-ego by rendering it complementary to the ego (the negation of the ego is the alter-ego, which means we can deduce one from the other)—and, on the other hand, the subject is other to the ego in the same radical sense that the otherness of language is other: so writing *S* in correlation to *A* is, first, not equivalent to *a* in correlation to *a*', and, second, the whole point is that these two correlations invert our expectations: the alter-ego is closer to the ego than the subject, who is in relation to what is most radically other to us (here, language *A*, rather than the *objet a*, which is later to be thought as that kernel which is other even to the Other). That is why the two axes are crossed, one as the imaginary (*a-a*', where the interiority of the ego overdetermines otherness) other as symbolic (*S-A*, where subjectivity is overdetermined by exteriority of language).

But the ingenuity of these relationships strikes when Lacan locates these four elements in a configuration such that the correlations cross each other. Therefore, instead of having two parallel relations (*S-A*, *a-a'*), we have two crossing relations (famously rendered as, see figure 3.1).

At the end of Seminars I & II, we find out what this twist really means: the imaginary relationship (the solid line) bars a complete symbolic relationship (the dashed line).[18]

Coming back to the question of "bootstrapping," one of the most important facets of subjectivization in phenomenology relates to the notion of recognition: the subject must be recognized by another subject. That is why a subject cannot recognize itself, because a subject does not know who it is. This brings the importance of the other in the process of subjectivization. Only an other can subjectivize an individual. During the session, the analyst receives this identity for the other.[19] Lacan later on discards phenomenological references: the subject's desire is to be recognized by the Other, and to this extent, the subject's desire fundamentally becomes what the Other desires. At every moment when the subject speaks, it is the Other who speaks: the unconscious is the Other's discourse.[20] Lacan, therefore, transformed the Hegelian grid from the inside. By the end of *Seminar II*, he defined the Other no longer as another subject but as the locus of the unconscious.

By positioning speech in the place of the subject and language in the position of the Other in *Schema L*, Lacan transforms the dialectic between subject and Other for recognition into the dialectic between speech act and language. The subject is both the agency of speech and the one who speaks, while the Other is the authority on how speech is to be organized. Through speech, the analytical process determines the rules of the language of the unconscious. Hence, this relationship evokes the anthropological studies of

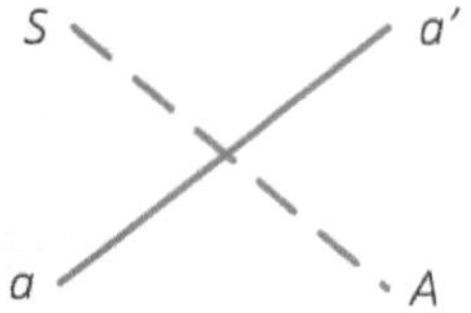

Figure 3.1 Schema L. *Source:* Author created.

extinct languages where one uncovers the rules of a language through fragments of its use.

Lacan's early itinerary demonstrates an engagement with many different schools of thought, in particular structuralist linguistics and phenomenology. Our takeaway is that Lacan saw no intrinsic or necessary entanglement between psychoanalysis and any of these schools. This is important to emphasize because, in light of the later commitments that Lacan made in his middle period, it is common to see Lacanian theory essentially as an extension of structuralist linguistics into the logic of the signifier. Yet Lacan borrowed from many different theoretical frameworks in order to capture something of his engagement in the analytical experience, especially when the paradoxical encounter between universality and particularity was at stake. Evidently, the more established a framework was, the further he engaged with its conceptual paradigms. It is evident that Lacan, using the logic of signifier, made advancements that would not have been possible otherwise, but it is reductive to equate Lacanian psychoanalysis with this logic. Lacan was definitely much more adventurous and flirtatious in his approach.[21]

This lesson is important for us from two perspectives: method and anecdote. In chapter 1, we saw an attempt to theorize the logic of signifier as the logic of all logics: of any discourses of determination, of any structuration that happens to include a reflexive element, a logic underlying number theory and that also separates sciences from ideologies. We saw how, in a brilliant rejoinder, Badiou showed that the program of the logic of all logic is itself an ideological attempt debunked by mathematics itself. As a rule, we should be wary of epistemological claims made about a discipline from the outside. Instead, every endeavor worthy of the name discipline has an interiority proper to it, and indeed at every decisive junction, it departs from such claims that may (temporarily) come to occupy it. Instead, disciplines are developed through a practice that has fidelity to its own content, formalism, and interiority. Therefore, the question becomes: Are such elements present in the field of Lacanian psychoanalysis? The answer cannot be found solely in the theory of signifier—a construct, separate from analytical practice, which is useful for conceptualizing the "elementary unit of speech" as an articulation between universality (of a system of differences) and particularity (of its local concatenation and enchainment), between what lacks (the referent) and what is present (the fact of speech) within the clinical setting. Part of the genius of Badiou, in our view, is manifested in his ability to discern this interiority in an obscure text from Lacan's early period, which pertains to a thought experiment: "Logical Time and the Assertion of Anticipated Certainty, A New Sophism" ("Logical Time").[22] For reasons that will be developed in chapter 6, this is one of the most important texts in the entire Lacanian corpus, avowed by Lacan himself on several occasions, even during his late period.[23] What makes this text doubly important for us is that it was written in the early period of Lacan's development, prior to his engagement with the logic of signifier. This acts as an affirmation of the interiority of psychoanalysis and Lacan's veritable fidelity.

Badiou's choice to select a text from the early period points to a unique Badiouian insight: "Logical Time" is a text that reveals psychoanalysis in its interiority and is free from certain allegiances that will become the bedrock of Lacan's logicism in his middle period.[24] The moment the logic of psychoanalysis gets sutured to the logic of signifier, even though there were evidently good reasons for it, this proper interiority is hardly retrievable. We see a similar isomorphism of theory and model underlying Frege's logicism when he took the theory of logical implications as a theory for the model of numeric succession.

By going back to Lacan, Badiou is not only trying to move beyond him—as we discuss in the next section—but also trying to de-suture Lacan from SD so that we can discover an alternative logic in HD.

Middle Lacan

The middle Lacan starts right after his epoch-making address in 1953 and at the beginning of his *Seminar I*.[25] What is important in this phase is his engagement with the relation of jouissance and signifier, as a particular expression of the singular and universal that we identified in the previous section. In a precise sense, we can track the passage from structural to historical dialectic in Lacan by surveying this relation between signifier and enjoyment in his teaching. This passage is relevant because it delineates the psychoanalytic form of thinking about the themes we have already developed, now cast as relations between structure and the real, determinate and indeterminate (or universal and singular). The signifier is that which is placed, in speech, in accordance with a given system of placements (a symbolic structure), while one's "mode of enjoyment" concerns what is singular and unrepeatable in one's experience. Speech is overdetermined by language, which tends toward the universal, but symptoms and unconscious formations can only be understood in regard to the singular history of a given subject. Hence the question: What is "real" in structured speech? This problem highlights what is at stake here is the in-mixing of the subject—as singular process and structure—as a general system of placement, which has interested us throughout this chapter. So, when we track the different relations proposed by Lacan between signifier and enjoyment, we are tracing the different relations he proposed between structure (signifier) and real (enjoyment), between determination (or general symbolic consistency) and indetermination (or local consistency), and so on. We should also make clear that there is a reason why we are following this periodization of Lacan via a text by Miller—whom we have previously criticized. Using Miller's recapture will not only allow us to track Lacan's development but also reinforce our point regarding Miller as a thinker of SD.[26]

Lacan, having departed from "objective" psychoanalysis and "subjective" phenomenology, and having established a structural basis for the concept of subject and the analytical experience, begins to import more of the Freudian constructs into a new edifice while attempting to untie these concepts from their genesis in "biologism" and to reconstruct them according to the structuralist orientation, with an emphasis on the linguistic theory of signifier. The genesis and causal force that Freud attributed to the libido is at the center of this reconstruction. In some sense, all of Lacan's developments in the middle and late stages of his intellectual life can be read as an attempt to re-conceptualize the relation between jouissance and signifier. In this relation, Miller captures these re-conceptualizations in the form of six successive paradigms, four of which belong to what we have referred to as the middle Lacan. In what follows, we shall use this paradigmatic approach.

We saw how Lacan, since his time as an analysand in 1936, emphasized the importance of speech and language. But it was not until his seminars that Lacan's teachings spring to life with the introduction of the symbolic as a dimension distinct from analytical experience. Speech is the giver of meaning, with language backing its structure. Communication is intersubjective and dialectical, which takes shape during the analytical cure. We saw how, as early as the Rome Discourse, Lacan saw in intersubjectivity an element of dissymmetry in the subject-to-subject rapport: the two subjects are never equal—in the case of the analytical process, the analyst is the one who listens, punctuates, interprets, and thereby decides on meaning. This is where the concept of Other is introduced and where the speech finds meaning: it is the place of language, a place where the subject is determined. This appeared to be Lacan's capital contribution and his innovation in the field: the inscription of the subject (*S*) within the field designated as Other (*A*), such that the Other, rather than the other-subject, is said to determine or constrain the meaning of what is said by the subject. The conceptualization in this phase begins by the framing of subject-Other relation in terms of two poles of speech and language. Lacan places the intersubjective on the side of speech and the autonomy of the symbolic chain on the side of language: the signifying chain that operates in the Other has its own demands, a logic of its own. In addition, this theorization of the symbolic for Lacan accounts for what is decipherable of the unconscious in the analytical process. During this early phase, the unconscious is sometimes bound to language and sometimes to speech. Likewise, the stress is alternately placed on the structure or the utterance, and overall, Lacan characterizes the unconscious as the subject.[27]

From the beginning, we witness Lacan's direct intervention into the Freudian corpus. This intervention marks a caesura that disentangles the technique of deciphering the unconscious and the theory of drives. In relation to the Freudian economical perspective, namely pleasure, Lacan proposes: "the essential satisfaction is to be found in communication itself, as it delivers meaning."[28] That is, within this first paradigm, Lacan has already conditioned pleasure on language and the process of signification. In his *Seminar V*, Lacan attaches satisfaction with the subject to the Other's fascination with the subject's (linguistic) formation. Understanding, on the side of Other, sometimes results in laughter and causes satisfaction in the subject in relation to the semantic structure used by the subject. On the side of the subject, satisfaction takes place when a certain meaning resurfaces.

Lacan's first attempt to link pleasure/satisfaction to the economic operations pertaining to consciousness is, therefore, symbolic satisfaction. This means that he already did not conceive of structure as something externally bound to singularity but proposed that the subject is produced, in her singularity, by this overdetermined system of placement, such that we experience ourselves mostly as "ourselves" precisely when the symbolic functions produce meaning. But, in addition to the symbolic satisfaction, there is also the satisfaction that pertains to the imaginary. According to Miller, in this phase, Lacan assigns to the libido and imaginary status. That is, the "singular" is something like a naturalization of an overdetermined effect of signification: we are determined by language, and one's mode of enjoyment is rather an "introjection" of something whose logic precedes and determines us anyway. But, at the same time, imaginary jouissance, which is jouissance proper, does not derive from speech or language; it does not belong to the symbolic register, and hence it does not originate in the subject. It belongs to the ego as an imaginary agency. Lacan symbolizes all that Freud characterized as a libidinal investment on the imaginary axis of *a-a'*, the place of jouissance as distinct from symbolic satisfaction.[29] The imaginary jouissance, distinct from symbolic satisfaction, is therefore not intersubjective and dialectical because it is intra-imaginary. The libidinal investment is, therefore, what ties me to my place, even though this place has a purely symbolic function.

In summary, in the first paradigm, Lacan stresses the disjunction between the field of the ego and the field of the unconscious, which is the strict distinction between signifier and jouissance, each having its distinct genesis and distinct course. What interests us the most is that Lacan is clearly within the paradigm of SD: the interplay between the system of placement and the place overdetermines the subject, which only appears as an autonomous and consistent instance due to the libidinal attachment which fixates his placing. Libido is here "topological" in the sense that it brings the imaginary continuity to cover up the "algebraic" logic of signification, which re-places us at every signifying point.

Using earlier Badiouian mathemes, we see Lacan developing the notion of language as the space of placement *P* (the symbolic), which fully determines the function of speech as placed identity (*A*). Due to singularities being cast off to the imaginary register, the placement of identity *A* (speech) is the full repetition of the space of placement *P*. This is an abstraction of the determination of scission under the operator of unity A_p (see figure 3.2):

In this matheme, speech makes its place in language, where it finds its meaning and relevance.

As we saw in *Schema L*, the crossing of *a* – *a'* creates a barrier for the symbolic establishment of *S* and *A*, causing a rupture in the rapport between the subject and the Other. But in Lacan's works, tension remains in this early interpretation regarding which one of these registers has the upper hand. Though he clearly mentions the autonomy of the imaginary from the language and speech, he frequently also refers to the symbolic ascendancy over the imaginary. Thus, early in this paradigm, we witness a non-rapport between the imaginary and the symbolic and the making of the imaginary as the place where Freud called libido.

This leads us to the second paradigm in Lacan's "middle" teaching, which does not chronologically succeed the first one, but blends with it and finally overtakes it.

During this paradigm, Lacan engaged in a genuine conceptual rewrite of all the terms that had been poured into the imaginary register, eventually making clear that they are thoroughly symbolic terms. Transference, drive, fantasy, and regression are all transposed from the imaginary to the symbolic. Phallus, an imaginary term, also becomes a concept of the symbolic. Its dominant image, which is already severed from the organ, is a sign of the privilege of the symbolic. Lacan now tries to conceive the topological aspect of the libido as itself somehow derived from the algebraic of the symbolic. The phallus is the signifier which marks within the structure the existence of something un-structured, which mediates between the place and what is placed, as something "transcendental." If the symbolic is transcendental in the sense that it is the general structuring principle, the phallus works as a local structuring principle; it is the signifier which binds language to enjoyment (hence being called

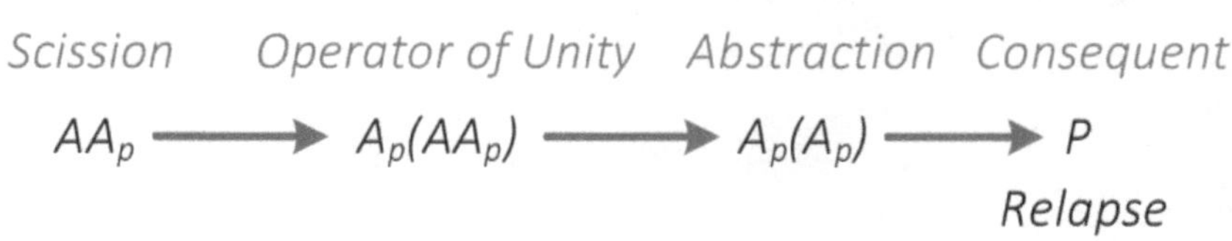

Figure 3.2 Matheme of Lacan's First Paradigm. *Source:* Author created.

later on "the signifier of enjoyment"), therefore binding the system of placement to the actual fixation of a place, without presupposing that the imaginary is somehow totally heterogeneous from the symbolic. So not only do we have a signifying chain, but we also have a signifier which, rather than representing the subject, stands for that in the subject which is un-representable.

What is said about satisfaction in this phase is always uttered in terms of the symbolic. In this phase, Lacan sees satisfaction as attached to the signifiers that bear desire. Satisfaction flows like the signified under the signifier—the pure satisfaction of metonymy.[30] But this does not answer the Freudian question regarding the drives' satisfaction because here, drive is reduced to a signifying chain. This looks like the obliteration of jouissance by the signifier. The signifier annuls jouissance, resituating it as signified desire. In this phase, jouissance is a dead desire.[31] Transposing this development to the machinery of SD, we can surmise that Lacan is attempting to render this machinery consistent by finding, in the system of placement itself, the cause of the fixation of a place and its identity without having to presuppose a substance that "tends" toward fixation.

In the dialectic of place and identity, this paradigm can be captured by the concept of determination (see figure 3.3). This takes place by importing the singularities from the imaginary to the symbolic register, such that the placed identity is no longer a relapse to the space of placement: we have a proper determination—a structural relationship of the whole (*P*—system of signifiers, desire) and part (*A*—unconscious demand).

As a result of Lacan pursuing the symbolization of the jouissance to its very conclusion, there arose a need for a third paradigm, introduced for the first time in *Seminar VII*, the paradigm of the impossible/real jouissance. Lacan considered this seminar as effecting a scission with the rest of his work. This is essential for us: this is the first time that the singular will be posited beyond the structure without being totally antithetical to it—this is the "germ" of HD, even though totally overdetermined by the grammar of algebraic structuralism, which is why the "real of enjoyment" can but appear here as the somewhat mystical "beyond." The real of enjoyment is now a singular consistency that is opposed to the imaginary consistency of the "I"—but since we do not have the means to think this real consistently, what we get is rather a Heideggerian privileging of the poetic as a rupture in structure, and *das Ding* as a sort of violent transgression of order (rather than the emergence

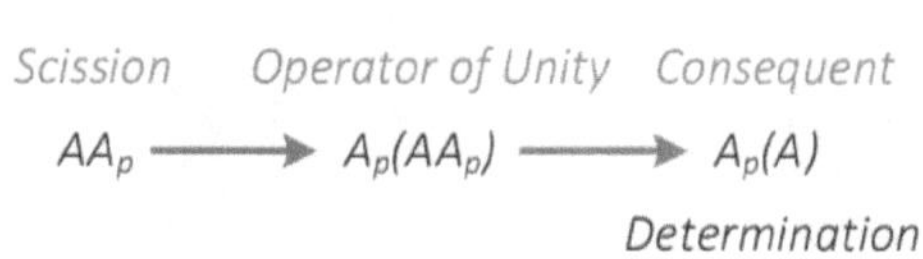

Figure 3.3 Matheme of Lacan's Second Paradigm. *Source:* Author created.

of another order). The outplace enters the scene as the inaccessible point of a given series, the "beyond" of the finite.

This is the meaning of the *das Ding*, which Lacan brings in from Freud's text as a sort of joke to signal that you are not dealing with a symbolic term. *Das Ding* denotes that true satisfaction belongs neither to the register of the imaginary (first paradigm) nor to the register of the symbolic (second paradigm) but to the real.[32]

Whereas the shift from the first paradigm to the second manifests a continuity in the transposition of terms (from symbolic to imaginary), the shift from the second to the third paradigm consists of a total rupture, a discontinuity. The signifier in the previous paradigm obliterated jouissance. By the movement of jouissance to castration, the signifier annuls jouissance and revives it as desire, as a shifting signified—the satisfaction was indeed in the shift itself, invested in the metonymy of the signifier. In the third paradigm, we witness a shift of perspective. By borrowing the concept of the vase from Heidegger, jouissance is exemplified as an empty place and can be written as equivalent to the barred subject, yet it will introduce the possibility of filling it up and the notion of a forever inadequate supplement.[33]

There is a sinister sense of jouissance in the register of the real. The symptom is the way in which the subject enunciates that jouissance is bad, positioning itself as a barrier between signifier and jouissance, reverberating the disharmony of jouissance and subject. Desire and fantasy together expel jouissance from the two respective registers of symbolic and imaginary, which ensure jouissance stays in the register of the real.[34]

In this paradigm, we are still within SD, but Lacan now has seen that in order to think of the singularity of the speaking being, he cannot think of the structure as containing within its system of placement all the determinations of subjectivity—something must take place "beyond" the structure that is not merely the illusion of a beyond, produced by the fixation of the imaginary. But until paradigm five, he still tries to conceive this within the algebraic grammar of SD.

In this paradigm, then we can recognize the limitations of symbolization, and this conception can again be captured through the matheme of place and identity. In this case, singularity cannot be reduced to the "imaginarization" of the symbolic; it is a *real* transgression of the symbolic. As such, the third paradigm works out a theory of outplace overdetermined by the space of placement of *P*, that is, a theory of the limit. This can be captured by an abstraction of the determination, similar to what we have seen before (see figure 3.4).

The fourth paradigm refers to *Seminar XI*, which poses an extraordinary antithesis to *Seminar VII*. In *Seminar XI*, Lacan works out a covenant between the symbolic and jouissance. In *Seminar VII*, jouissance has a massive dimension that puts it beyond the grasp of the symbolic and thus out of reach. It calls for transgression and force, in an abyssal place, as the only way

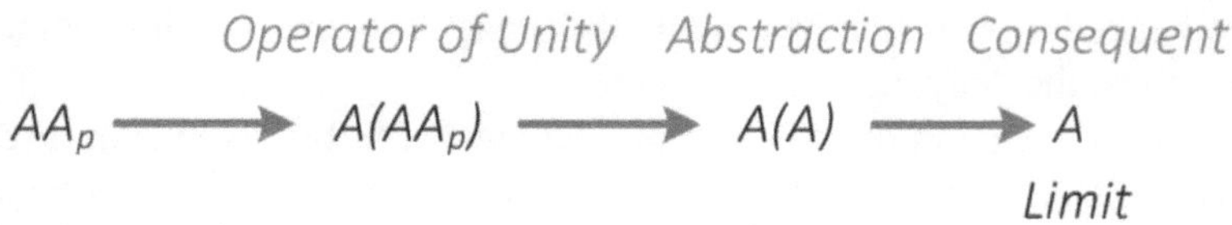

Figure 3.4 Matheme of Lacan's Third Period. *Source:* Author created.

to access it. In contrast to this, in *Seminar XI objet a* is in a little hollow, or it is simply the presence of a little hollow, a void.[35]

There is no harmony between these two seminars. In *Seminar VII*, jouissance is connected to horror, whereas in *Seminar XI*, there is a rapport with jouissance through art and peaceful contemplation of the art object. The work of art calms people and comforts them. There is an inverse itinerary between the two seminars. In the model of paradigm three, jouissance fragments the whole of the symbolic. In the model of paradigm four, it makes the fragments of the partial drive a whole. In *Seminar VII*, there is a cleavage between the signifier and jouissance. In *Seminar XI*, there is a covenant. *Seminar XI* revises the foundation on which his Rome Discourse was based. In *Seminar XI*, Lacan introduces two key concepts of alienation and separation and uses set theory in describing them. This precisely relates to the tight articulation that he is creating between the symbolic and jouissance. Not only is jouissance not a surplus over the signifier, it rather fits its function; it is akin to it.[36] This marks the beginning of the end of Lacan's SD period. The thinking of the consistency of the symbolic somehow correlates to the singular consistency of the speaking subject already looks more like the Hegelian logic of force than the logic of whole and parts.

In this paradigm, jouissance persists within the boundary of a mechanism with the operators of alienation and separation, whereas in the third paradigm, it is beyond the reach of any operation, symbolic or imaginary. The fourth paradigm is the revival of Lacan's ambition for the domination of the symbolic. We could even say it's a sort of impotent last attempt to save SD from Lacan's own orientation toward the real or singular dimension of the speaking being's consistency. Jouissance, instead of being irreducible to the symbolic, or being reduced to a signifier, is singularized as such and inscribed in the function of a system through the operators of alienation and separation. Jouissance is then appointed under the guise of *objet a*, a modest destination in comparison to the menacing *das Ding*. This paradigm makes *objet a* an element of jouissance.

In this paradigm, we have a singularity that cannot be "beyond" the symbolic. Hence a new relation between place *P* and singular enjoyment *A* is needed based on the topology of the erotization of the body. This, in Badiou's terms, is the first attempt on the side of Lacan to think about outplace, not in the grammar of whole/part, but as force, which is overdetermined by the "algebraization" of outplace as "absolute difference."

In fact, the very consistency of the *objet a* as an element which orients the analyst towards "absolute difference" depends on a certain oscillation between the epistemological and ontological uses of the grammar of SD.[37]

The philosophical problem which haunts the concept of the *objet a* is its being-other to the One. How can there be a conceptual treatment of that which is absolutely removed from the One? This problem will gain center stage later on in our investigation, but it is worth noticing that the *objet a* has its "elementary" character, its symbolic inscription or place, not due to its ontological unity, but rather due to the limits of the symbolic now rendered positively. If absolute difference is not differentiable—if there were "different differences," then they would not be absolutely different, but only relatively so—then the unity of the *objet a* is not guaranteed by its internal consistency but by the weakness of the symbolic in differentiating it. The limit of the structuring principle is itself taken for a structural contour of the object. When dealing with Miller's formalization of the relation between the All and the Nothing, we noted that the unity of Nothingness was guaranteed by the non-indexing of its iteration: qualitatively, nothingness cannot be indexed, for there is nothing to index, and if this is the case, then we can treat it, by in-distinction, as always-the-same Nothing—as "one" Nothing. The limit of the indexing system became the positive unifier of nothingness, a way to negatively inscribe it into the structure as a punctual and elementary nothingness.

Miller's little text called *Matrix*[38] may be the best condensation of this pure logical strategy of the algebraic treatment of the singular character of the *objet a*: the wager that we can place the un-place-able by turning the limit of the system of placement into the marker of this remainder. The *objet a* is not a part of the whole. But insofar as every placement leaves something outplaced, and insofar as we cannot distinguish between these outplaced instances—for that would require us to index them to some identity—these un-placed elements can be thought of as a part against the whole ("partial objects" with no corresponding body to them, as Freud would put it). This is the logical impasse behind the algebraic concept of the *objet a*: the means to conceptually form its unity are derived from the overdetermining of the unstructured by the structure. Were it possible to think of absolute difference more directly than by the paradox of distinguishing between different instances of pure otherness, this construction would no longer hold.

Late Lacan

Late Lacan, in our view, is the beginning of a division between the Lacan of consistency vs. the Lacan of cause, to use Badiou's terms. This is where

Lacan begins to realize that symbolic causality is frequently interrupted by another force, which is beyond symbolization. Lacan's *Seminar XVI* is the inauguration of these thoughts, which find its more elaborate articulation in *Seminar XVII* and the theory of discourse. Miller calls this the fifth paradigm. The theory of discourse unifies alienation and separation, and this grants the statement: "there is a primal relation from knowledge to jouissance,"[39] and thus the statement that there is a primal relation between signifier and jouissance. The theory of discourses consolidates this "embeddedness" which began in the previous paradigm: the structure of the unconscious, as previously conceived as language in opposition to concrete discourse, now appears as something conditioned by discourse itself.[40]

The statement "signifier represents the subject for another signifier" sums up symbolic alienation. But the theory of discourses introduces the idea that the signifier also presents jouissance to another signifier.[41] We know from Miller's *Suture* that this is the subject that makes signification work:[42] a signifier relates to another signifier through the subject. As Miller puts it: no identifying representation is ever complete, and that is why the representation repeats itself. It is from the standpoint of the "presentation of enjoyment" that repetition, as the becoming-matter of the signifier, makes sense. The negativity of a Cartesian subject is implicated in the process of signification. The signifier presents enjoyment to another signifier: singularity emerges as an impasse in the process of signification, as repetition renders sense senseless, getting in the way of the process that allows the subject to be negatively implicated. To make things worse: the two sides of this process, albeit connected, have no relation to each other: what renders us singular is not subjective, it's "objectal."[43]

Insofar as the subject is represented, it is never presented; it is never in the present. Yet, to represent the non-representable leads the signifier towards repetition. Repetition is a sign of impossibility—as we have earlier discussed in this work—and comports with a failed representation.

So, what of jouissance and unconscious in this model? Just prior to this paradigm, in *Écrits*, Lacan tries to create a relation between the signifier and jouissance, while maintaining that they are of different registers. For example, in the previous interpretation, *S* couldn't inscribe jouissance as such, only as emotion and affection of the body. Miller discusses Lacan's interpretation from another prominent article in *Écrits*: "The Subversion of the Subject and the Dialectic of Desire in the Freudian Unconscious."[44] In this article, according to Miller, Lacan attempts for the last time to chart the statute of jouissance in terms of signifier and signified, asserting jouissance as lacking in the Other. He attempts to exert the scheme of signifier and signified on jouissance as phallus, and through a complex signification of jouissance as forbidden, lacking, barred, mortified, with the signifier of jouissance that cannot be annulled. He discerns between two signifiers in this regard: $-\phi$ and Φ, the former as the signification of jouissance and the latter as the signifier of jouissance. This scheme still attempts to transcribe libido in terms of desire. For example, the death drive is the desire of the kind that "creeps everywhere and intrudes its bizarreness and oddities to everyone."[45] Jouissance, the leftover of libido, is the part that is impossible to transcribe as being outside of symbolic, but nevertheless inscribed as Φ. This notion of signifier of jouissance "somehow turns Φ into the symbol of *das Ding*, as a signifier rendered absolute."[46] The difference here—as one could glean from Miller's interpretation—from the second paradigm, in which Lacan's absolutist approach to jouissance had made it mortified even beyond death, is that, although "jouissance can be termed forbidden, it can however be read in-between lines"—that is, we are able to pin down "the outlining of the metonymy of jouissance; perhaps not only the barred subject, the subject-in-lack, is conveyed by the signifier, but also jouissance as lost object."[47]

Therefore, to sum up, the being that is alienated is a being that loses something. Prior to alienation, or launching of the signifying system, the being in question is the being of jouissance—a body affected by this jouissance which is the insertion point of the signifying apparatus. So, what the subject loses in the process of signification is the lost jouissance. In that sense, the signifier becomes the apparatus of jouissance. This means the signifier is a localized means to render the materiality of a symbolic machinery compossible with a subject. The signification process is no longer seen as oriented by meaning—meaning is a side product like libido previously was—but by the production of enjoyment, that is, oriented toward the sedimentation of this material dimension of signifiers, which presents something of the speaker.[48]

In the fifth paradigm, what was conceived as a loss of natural life as the result of alienation is interpreted as the loss of jouissance by the signifier: signifier comes to replace the lost jouissance, and that provides the inherent relation of signifier and jouissance: *objet a* becomes the signifier of the loss of jouissance. Therefore, in contrast to the second paradigm, jouissance, represented by *objet a* in the signifying chain, is no longer interpreted as forcing or transgression but rather repetition. The inscription of lost jouissance as *objet a* accounts for the repetition of jouissance.[49]

In *Seminar XVII*, the emphasis is on the signifier as the mark of jouissance. "The master signifier celebrates an irruption of *jouissance*—yet it also introduces a loss of *jouissance* and brings about a supplement of *jouissance*."[50]

Miller refers to a function that Lacan uses to theorize the accretion of jouissance, that of entropy. In *Seminar XVII*, Lacan equates the effect of the signifier to the thermodynamic law as it implicates physical systems. Every physical system follows a formula that can roughly be expressed in the following way:

The energy that a system takes in (Energy) is always greater than the energy required to maintain the working of the system (Synergy) plus the work that the system produces (Exergy) (see figure 3.5). This excess, usually produced as heat, is referred to as Entropy. Lacan's brilliance is to see the terms of this relationship in the working of the signifier.[51]

Adopting this model, Lacan proposes that the signification of jouissance is thermodynamically equivalent to the unavailability of (a system's) jouissance (thermal energy) for inscription (conversion) into the signifying chain (mechanical work), which allows Lacan to interpret it as surplus jouissance (the degree of disorder or randomness in the system). This model explains that the more signification persists (resulting from the impossibility that causes repetition in the signification chain), the greater excess produced, causing the accretion of the surplus jouissance.[52]

It is then rather straightforward to see how Lacan's formalization of jouissance leads to a definition of surplus jouissance, not as transgression, but as the effect of the signifier itself. This resonates with the lecture title we mentioned before: knowledge as means of jouissance. The next chapter of the same seminar is titled "Truth, the sister of jouissance,"[53] which essentially states that truth is inseparable from the effects of language and therefore bound to surplus, barred, or forbidden jouissance. Truth is the sister of forbidden jouissance, represented by the phallic signifier *Φ*, the perfect, paradigmatic, and exemplary jouissance.

Our underlying thesis has been that the middle Lacan was a structural dialectician. In other words, if the early Lacan is determined by his interlocution with different models, the middle Lacan comes with a specific linguistic and structuralist commitment. In light of what we developed in chapter 1, for

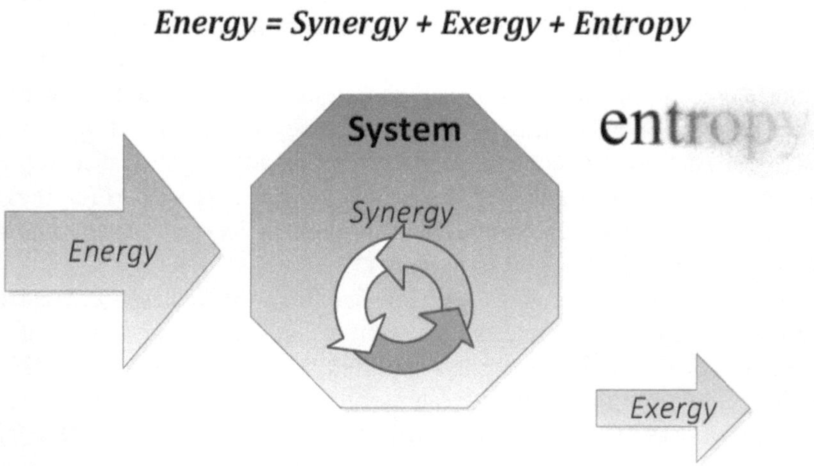

Figure 3.5 Energy, Synergy, Exergy, and Entropy. *Source:* Author created.

the middle Lacan, the relation of the signifier to psychoanalytical theory is analogous to the relation that logic has to number theory for Frege. Therefore, what Miller names the "passage" from the fourth (middle) to fifth (late) paradigms, is, for us, a rupture. This break is not so much with the economy of the symbolic (as Miller seems to be concluding in his work), but rather with the idea that there exists within structuralist linguistics a fundamental, non-symbolizable residue that lends itself to the metaphor of entropy. Earlier in the middle Lacan (in paradigm three), there was something singular about the subject, which could not be captured by language. Later in the same middle period (paradigm four), this "beyond" is transformed into a deformation: it is not because the "beyond" is too far away, but rather because it is too close, that we cannot capture it in language—due to a deformation inherent in the materiality of language itself, because language is in the world. That is why *objet a* finds a paramount role in paradigm four.

The late Lacan completely parts ways with this. Language is embodied; it is material and contingent. Whatever symbolic determination we ascribe to language from the outset must be conditioned by the fact that it is in history. The relation between signifier and signified is not merely the relation between whole and parts; words are now equivocal, not transcendentally but immanently, and there is no longer a third position from where one could arbitrate and intervene into this univocity of the word. A statement can be only partially fixated by whatever is contextually decided. But even the contextuality is submitted to the contingency of speech being material. This is why ideas of repetition and entropy are so important here: they are indices of how a particular informational transit is submitted to some non-informational noise due to the materiality of signification itself, and not due to some extrinsic interference. This new posture inverts everything for us.

Lacan will say now that the Other is a reactive formation to the fact that communication is impossible; there is no pure conduit of information, and therefore noise appears. Though noise intrinsically does not make sense, we nevertheless invent the Other for whom that noise is sense. This is an inversion because there is no longer a symbolic determination that, like a machine, works behind our backs. Rather, our speech simultaneously produces both sense and the lack of sense. The lack of sense is then retroactively inscribed in the figure of the Other as an autonomous field—the figure that is instrumental to the signification itself. So, the lack of sense comes before the signification. This means that ultimately, we begin from the standpoint of embeddedness—we do not arrive at it.

This is in contrast to the previous formulation, where the machinery of sense is always imaginary, and if we take it to be consistent, it is because of the retroactive effect of suturing an element to its particular place in the structure. In the middle Lacan, the autonomy of the symbolic under structural linguistics is thought under the guise of a logical machine. However, the moment such a mechanical-logical model is assumed, we have to deal with the way the singular effects of the experience of the subject can be grounded and thought. Now, the inversion of the late Lacan shows that it is the impression of an autonomous machine itself that is an effect of how language functions.

This, being a classical problem of the dialectic of whole and part, is compounded by its very history: How can we explain that totally different registers come to inhibit the subject from the get-go? If the general model operative in the late Lacan—by bringing this singularity into the realm of the symbolic itself, which changes or enriches the order of the efficacious interaction between singularity and regularity, the latter being the retroactive effect of the impasse of the singularity—looks a lot like Hegelian sublation, this is no coincidence: it is the sublation of the dynamic of whole and part to the mutually efficacious dynamic of forces, what Badiou calls HD.

With this, we now have a better appreciation for what Badiou said about his relation to Lacan: "Like Hegel for Marx, Lacan for us is essential and divisible."[54] Lacan is the major dialectician after Hegel, who advanced the Dialectic of Four: the two dialectical matrixes of SD and HD, to the point where their inflection point became clear and visible—the algebra of the symbolic and the topology of the real. There are two, successive "Lacans": "the one of the lack of being" and the other one "of the being of lack."[55]

LACAN IS IDEALIST

In order to understand the claim that Lacan is an idealist we must know two things:

- What does idealism mean to Badiou at that time?
- In what sense does Lacan fit this definition?

In this section, we will try to answer the above two questions.

The Idealism of SD versus Materialism of HD

What does idealism mean to Badiou in the context of *Theory of the Subject* and the formulation of the two dialectical sequences we've just investigated?

a) The structural dialectic is certainly a form of dialectical thought (this is its materialist side) in the sense that, broadly speaking, it stems from two crucial ontological principles:

- the primacy of process over equilibrium, of the movement of transformation over the affirmation of identity;
- the primacy of the Two over One (contradiction).

b) The structural dialectic has a tendency (this is its idealist side), first, to make the structural aspect of the dialectic prevail over its historical aspect, that is, place over force; and, second, within this very same primacy of the structural foundation, to make the theory of the splace, on the basis of its regulated universe, predominate over the emergence of the outplace.[56]

Hence, for Badiou, idealism is defined as the tendency within SD to give primacy of place over force, and in that primacy, splace over outplace. Another important division of SD and HD resides in their encounter with the real. SD and HD both share the encounter with the real. However, while the real is the "effective thought" of HD, in SD it is a "nuisance" and thereby subordinated to the combinatory that constitutes the structure. The tendency of SD is to dissolve the real into the authority of the structure.[57]

In this relation, the project of SD has to solve three main problems:

1. How to reduce strong (qualitative) difference to a weak difference and effectively reduce it to the difference in position. This is the problem of the schematization of the un-schematize-able.
2. How to get rid of force so that the above reduction is possible. This is the problem of the vanishing term.
3. How this vanishing force or term can cause the movement of place and make the entire structure a whole? This is the problem of the causality of the lack.

From the above, we can construct a relation to the three registers of contradiction: difference, correlation, and position. The first problem is the problem of weak (difference of position) vs. strong (qualitative difference) difference of force/identity. The second is the problem of reducing the struggle of force to a mere scission, opposition. The third is the problem of structure and how the elements of the structure are affected by the structure as a whole.

Thus, in order to understand idealism, we only need to understand the position of thought in relation to the three registers of contradictions: idealism is the thought of weak difference, correlation as scission and invariant position. These three thoughts in the dialectic of whole and its parts result in the domination of the whole over its parts or place over force. It is this domination, as the core operative within the idealist thought, that gives ground to other and perhaps more visible tendencies that define idealism, such as the primacy of spirit over matter (nature). The idea here is that idealism is the ideology of the ruling class, the class that is the guardian of the place, and its functioning over and against the force. This idea is intuitive enough: for the conservative forces, the ideal is the law, which represents the embodiment of the functioning of the place.

What is interesting here is that idealism may accept the dethroning of the law, but not the splitting of the law. Because in the former case, one ideal replaces another: this operation is conceived and legitimated by the lexicon of the structure. But splitting is beyond the comprehension of idealist thought and, therefore, not legitimate. To seal the indivisibility of the law and therefore establish its radical authority, the law is believed to belong to the real. Idealism is, therefore, the doctrine that assigns to the register of the real laws of the State.

It's then rather abundantly clear how HD appears as the renegade orientation, which attempts to access the real that is veiled by the idealism of SD. Likewise, it is straightforward to conceive HD as materialism, which is the orientation of uprising rather than ruling. Once the place is reorganized, you need an ideology that preserves the place. But given that idealism never uproots what is there to replace it, we must conclude that what idealism preserves is what needs preservation, that is, something impossible to perpetuate on its own—something marked by an impasse. In that regard, we must conclude two things. First, it is HD, and the philosophical orientation that it fosters, which allows us to think change beyond the boundaries of the place, and second, SD protects the place that has come to dominate because HD destructed the old, dominant place. Badiou recites from the modern history of Europe sequences that illustrate the latter dynamics. First came the idealism that, with its three rules, turned the Enlightenment to a regime which preserves the new place:

1. Reduce materialism to scientism
2. Reorganize idealism around a symbol: the subordination of the manual to the intellectual in the hierarchical division of labour
3. Make the maintenance of the place and assignation of repetition a juridical and moral order

> The product of all this is a specific idealism, centred on Man, and no longer on God. Consciousness as the focal point of experience, the subject as guarantee of truth, morality as atemporal formalism: this average Kantianism lasts, on a massive scale, to this very day.[58]

After this assault against the Enlightenment, the second materialism comes into conflict with the above idealism. Lenin was the Robespierre of this revolution: Robespierre's philosophical ally against the Aristotelian core of oligarchy was Descartes, and Lenin's was Hegel against the Kantian moral religion.

The third revolution, the one that Badiou hopes for our era, is the revolution of Lacan against Althusser and his "process without a subject." In these sequences, natural materialism first kills God and replaces it with Man. Then, historical materialism kills Man and replaces it with a sociohistorical structure called class. The third materialism, dialectical materialism, kills the structure and its various representations, such as social class or language, and replaces it with a new understanding of the subject. The idealist addendum of the ruling order with the objective of the preservation of what is there is thus to replace the fragile figure of the Man with a more robust representation of a structure: Althusser, Foucault, and Lacan each contributed to this.

Germane to this progression are three idealisms: the religious, the humanist, and the linguistic. Today, materialism targets idealism in the contemporary linguistic orientation, whose central axiom is that language "is that of which experience is the effect, and it is from that which makes his speech possible that Man derives the power not to exist."[59] Linguistic idealism does believe in a subject: the individual consumer. Individuated and selfish, this subject does not take on any project that goes beyond the satisfaction of ersatz-enjoyment. Nothing is more far-fetched and irrelevant to it than the eradication of the place. Language not only proceeds but also founds the world.[60]

What is this new revolution to do? Marx invoked an inversion of Hegel. Should materialism also invoke the inversion of the dictum: the world is made of language, whose opposite is that language is the stuff of the world, from the language being the cause of the subject to the real being the cause of the subject? For this materialism, a simple inversion like what Marx had envisioned is not sufficient. The fate of the subject is tied to the effect of the chain! The reversal cannot take us back to the concept of Man, whose philosophical edifice has been worked out in transcendental idealism. The path is the redemption of Descartes, and in particular, the Cartesian subjectivity in the way that it has been revitalized and enriched through the psychoanalytical theory of the split subject. The third materialist inversion is, therefore, the culmination of a new, materialist theory of the subject.

This means that Badiou is drawing a line between the materialist discourse of our time with the materialist discourse of the French or October Revolutions. The materialism of the French Revolution was the materialism of nature against the idealism of religion. This revolution was successful, and its ideas of the primacy of Nature in the cosmos, and Man in ethical law, have held sway ever since. The second materialist discourse was tied to the communist revolution: it was the confrontation of the idealism of the division of labor to the materialism of the masses. This revolution, despite succeeding in seizing power, fell short of its communist goals, although its impact on the dominant ideology of capitalism was profound. In the aftermath of this second revolution, the idealist core of Nature + Man shifted to a new idealism that represents a new form of dominance, the dominance of a system with no subjects—an abstract, non-existent subjectivity that allows for no essential or discernible difference in the functioning of the system based on race, division of labor, gender, and so on.

The ideological narrative of the latter form of idealism is what Badiou refers to as "idealinguistry," or in short, linguistics, whose philosophical core calls for the dismantlement of the concept of the cause (vis-à-vis the law) and of the subject—concepts that were so central to Kantian humanism. What Badiou finds in Lacan is nothing short of the required theoretical framework for confronting this idealism.

The diagram in figure 3.6 summarizes the foregoing shifts.[61] On the left side of the diagram, we have three dominant idealisms from the beginning of modern Europe: religious, humanist and discursive. But, as the diagram shows, the movement of a dominant idealism toward the next is not through a direct path that runs between them, along a vertical axis on the left, but through a movement to an associated materialism on the right side of the diagram. The solid arrow from an idealism on the left and its associated materialism on the right pertains to materialist revolutions we discussed earlier. For example, the humanist materialism of the Enlightenment was a revolution against religious idealism. Each solid arrow between an idealism-materialism pair has a specific operator: inversion is the materialist operator of the revolution against the religious idealism. Finally, the dashed arrow illustrates the

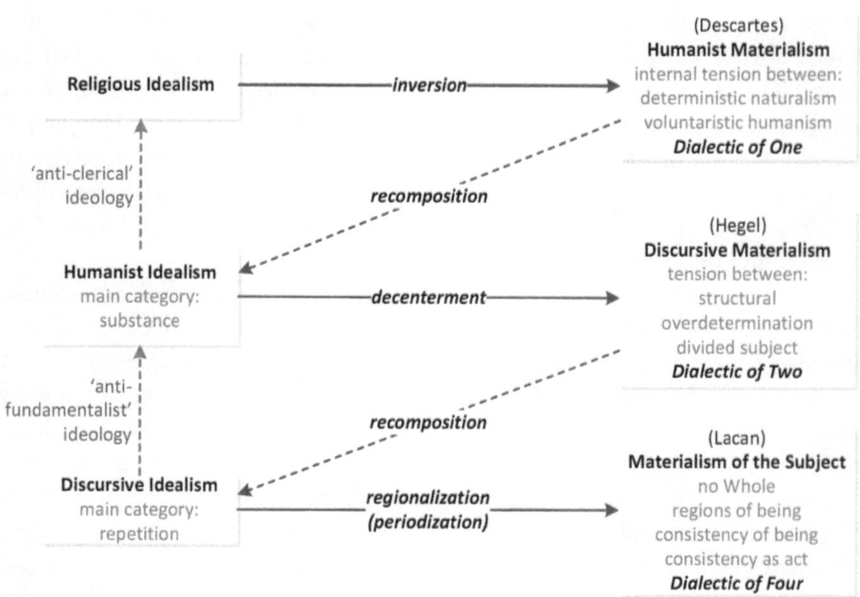

Figure 3.6 Historical Schema for Materialism. *Source:* Author created.

recomposition of a form of materialism to an ensuing dominant idealism: humanist or transcendental idealism is an idealist recomposition of the Enlightenment thought. The relation between these two successive forms of idealism, religious and transcendental, is mediated by a particular materialist thought. Furthermore, each of the idealist thoughts holds a specific ideological stance toward the previous dominant idealism. This is the residue of the materialism that overthrew the previous idealism and is recomposed within a current form of idealism. For example, in this diagram, the deconstructive and anti-fundamentalist attitude of discursive idealism against the humanist, ethical idealism of modernity is a recomposition of the materialist operator of decenterment (see below).

Each of the materialist philosophies makes use of a particular idealist philosopher (Descartes—for the authors of the Encyclopaedia, Hegel for Marx and Lacan for Badiou). The materialist operator of the second revolution is decenterment, where the anthropomorphic references to reason, ethics, and ego are decentered via the socio-historical-centric subjectivity and the conception of the divided subject, of which the idealist recomposition is the discursive idealism (what Badiou calls idealinguistry) dominant in our time. It is crucial to point out that, although in the diagram we placed Lacan in an analogous position to Descartes' and Hegel's, his placement according to *Theory of the Subject* is much more complex, and comprises all of the following recognizable positions: Lacan as a "discursive materialist" against the idealism of Man, Lacan as one of the founders of "discursive idealism," and Lacan as the philosopher whose thoughts are central for theorizing HD and a new materialism of the subject. In what follows, we will trace some of these placements more closely.

Each phase of materialism is designated with a dialectical thought: Dialectics of One, Two, and Four respectively. Although the split substances of extension and mind emerge in the humanist era, the operation of thought still surrounds one essence: either extension or mind. That is why the operator for this materialism is inversion, because it simply inverts the one of God to the one of Man. The dialectic of discursive materialism is Dialectic of Two, where we not only posit two but also define operations that work between the two posited principles. That is why the operation of this materialism is decenterment: we are decentering from a single principle per region of being, to two principles and their interrelations. The materialism of the subject is designated Dialectic of Four: the main thought in this materialism is the absence of the thought of return, but the thought of periodization, or stratification as we saw developed in the previous chapter. That is why the materialist operator of the shift in our time is "periodization," whose core operation revolves around the following basic ideas (which will be discussed in more detail in the next section of this chapter):

- structuralist dialectic is unable to compose a structure of all structures: one failed attempt was Miller's logic of signifier, which we covered in the previous chapter
- the real of the structure is beyond theorization only relatively and not absolutely: a consequence of the above point, this point attests to historical, that is, situated, effect of all structures, and the fact that every occurrence of SD has a complementary and necessary occurrence of HD
- the real of the structure is not only inscribed as cause in the structure but also as consistency: a consequence of the above point, this point speaks to the capacity for an internal critique of the structure, as opposed to an external critique (also discussed in the previous chapter)
- consistency of the real is mediated through act: thereby the necessity of the category of the subject as the requirement for the development of a materialist theory

The above properties will form the capacity for constructing a concept of discipline in contrast to the theory of structure, which we introduced—although merely in a negative capacity[62]—in the previous chapter. Discipline avows the structuration effect of SD but only within a horizon that HD sketches. We showed how moving to Hegel through a particular reading of *Logic* allows the development of the theoretical tools to critique structuralism without committing to metaphysics. The operator of the periodization will then perform the following two tasks:

- develops a critique of structuralism and shows its limitations and weaknesses from a non-metaphysical standpoint
- develops a positive theory of the historical, that is, situated structures, that are subject to both SD and HD that involve the subjective act

For reasons that will become clearer later in this chapter and more so in the next, we have preferred to name this operator "regionalization" instead of "periodization." Although the latter is closer to the project of *Theory of the Subject*, the term has a suturing to historicity and politics, something from which Badiou moves away in his mature work (*Being and Event* and beyond).[63]

Badiou reminds us that, under the name materialism, we understand two different theses that completely contradict each other: there is the One, and the One precedes the Other, so there are two. Monism in materialism is the conviction that being is a unity. The signifier of that unity in materialism is the name "Matter." This monism is very similar to idealistic monism, with the subtraction of the anthropomorphic properties from the signifier God. In his *Saint Paul*, Badiou does mention such a God, especially under the Aristotelian articulation of unmoved mover. A non-anthropomorphic, idealist signifier evokes matter itself, something for whom/which the whole world is made, and who/which is very indifferent to man. This is why several modern commentators find Spinoza's monism materialistic, as his signifier, substance, behaves very similarly to matter.[64] That there is One and nothing else is the monist thesis. The thesis of the One will either lead to a mystic thought or to negative theology, which objects to anything predicated on God.

The Two, however, are the inherent dualism of unity vs. multiplicity, intellect and perception, infinite and finite, the creator and creation, the form and matter, body and soul, etc. There are always two, and there is always a knot of the two. The knot ties all three together: in Platonism, the idea and Presence through the knot of participation; in Judaism, the split of the Creator and creatures through the knot of creation; in Christianity, the split of the infinite and finite through the knot of incarnation, in Islam the split of God and Cosmos through the ascension of the Perfect Man. Such a split and knot is the two of the idealism of theology.[65] After the idealism of theology, as we saw, we had at least two more idealisms: of God and of Man. The same split and knot are operative through both: for example, the idealism of Man produced noumena and the phenomena with the knot of the Transcendental Unity of Apperception.

Hence, we have the One and the Other. It is inherent to the idealisms we have reviewed to affirm the Two and not just One. There is no ideal in the being of the One alone; not having a schism undermines idealism altogether. The ideal depends on schism, a minimal otherness and distance between what is and the ideal. As we also saw, the ideal needs protection and law. With the One alone, there is no such requirement. From the outset, the idealist project has no issues with admitting the Two, the One and the Other. However, idealism gets the precedence wrong: in the relation of creation and created, infinite and finite, form and matter, and so on, idealism protects the wrong side. The Ideal is initially derived from the Other and then subsequently posited as the One, and therefore, it is a representation of the Other, not the One. Now, if materialism also admits the two of the One and the Other, it is the *only* orientation that gets the precedence of One over the Other correctly.

But in what sense does materialism even admit two and not just the One, and why?

> In order to distinguish itself from idealism, materialism is forced to abdicate its essential axiom, which is monism, and to posit the thesis of all major idealisms, namely, that there are indeed two regions of being.
>
> However, it does so only with the aim of annulling this thesis. For in truth there is only one region of being for materialism.
>
> What does this mean, if not that thought for materialism is the vanishing term from which it follows that there is only matter?

> Indeed, in order to name the One as such-or-such-a-One (in this case, matter), what is needed is the real of the Two. Two signifiers, in any case (matter and the idea).
>
> This is what Marx summarizes as follows: "Thinking and being are thus certainly distinct, but at the same time they are in unity with each other."[66]

Thinking, having been posited as a different region of being by idealism, is eventually brought into being, by realizing that idealism did not correctly understand the function of thinking, which is a vanishing term and not a substance that renders or founds a different region of being, other than being itself. Thinking is the vanishing term exactly in the way that it affirms the oneness of being, in other words, the essence of all dialectic—and for sure, the Hegelian dialectic.

In the end, we have two regions of being, matter and thought, for which we can articulate two materialist theses: identity and primacy, with two metaphors of reflection and asymptote:

1. *Identity:* admitting that there are two regions of being, but on every occasion, they speak of the same being, not two beings, where one of them is the reflection of the other.
2. *Primacy:* Primacy is of being, so it is thought that it is the reflection of being—that is, it says the same thing as the being in the real. But this reflection runs side by side with the thing in the real, like two asymptotes, always approaching each other yet never meeting.

So now we have a double occurrence for materialism: structure and history. Badiou proposes the name "algebra" for the structural dialectic of materialism under the metaphorical law of the reflection, as the logic of the thesis of identity and "topology" for the historical dialectic of materialism with the metaphor of the asymptote, as the logic of the thesis of primacy. These two maps what Lacan refers to as causality and remainder.

In What Sense Is Lacan an Idealist?

The essence of the foregoing discussion, therefore, resides in the question of the relation between being and thought, to which the materialist thesis asserts that thought is the reflection of being (postulate of identity). In addition, the thesis also asserts that thought always attempts to approach being but never reaches it—that there is always a remainder, in the real, that cannot be thought (postulate of primacy):

being = thought + unthinkable remainder

The mathematical representation that captures this relation is through metaphors of algebra (identity) and topology (primacy), which were used by Lacan originally in his theory of the subject (and from where Badiou, as he openly avows, fashioned the same usage):

jouissance = desire + objet a

In late Lacan, the signification of jouissance always leaves an un-signifiable residue.

The similarity of the two formulae is obvious. Why is Lacan an idealist, then?

In order to answer this question fairly and accurately, we have to first recall the placements of Lacan within the historical schema we saw before (figure 3.6), which is much more complex than what the question of Lacan's idealism can afford. We could precisely enumerate three placements, which roughly map to three different Lacan's that we developed earlier in this chapter.

We first have the discursive materialist Lacan who maps to the early Lacan. In that position, Lacan is trying to establish a theory of the subject that has non-humanist and non-substantialist commitments in order to enrich the structural theory that is without the subject, so that we finally have a theory that can account for the Freudian split subject that works with both scientific structuralism and linguistics. This effort squarely puts Lacan contra the humanist idealism of phenomenology and existentialism and is one of the pioneers of discursive materialism.

We then have the Lacan, who attempts to come up with the ontology of the entire field of experience through a unified theory of the signifier. This maps to the middle Lacan. During this period, Lacan's emphasis on the primacy of the signifier, and his successive efforts (in different paradigms, as we discussed) of subordinating the indeterminate (the jouissance) under what determines (the signifier), cast him as a discursive idealist.

And we finally have the Lacan who is designated in the diagram as a materialist thinker (like Descartes and Hegel) whose thought represents an upheaval against the dominant form of idealism. We think this marks the late Lacan, the so-called Lacan of the "primacy" of jouissance over signifier.

So, we have three Lacan's (early, middle, and late) and three positions in the history of materialism. This is partly why the materialist operation of our time is regionalization (as opposed to inversion or decenterment). Part of the reason is the unique role that Lacan has had in the history of contemporary thought: first, as a materialist, then as an idealist, and then as a critic of his own idealism.

We should then ask in what sense is Lacan's idealism important to us. Again, precisely because Lacan, during his middle period, while producing an accurate portrayal of the dominant idealism of his time, is impregnated with materialist commitments that he had developed during his early period as well as traces that lead to his late period. These materialist commitments provide the possibility for Badiou to drive a wedge inside Lacan's idealism and to attempt a division that allows the materialism of the subject to find a proper and relevant expression. The relationship between Lacan and Badiou is thereby analogous to the relationship between Hegel and Marx: while Marx divided Hegel to unearth the rational core of his idealism, Badiou divided Lacan's thinking to unearth a historical dialectical matrix.

Therefore, according to what we have presented previously, the middle Lacan should have the clearest expression of the idealist Lacan, and although Lacan during his late phase did have a conception of the topological and asymptotic relation of two regions of being, jouissance and desire, he remained committed to privileging and thereby preserving the structure—or an ideal of the structure and hence the ensuing idealism (idealism = primacy of place over force).

As a "discursive materialist," Lacan suggested important adjustments to linguistics and structuralism so that there would be room for a new theory of the subject and subjective cause.[67] We briefly take note of two of the most important advancements that Lacan made to structuralism. The first is Lacan's changes in the field of structural linguistics to the concept of identification changes that result in a constant shifting of meaning. Lacan asserted a theory where the signifier has dominance over the signified, an inversion of Saussure. This dominance is registered in two ways:

1. The signifier does not ever completely coincide with the signified; there is always a gap between the signifier and the signified—the metaphoric effect.
2. The signifier always shifts other signifiers, in turn creating an indefinite chain of signification, with none of the signifiers therein ever completely exhausting the signified—the metonymic effect.

The point is that this scheme in language makes it so perfectly suitable for a structural representation of repression and displacement, originally thought of by Freud as two mechanisms responsible for creating the unconscious mind, thereby making language the paradigm in which the unconscious structure can be thought and formalized. The famous Lacanian dictum that the unconscious is structured like a language leverages precisely this interpretation of language to confront the ego- and depth-psychology of American and Jungian breeds.

The second change pertains to the theory of structure in which Lacan reveals an important characteristic inherent to the consistency of all structures having to do with the notion of stability: unlike what structuralism had thought, the structure is not the home of stability—but precisely instability. What holds the structure together is its inherent law of exception. This was Lacan's insight: that aside from the minute and detail-oriented effects of the algebraic operations; there resides something that stands outside. That is why Lacan was so ahead of his time: he said something that no one—in the structuralist milieu—could see at the time, which made him the leading figure of "discursive materialism." The structure—and hence the algebra—is not what maintains stability. It is a place of instability. The lack causes Metonymy, and algebra stands for the mechanism of motion that is caused by this lack. Stability is not a consequence of structure, which is what the classical structuralist thought had postulated.[68]

These innovations allowed Lacan during his middle period to come up with successive formulations in which jouissance is conditioned on the process of signification. This, as we have seen in the previous section, could be reconstructed with the matheme of placement of identity (*A*: speech or discourse) placed in a system of placement (*P*: language). During the first two paradigms of the middle period, we saw how the space of language completely determined the identity of speech, while the singularities of identity were pushed to the imaginary register. According to the schema developed earlier in this chapter, we have the determination notified as $A_p(AA_p) = A_p(A)$, where what is left of speech in the symbolic register is completely determined according to this matheme. During the third and fourth periods, we begin to see how the symbolic and real registers re-appropriate the imaginary singularities. The singularities of identity leave traces within the symbolic, which we captured using the other branch of Badiou's matheme as $A(AA_p) = A(A_p)$, which we had mapped to the semantic of identity determining the determination: this was how Badiou captured the category of Limit in *Logic*. Based on this, as we have seen before, we are still operating within the Dialectic of Two. We have the establishment of a scission against the One, but we are still missing the operator of the periodization. This operator is only available in the Dialectic of Four. Although Lacan begins to tinker with this Dialectic, he never fully surpasses it, and that is what leaves his formulation bound to SD rather than HD, and this makes Lacan an idealist thinker: the thinker of the Dialectic of Two.

Therefore, returning to the question of this section, Lacan, in his middle period, was an idealist because he was a thinker of the Dialectic of Two, of SD and the operator of return, instead of the Dialectic of Four (HD = SD + the operator of the periodization).

The late Lacan, however, is a divisible thinker. We will see in the next section precisely in what way Badiou performs this division, and what exactly this offers us the necessary tools to realize the materialist shift in our era. But in the remainder of this section, we would like to clarify why, even in this late period, Lacan for the most part, remained within idealist commitments that he had developed during the middle period. We do that by focusing on two main topics that remain of interest and importance to Badiou's presentation in *Theory of the Subject*.

We noticed how Lacan's formalization of language, as a system of signifiers that always falls short of "hitting the spot" and allowing for the shift of meaning and surplus enjoyment, is for Badiou the same formalization that materialism can leverage today on several fronts:

1. To help to build a narrative against "idealingusitry" as the dominant ideology of the global capitalism. This narrative helps to isolate the outplace—the real—as the force incommensurable to the linguistic and metonymic manipulation in the system, which will serve as the dialectical counterpart to the place.
2. To articulate a theory of change as the outcome of splitting the place of the symbolic through the force of the real.
3. To come up with a theory of the subject where the above dialectic plays out in a (political) subjective act.

This formalization also allows for building the dialectic of thought and being, knowledge and known, identity and primacy, monism and dualism, and so on. But this formalization for Lacan stops at the concept of cause as a category of the subject. This poses a limitation for the materialist agenda, as this knowledge is the locus for both thought and the real. One way to understand this is to put it in the context of point 1 above. If the relation of thought with respect to being always has a remainder or surplus, inexorably exceeding the thought, then that remainder is like the force operating in the system of placement, or the system of representation of being. This force will eventually split the place (the representational space) into a new place that includes what was once a remainder. For Lacan, such a break is inadmissible. There is absolutely no crossing to the real. But for the new materialism of the subject, due to the periodized nature of all systems of representation, such a crossing is not only admissible, but it is also of historical necessity.

As we shall see in more detail later, this is how the "knot" is ultimately supposed to function: the guarantee of identity of being and thought. This identity is in the real and not something in the symbolic or imaginary registers. It is of utmost importance in Badiou's system for being and thought to be the same thing. The real is essentially thinkable—when we cannot think it, it is not because of an unsurpassable chasm, like the one that is between the conscious and the unconscious minds, but only because the place, the system of representation, does not have the ability to name the remainder. Once we augment the system with that ability, we can make it name the remainder. This shows what a great discovery it was for Badiou to encounter Cohen's invention, the method of forcing. It was a vindication of his insight that materialism must periodize/regionalize Lacan in order to go beyond him, and, subsequently, in the process, to also go beyond idealism.

This doctrine of remainder points to the gap that exists between reflection and asymptote. It should be obvious now why this doctrine, according to which the new materialism, is not Kantian. The postulate of the identity (of thinking and being) itself is anti-Kantian at its core. The doctrine of the remainder only adds a proviso to this postulate: being is thinking plus a remainder. To this, we should also add that the remainder is the real of knowledge, or the force in the system of representation—which splits the place. By virtue of having a remainder, thinking (and knowledge) is historical: or as Badiou says, the remainder is thus what periodizes knowledge.[69]

We've seen how Lacan, during his late phase, did observe that any systematic representation has an unsystematic remainder, metaphorically formulated by the law of entropy: every specific order has chaotic excrements. Badiou's attempt here is analogous to Marx's inversion of Hegel. Lacan, in seeing the new dynamics of order and disorder, turns inwardly towards order. Badiou's regionalization of Lacan, which is much more than a Marxist inversion of Hegel, consists of three moments. In the first moment, Badiou generalizes this problematic to a new orientation in thought—not of just of jouissance and language, but of being and thinking, In the second moment, Badiou turns outwardly and towards the so-called disorder. In the third moment, Badiou relativizes the relationship of system and its remainder to a localized relationship. This latter moment is akin to the character of "unworkable" entropy, which is always relative to a given system—an aspect of this model that Lacan did not explore in his own use of it.

There is a brilliant historical anecdote that Badiou provides that we shall recite here. This anecdote comes from the development of number theory, from the Pythagorean era to Rafael Bombelli in the early Renaissance, and is an illumination of what such regionalization entails. In the knowledge of numbers, Pythagoreans only admitted whole numbers and their relations. That doesn't mean that the relation of a diagonal of a square to one of its sides or the circumference of a circle to its diameter was not known to them: they simply did not call these numbers. In only admitting whole numbers and their relations, Pythagorean knowledge was the reflection of numbers with an inadmissible remainder. As Badiou remarks, following the famous Lacanian "instance," the symbolic in this case posits as a law that the denumerable consists only of whole numbers and the impossible real the non-denumerable. In

Badiou's own terms, Pythagoreans were forcing the law of the place, which prescribes no place whatsoever for a number that is not whole. So, what happens? The force of the remainder of knowledge, the ratios known to Greeks, changed the place. By naming the remainder within the field where the lack is sustained, they were able to able to create a reflection out of an asymptotic segment. This is what Eudoxus achieved: he created a new classificatory arrangement that included the irrational numbers—"it forces the impossible, it symbolizes the real."[70]

Asymptote was sutured to the reflection, the remainder to the place, and the impossible real to the legality of what is possible. A similar move came to pass almost a millennium later, which brought the imaginary numbers into the fold of the reflection, which itself was an inadmissible real within the domain of the real numbers. Why are these examples crucial? Because they show the actual process in the field of science, which in Badiou's later development, is one of the four conditions of truth.

We saw earlier the division in materialism: the division according to algebra and topology, the knotting of which is the supreme problem of materialism. In Lacan's theory of the subject, this scission finds a consistent formulation: the knot. For Badiou also, the theory of the subject seems to fulfill the same fusion of the place and the outplace—but as formerly mentioned, not just within the subject, but also within being itself. This is why the theory of the subject must address the supreme problem of materialism: it is only in a particular situation of being—the subject—that the scission of two orientations of materialism can be fused together.

To compare these two models, Badiou provides the example of the masses as the real. Masses do not constitute a substance, and as such, they don't belong to the register of the symbolic: in other words, the State does not represent the masses. This means that masses are neither thought nor thinkable, which means the "there is" of the masses is the vanishing mode of the historical real. Masses do register within the symbolic, the State, only asymptotically. That is, the State can never fully represent the masses, and there is always an unrepresented/unsymbolizable remainder. At the same time, masses are not an imaginary construct, as when they are encountered, they are a real force. They can be neither grasped by the State nor by the party. This establishes masses as fulfilling the character of a register on its own.

HOW DO MASSES EXACTLY MAP TO THE OTHER TWO SENSES OF REAL: AS CAUSE AND AS CONSISTENCY?

Masses make history by registering a remainder within what the State represents as a whole. In this function, masses cause the periodization/

historicization of the State, which in its falling short, has been historically forced to become a more inclusive representation of the masses. Masses are also the sole principle of consistency in politics. Representation of the masses in the symbolic is registered by the consent that the State proclaims to have from them, either snatched up or peacefully acquired. At the same time, in the register of the imaginary, the masses are represented as social classes. In this sense, masses are the real to which both State and classes are tied, and whose consistency is a guarantee for the consistency of both State and classes.[71]

The question remains: Why is consistency in the domain of topology and not algebra? Apropos this division, Badiou proposes two definitions of consistency: weak and strong—consistency of the symbolic, the algebraic consistency, and the consistency in the real, the topological consistency. The strong consistency, which overdetermines the algebraic, is anchored within the adherence and excess, which comes together in the consistency of neighborhoods. On the other hand, weak consistency is based on the concept of the lack: for Badiou Lacan's cause is a weak consistency. This dichotomy of two consistencies, weak and strong, is itself a problematic belonging to philosophy proper: Should being be conceptualized according to cause or consistency?[72] Badiou provides the following brilliant recapture:

> Consistency supports the question: what about the being of what is? Causality, the problem of the supreme being, by which the world is the form of what is.
>
> Ontological question, theological question.
>
> Heidegger intends to deconstruct metaphysics, previously defined as the concealment of the first question by the second.
>
> Heidegger would like to put an end to the philosophical idea of a *guarantee of consistency by the cause*.[73]

This is a brilliant capture of the idealistic core of Lacanian theory: it is a metaphysics of the cause, which seeks to establish its connection to the cause.

Apropos this dichotomy, being could be called in three ways. Cause and consistency are primitive concepts, whose combinatory, recognized in the asymmetry of dialectical divisions will give us the following:

- The primacy of cause over consistency (of the algebra over the topology): being-placed, in a position of vanishing as to the object, and of dispersion-repetition as to the subject.
- The equilibrium of the cause and of consistency: being-structured gives the position of variation to the object and of similitude to the subject.
- The primacy of consistency over the cause: being-forced in a position of destruction to the object and of excess to the subject.

The above maps to three visions of the world as to how people live and think: the ordinary, the trade unionist, and the political and accordingly, one can infer two definitions of the subject, which will oppose one another:

- the subject is a consistent repetition in which the real ex-sists; (Lacan)
- the subject is a destructive consistency, in which the real ex-ceeds. (Badiou)

To conclude, the regionalization of Lacan is the path to go beyond the idealism of SD, which gives dominance to place over force. This regionalization calls for a materialist doctrine of the force against the place, of asymptote against identity, and must constitute a theory of the subject. This is the inquiry on which our next section will focus.

NOTES

1. "Lacan, I said earlier is our Hegel, that is, he presents the (idealist) dialectic of our time" (Badiou, *Theory of the Subject*, 132).
2. The discussion of dialectic here is in the context of *Theory of the Subject*, and thereby pertains to the classical, that is, Hegelian sense of the term, not the sense in which later Badiou refers to this term. We will return to the issue of dialectic, and the sense in which its revision and revival is tied to the thought of affirmation and infinity, in the final chapter of this work.
3. Badiou, *Theory of the Subject*, 132.
4. Jacques Lacan, "The Function and Field of Speech and Language in Psychoanalysis" in *Écrits, the First Complete Edition in English*, New York: W. W. Norton & Company, Inc. 1999, 197–268.
5. Lacan, "The Function and Field of Speech and Language," 241–242.
6. Jacques Lacan, "Beyond the 'Reality Principle" in *Écrits, the First Complete Edition in English*, New York: W. W. Norton & Company, Inc. 1999, 58–74. This paper was written while Lacan was an analysand and had completed the fourth year of his analysis.
7. Jacques-Alain Miller, "An Introduction to Seminars I and II – Lacan's Orientation Prior to 1953 (Parts 1 to 3)," *Reading Seminars I and II Lacan's Return to Freud*, 3–35. Albany: State University of New York Press Albany 1996, 17.
8. It is not so much the thing referred to that is important but the other to whom one speaks. See Miller, "An Introduction to Seminars I and II," 18).
9. Miller, "An Introduction to Seminars I and II," 18.
10. See the section "Late Lacan" in this chapter.
11. Lacan, "Beyond the Reality Principle," 73.
12. Lacan, "Beyond the Reality Principle," 73.
13. Jacques Lacan, "Aggressiveness in Psychoanalysis," in *Écrits, the First Complete Edition in English*, New York: W. W. Norton & Company, Inc. 1999, 82–101.

14. Miller, "An Introduction to Seminars I and II," 21.
15. Lacan, "Aggressiveness," 83.
16. Miller, "An Introduction to Seminars I and II," 28.
17. Jean-Paul Sartre, *Transcendence of the Ego*, New York: Hill and Wang 1991, 88–91.
18. Miller, "An Introduction to Seminars I and II," 30. As an aside, we should mention that the resistance referred to above does not reside in the daily life of an individual—the notion of the subject does not apply to those currents. In psychoanalysis, an individual becomes a subject only on the couch. But it is crucial to realize that the "handle to turn" the axis a-a' to *A-S* is not on the side of the analysand, but rather on the side of the analyst: it is the analyst in whose position a', and A are supposed to coincide (the alter-ego being the source of speech's constraints), hence the resistance, which prevents the shifting of one axis to another, concerns the analytic technique and implicates the analyst more than it does the analysand: when this shift is not possible, one must ask what on the side of analyst has prevented it from happening—Lacan would call this "counter-transference."
19. Miller, "An Introduction to Seminars I and II," 31–32.
20. Miller, "An Introduction to Seminars I and II," 32.
21. Lacan's involvement with structuralism was not straightforward by any means; as we saw in chapter 1, structuralists attempt to explain everything in terms of the first level, that is, in terms of a more or less mathematically determinate combinatory which plays itself out without any reference whatsoever to subjects or objects. While structure plays a very important role in Lacan's work, it is never the whole story at any point in Lacan's development. In *Seminar X*, Lacan associates the supposed progress of science (and structuralism's scientific pretensions were rarely kept secret) with our increasing inability to think about the category of cause. Continually filling in the gap between cause and effect, science progressively eliminates the content of the concept of cause, events being viewed as leading smoothly, in accordance with well-known laws, to other events. Science, attempting to suture the subject—that is, trying to evict subjectivity from its field—also tends toward a suturing of the cause.
22. Jacques Lacan, "Logical Time and the Assertion of Anticipated Certainty, A New Sophism" in *Écrits, the First Complete Edition in English*, New York: W. W. Norton & Company, Inc. 1999,161–175
23. Lacan repeatedly refers to this text as the purest representation of his method. For example, the following is from his *Seminar XX*: "If there is something in my *Écrits* that shows that my fine orientation, since it is of that fine orientation that I try to convince you, is not such a recent development, it is the fact that right after a war, where nothing obviously seemed to promise a pretty future, I wrote 'Logical Time and the Assertion of Anticipated Certainty'" (Jacques Lacan, *Lacan Book XX, 1972 – 1973*, New York: W. W. Norton & Company, Inc. 1999, 48–49).
24. This text shows the internal tension within Lacan's logicism. We will show that in this text, Lacan is "trapped" within the algebraic paradigm, but still, the topological is at least presupposed, whereas later, the topological would disappear when he is submerged in the modeling of the analytic experience in terms of the signifier during his middle period.

25. So technically, what was just presented in the previous section, especially the bit related to Schema *L*, belongs to this middle phase, that in order to preserve the coherence of the presentation, we included as a part of the early Lacan.

26. Through the texts we read from Miller, we can safely assume that SD is the only logic Miller would require: for him, it is indeed the logic of all logics. This fact proves the insight that Badiou had to go back to Hegel, because with Hegel, even though only implicitly, we get both SD and HD. De-suturing Lacan from SD allowed Badiou to recognize Lacan as the only thinker since Hegel who has seriously thought about a logic that thinks of indetermination within the general space of determination.

27. Jacques-Alain Miller, "Paradigms of Jouissance," *Lacanian Ink 17*, 10–46, Wooster: The Wooster Press 2000, 11.

28. Miller, "Paradigms," 12.

29. Miller, "Paradigms," 13.

30. Miller, "Paradigms," 17.

31. Miller, "Paradigms," 18.

32. Miller, "Paradigms," 19.

33. Miller, "Paradigms," 20.

34. Miller, "Paradigms," 21.

35. Miller, "Paradigms," 21.

36. Miller, "Paradigms," 21.

37. "The analyst's desire is not a pure desire. It is a desire to obtain absolute difference, a desire which intervenes when, confronted with the primary signifier, the subject is, for the first time, in a position to subject himself to it. There only may the signification of a limitless love emerge, because it is outside the limits of the law, where alone it may live" (Jacques Lacan, *The Four Fundamental Concepts of Psychoanalysis*, *The Seminar of Jacques Lacan Book XI*, New York: W. W. Norton & Company, Inc. 1998, 276).

38. Jacques-Alain Miller, "Matrix," *The Symptom 13, Summer 2012*, Edited by Jacques-Alain Miller.

39. The title of the third chapter in *Seminar XVII* is "Knowledge, a means of jouissance." See Jacques Lacan, *The Other Side of Psychoanalysis, The Seminar of Jacques Lacan Book XVII*, New York: W. W. Norton & Company, Inc. 2007, 39–53.

40. Miller, "Paradigms," 28.

41. It is interesting that Miller's text phrase this slightly differently: "However what discourses introduce is that the signifier *represents* a jouissance for another signifier" (Miller, "Paradigms," 29, our emphasis). But the whole spirit of this paradigm shift, in Miller's own words, is the "primal" relationship of signifier and jouissance—the fact that jouissance is embedded in the signifier itself. In that level, signifier is no longer a representative, nor is jouissance represented. They are interconnected in a primal sense. Without reading too much into this "slip," we nevertheless think the Badiouian presentation is a much better way to capture how the relation of jouissance to two signifiers could be played out.

42. See Chapter 1.

43. Miller, "Paradigms," 30.

44. Jacques Lacan, "The Subversion of the Subject and the Dialectic of Desire in the Freudian Unconscious," in *Écrits, the First Complete Edition in English*, New York: W. W. Norton & Company, Inc. 1999, 671–702.

45. Miller, "Paradigms," 33.

46. Miller, "Paradigms," 33.

47. Miller, "Paradigms," 33.

48. Miller, "Paradigms," 33.

49. Miller, "Paradigms," 34.

50. Miller, "Paradigms," 34.

51. Jacques Lacan, *The Other Side of Psychoanalysis, The Seminar of Jacques Lacan Book XVII*, New York: W. W. Norton & Company, Inc. 2007, 48.

52. One brief observation may be illuminating here. The ground for theory of the entropy is the unavailability of the thermal energy for conversion to work. In this sense, it is impossible to capture the thermal energy represented by the signifier of entropy and to re-inject it into the working system: this thermal energy is barred from signification, which stands for the working system. This clearly contravenes what Badiou is after, both in *Theory of the Subject* and *Being and Event*, in both of which there is an operator for converting excess into supplement (the operator we know after Paul Cohen as the operator of forcing). It can be conceived that a model based on thermodynamics may only breed regional theories of justice: a system of justice causes more injustice outside of the system—the welfare of one causes the misery of many. This can explain why Badiou had to go to an order of thought "lower" than the physical theories in order to find the operators that can support the category of justice: the operators of consistency and forcing in *Theory of the Subject* and *Being and Event* respectively

53. Jacques Lacan, "Truth, the sister of jouissance," in *The Other Side of Psychoanalysis, The Seminar of Jacques Lacan Book XVII*, New York: W. W. Norton & Company, Inc. 2007, 54–68.

54. Badiou, *Theory of the Subject*, 133.

55. Badiou, *Theory of the Subject*, 133.

56. Badiou, *Theory of the Subject*, 53–54.

57. Badiou, *Theory of the Subject*, 54.

58. Badiou, *Theory of the Subject*, 187.

59. Badiou, *Theory of the Subject*, 187–188.

60. "Neither God nor Man, in modem idealism, has the function of the organizer of being. The constituent function of language, which excentres every subject-effect, deactivates the materialist operator of the inversion of the inversion in the sense in which Marx spoke of putting Hegel back on his feet" (Badiou, *Theory of the Subject*, 189).

61. We thank Gabriel Tupinambá for providing the original. The above is but a slight modification of Gabriel's diagram.

62. The subjective aspect of discipline, as was alluded to in "Mark and Lack," involved the operator of fidelity on the side of the working scientists.

63. Badiou, in *Manifesto for Philosophy*, explains how philosophy is a space for the compossibility of all four truth procedures—science, politics, art, and love—and

as such, it should not be sutured to any of them. (See Alain Badiou, *Manifesto for Philosophy*, Albany: State University of New York Press 1999.)

64. For instance, see Steven Nadler, *Spinoza's Heresy: Immortality and the Jewish Mind*, Oxford: Oxford University Press, 2004.

65. Badiou, *Theory of the Subject*, 191.

66. Badiou, *Theory of the Subject*, 192–193.

67. See Chapter 1 for Lacan's conception of a subjective cause.

68. This goes with the other of Lacan deep insights we discussed earlier in this section that idealism is the conservation of the Other, not of being—or rather the conservation of representation against presentation. The common belief is that what needs conserving must be of the order of being, and therefore conservatism is defined by its perseverance in being. But as Lacan teaches us, it is exactly the opposite: conservatism, and its perseverance, is due to the lack in representation, and not being.

69. Badiou, *Theory of the Subject*, 202.

70. Badiou, *Theory of the Subject*, 203.

71. Badiou, *Theory of the Subject*, 228–229.

72. As a side, Lacan's categorization of philosophy in the register of the imaginary becomes uninformed, and itself is the result of him already making a philosophical commitment to categorize being under the concept of cause.

73. Badiou, *Theory of the Subject*, 234–235.

Chapter 4

Lacan Is Divisible in Two

We have come to a point where Badiou, by a splitting operation, parts ways with the idealist Lacan. The question of consistency, and ultimately the materialist disposition, puts the final wedge between the idealist and materialist Lacan.

The question of consistency for Badiou is both a philosophical and a political one: it is philosophical insofar as Badiou's project regionalizes Lacan. It is a political question, as this inversion will finally anchor consistency within the political subject. In this relation, what is important for Badiou is to untie consistency from its Lacanian roots: the knot and the imaginary.

subject (in the real) = representation of subject (reflection) + the part that remains outside of this representation (asymptote)

For example, a political subject such as the class party should be taken as an existent in the real and thus has a materialist capture. Therefore, there must be both an identity (reflection) and a remainder (asymptote). How can we identify these two sides of the political subject? Empirically, we see two manifestations of the political party: during insurrection and when it takes the State power. Badiou calls these empirical observations of the party the "two circumstances" and claims that, in each, the knot is different.[1] The knot and the cutting operate over the set of Lacanian loops superimposed by the political situation:

class = imaginary State = symbolic mass = real

In these registers, the knot is holding the three together. In the first situation (insurrection), the class that was tied to the State (imaginary to symbolic) is untied and, through this process, it gives its place to the real of the masses, while vanishing (the party vanishes in bringing to the fore the masses). The State falls to ruins, and class takes hold of the mass—the proletariat over the old dominant class. The second operation ties the knot in a different place: the new class sutured to the State represents a consenting mass. So, what we have is an operation of cut and an operation of linkage. What ties the knot depends either on its cut or on its linkage. The knot cannot be only cut; it has to be tied again. The subject crosses these two operations.

Here we should take note of a slightly vulgar, but totally warranted fact: the political procedures are knotted out of material elements which could endure even if the knot is cut. Such a luxury is absent in analysis. This is a simple way to ground why it is that Badiou's concept of the real is *relatively* inaccessible while Lacan's is *absolutely* so: the division (i.e., the cutting of the knot) of the individual at the clinical level leaves us with less than an individual, with residues that do not endure. The division and cutting of the knot in politics leave us with the individuated mass, cut from their class designations and thereby from the State, but perfectly accessible for thinking: to decouple the masses from their class status is not to fall on nothingness, as in the case of the decoupling of the "imaginarized" subject from its symbolic status (which leaves us in nothingness). It leaves us with an enduring material instance. In short, if we transpose the same model into a political theory, the philosophical commitment to the real as absolute appears as an ideological side effect of clinical psychoanalysis (which itself appears to psychoanalysts as a commitment to an imaginary character of philosophy). So, just as Badiou will criticize the logic of the signifier because of his trust that logic does not "disperse" when inconsistency is found *in a given logical system*, he can criticize the theory of the knot (as Lacan's second attempt to ground the real of the structure ontologically) on the basis that politics does not disperse when inconsistency is found *in a given political state*—what we get is rather a situation in which what functioned as the inconsistent exception to "a" consistency is now available for a new (constrained, but real) consistency. The abolition of the State does not lead us to total chaos; it leads us to relative chaos, in which a new orientation is hard but still possible. So, the transposition from clinic to politics, from the real of the individual to the real of the masses, allows Badiou to un-suture the theory of the knot from the ontological backing that Lacan gave it—a sort of "proof" that, even if the splace/outplace contradiction has been accepted as primary, consistency remains an effect of the symbolic and not a primitive term.

With this in mind, let us dig deeper into Badiou's surgical analysis of the operations of cut and linkage. The following figure shows an image of the Borromean knot (see figure 4.1). In the example above, if real/mass = red, symbolic/State = blue, and imaginary/class = green, the first cut undoes the link between blue and green (the cut that ruins the link that ties the old class to the old State.), it amounts to two moments/processes: the moment of separation and the moment of linkage. Badiou does observe that Lacan was aware of these two moments, and in fact, Lacan adjusted his teaching in recognition of these two moments.

subjectivization = algebra: algorithmic explorations of the chain
subjective-process = topology: the localization by neighborhoods.[2]

Therefore, if Badiou's analysis holds, in Lacan's mind, these two moments are still clearly different, yet his treatment of the second is still through algebra rather than topology (although the knot is a topological object!). This is so because, for Lacan, the existence of the knot is only provable by a cut—and cut is an algebraic function. In a sense, Lacan assumes proof by the existence of the opposite:

knot = cut, and ¬cut = knot
(if we cut, we are proving that there was a knot)

To Badiou (and, as we shall see, to the materialist disposition of consistency), this is a wrong assumption. The cut is not the opposite of the knot. It is the *destruction* (apropos communism that destroys the state and classes) that is the opposite of the knot.[3] In communism, there is no knot, since the

Figure 4.1 Borromean Knot. *Source:* Author created.

imaginary classes do not exist, nor does the State. In fact, the correlate of two out of the three registers do not exist in communism, classes and State (it is totally a different matter what happens to the correlate of the register of real, that is, masses—apropos the uninterrupted subjective process that we will see later). Destruction destroys the knot altogether. In contrast, cut not only does not destroy anything, it assumes the existence of the three correlates, each mapping to a register. Within the paradigm of SD (that is, idealism), where consistency is an imaginary category, a cut is analogous to dispersion. However, within the paradigm of HD (that is, the materialism of the dialectic of four), consistency is a category of the real, and the destruction of the knot does not result in dispersion—unless HD relapses to SD. Therefore, it is completely legitimate for us, in the context of the paradigm of HD, to talk about the operation of recomposition, since the cut did not annihilate anything.

This different treatment of SD and HD of the category of consistency is echoed by Lacan when he says the knot is not real, nor is consistency, but they "ex-sist." Lacan differentiates between existence and ex-sistence. For Lacan, existence pertains to the symbolic. Only what is integrated in the symbolic order "exists." In that sense, "reality," as what is comprised of existents, is also what pertains to discursivity.[4] In addition, everything that exists in the symbolic only exists by virtue of its difference from others, including its own absence.[5] This is Saussure's differential description of language.

In contrast to the above sense of "symbolic" existence, there is an existence that a signifier cannot register. In this sense, it is only that which is impossible to symbolize that exists: such as the impossible thing at the heart of the subject.[6] This is the existence of the subject of the unconscious.[7] This second sense of the term is exactly the opposite of its first sense. Lacan uses the term "ex-sistence" to refer to this second sense. He borrows this term from Heidegger to say that, while reality exists (through the symbolic), the real "ex-sists." The neologism ex-sistence expresses the idea that the heart of our being is also radically Other, strange, outside; the subject is decentered, its center is outside of itself, it is ex-centric. Lacan also speaks of the ex-sistence of desire in the dream, since the dream cannot represent desire except by distorting it.[8]

There is no temporal distinction between existence and ex-sistence. Although, for illustration purposes, we may indicate that the real is "prior" to the symbolic, the actual temporal relationship is not necessarily determined. The real ex-sists alongside the symbolic. It can be what is not "yet" symbolized or resists symbolization. For example, psychoanalysis enables the analysand to verbalize an experience that was never symbolized. Ex-sistence, in this case, is warranted by not having a symbol for that which nevertheless asserts itself, in a relation which the Lacan of the 1950s would have called "extimate."

Lacan's debt to Heidegger regarding the concept of ex-sistence carries with it a certain commitment to think of being in opposition to logos, posing consistency as a category of the latter (the logos, the idea), while Badiou seeks precisely to surpass Heidegger, by placing consistency at the point of contact of being and thinking (consistency as the very force which passes through the impasse of formalization, that is, it binds the disjunction of being and thinking without suturing them). Lacan shows his indebtedness to Heidegger by thinking of consistency as a property of the global logos, in opposition to the opening of being, which he thinks of in terms of dispersion.

In a passage that Badiou quotes from Lacan, we find evidence for a similar understanding of the category of consistency. We can extract the following important assertions from this passage:[9]

1) What causes consistency ex-sists in the real, and like everything in the real, it withdraws from the inscription by the signifier.
2) This cause has an effect on what consists.
3) What consists turns the aforementioned absence (from reality) into the representation of consistency in the imaginary.
4) So, consistency has an imaginary *inscription* and a real *ascription.*

Next, Badiou focuses on Lacan's passage and makes important observations as follows:

1) The Knot is of the real—aside from the real that is knotted to the other registers, the knot is also *ascribed to* the real. That is, what we see between the three loops, the one-of-the-knot, is the trace of the real of the knot that causes what we have symbolized as the topological knot. except that,
2) The one-of-the-knot would find the guarantee of its consistency in the imaginary rather than in the real, except that,
3) The real causes the imaginary consistency to ex-sist, "whereby it touches upon the scission of the very being of consistency into object-cause (object-knot), on one hand, and ex-sistence, on the other."[10]

Badiou ties the above reading together with his own developments in the following: "You must not think for a moment that this inconceivable circle is pure sophistry. To tell the truth, it is what we witness everyday."[11] In other words, Badiou sees in Lacan's "there is something of the One" the operation of count-as-one, as Badiou develops in *Being and Event.* This "everyday" operation shows that consistency (the count) is not yet the formation into One which consolidates the state; it is possible to subvert the count (the generic subset). There is something of the One in the real of the situation (in the site, which is un-structured or un-situated only relative to that situation): once the situation is "destroyed" from the standpoint of its state, it is still possible to produce out of the 'ultra-one' of an event (that is, out of the weakness of the structure) the basis for a situated and indeterminate counting process (a real—since it is situated and spatially/temporally bound—consistency). We shall return to this reading in the next section.

If consistency is largely imaginary, as Lacan says, and if the only way for this "ideal" of imaginary consistency to stay alive is to prevent its transformation into a utopian demand, of the plebs that come to disturb society, for example, consistency must keep its footing in the real; it must keep its vanishing effect by having an ex-sistence in the real. Imagine what will happen if an impulse of the masses cuts the knot: it will let loose the imaginary ideal as a way to compensate for the knot's dispersion. Such is what gives impetus to the anti-revolutionary rhetoric and what constitutes the essence of conservatism—the fear of the terror of the imaginary ideal. Consistency is imaginary: How did Lacan "suddenly" turn so an anti-Hegelian? Lacan departs the realm of philosophy when he asserts that philosophy itself belongs to the register of the imaginary. The core of this proposition is this: the knot, a coefficient of the real, cannot be replaced by the ideal of consistency that is from the imaginary register. Therefore, if the real of the knot is cut, by another force equally in the real, the consequence is dispersion, and, subsequently, terror.

WHAT IS THE REMEDY TO THIS?

Lacan taught us that the real anchors causality: the real is the vanishing cause that puts the entire symbolic order in motion. To bring home this novelty, to a home where thoughts of change can nourish and grow, we must make sure that the real also anchors consistency as much as causality: "we must determine the subject in the subjectivizing differential *and* in the being of the subjective process."[12]

Let's have a quick recap in order to zero in on the novelty that Lacan inaugurates.

We started with the idea of two dialectical matrixes in Hegel. Having two dialectical matrixes is the only way to maintain the idea of periodization, and for Badiou, the idea of periodization, as opposed to circularity, is the only acceptable "outcome" of dialectical thought. But we should also recall the motivation behind the pursuit of such a dialecticity—dialectics is the logic proper to the thought of interiority, a logic independent to what lies outside—what underlies the domains of thought that are uniquely circumscribed to produce truth: science and politics. We saw how, through Lacan, the dialectical matrixes of SD and HD find a full articulation using the materialist principles of identity and primacy. We saw Lacan's articulation of these two principles in the formula that states jouissance (real) is never exhausted through the representation of jouissance as desire (signifier); there is always an un-symbolizable remainder.

This formulation also relates to what we call the Dialectic of Four: comprising not only two premises (splace/outplace) but also two processes (SD/HD). Lacan teaches that under the law of the whole over parts, we never find consistency within the whole itself, which is equivocal and historical. Where does this consistency that we experience come from? While to the middle Lacan, apropos his structuralist orientation, it still comes from the deterministic laws of the whole, the causality of the structure, but with the late Lacan, this determinism of the whole is just a smokescreen, that is, in fact, externally caused by the real, and the whole itself—the system of representation in language—is contingent and impossible. The Other is the retroactive positing of this impossibility which manifests itself within the insistence of the signification.

The novelty is thus this: we discover the relation of the whole to its parts to be just a brittle and weak constitution; the structure could never be determining, as it is always already anchored in contingency, and it is, by signification of this impossibility, through symbolic operations as well as imaginary constructs that it could buy the consent of its constituents—the masses—of their domination; therefore, this domination could never endure without thick protective layers of ideological reversal, conservative protectiveness, and idealist capture.

Lacan's proposal for where the (historical) consistency that we experience comes from is thus not the symbolic but the imaginary register. As said earlier, this must have been discouraging for Badiou to witness how Lacan, after having taken us so far, just expels consistency from the symbolic in order to re-ascribe it to the imaginary register.

Badiou proposes something different. For him, consistency is in the real itself, and the real is its guarantor. In our view, it is rather straightforward to see how this difference in position ties Lacan and Badiou back to their founding philosophical commitments—that is, how Lacan, by finally depositing consistency in the imaginary in order to preserve the knot, as the sole index to consistency, stands to be an idealist, and how Badiou, by taking consistency as real, makes him a materialist. If one believed (as Lacan did) that the knot is what is in the real, whose effect—the keeping of the three registers together—is what the imaginary idea (or ideal) of consistency captures, couldn't we say that this exactly performs the aforementioned idealist reversal to sell a capture of what *there is* as what *is*, as selling representation for presentation? The imaginary status of consistency, as a historical ideal, serves as the perfect counterargument against the need for regulatory interventions against unfettered financial markets—this is just a childish dream of uninformed plebs to think economy and consistency together, a dream whose actualization will no doubt undo the knot that holds their contemptible lives together! In response to this, if we assert that the knot itself stands for the retroactive capture of the impossibility of inscribing the real of consistency within our imaginary and symbolic apparatuses, don't we sound a lot closer to the late Lacan? If so, Badiou is more Lacanian than Lacan was (the same way that we saw how he was more Christian—or at least Pauline—than Hegel).

Now, for Badiou, it would be just an empty—materialist, but nevertheless still empty—claim to say that consistency is anchored in the real because this does not result in any specific political theory, let alone a political theory that is committed to politics as an interiority, a discipline. But Badiou did not stop there: he identifies exactly, insofar as it concerns politics, the "location" of the real, where we shall find the anchorage for the consistency of the practice of politics, as "the real of the subject," of the political subject, as the guarantor of the consistency, "without the mediation of the imaginary." The second important question is: How is this real "reachable" and not just a vanishing moment? For Badiou, the real of the political subject is not merely a vanishing *cause* (as it is also a consistency) but also not a *vanishing* cause (as it endures in a process). The answer for Badiou, fashioned after Lacan's *analysis situs*, is also the subjective-process: you allowed the beast to come out of the cave, or the chained to emerge to the light (subjectivization), and you must now maintain its presence by what is analogous to the "analytical situation," that is, the "uninterrupted" subjective process.

That is how Badiou proposes to proceed, and proceeds he does, by honing onto "the infamous gap through which the same passes over into the other."[13] This is where his reading of Lacan's "Logical Time" takes over, a reading that is supposed to clarify the fate of his proposal, and "identify his future," the same way that in 1843 a reading of Hegel's philosophy of right identified Marx's future. Why "Logical Time"? Because the peculiarity of this text, and we shall see that shortly, had allowed Lacan, and now Badiou, to elicit, in a purely logical form, the notions of subjectivization and subjective process.

"Logical Time": Some Introductory Remarks

Badiou's presentation of subjectivization and subjective process hovers around yet another encounter with Lacan: the obscure text "Logical Time." This is a very special early text, written during the Second World War and first published in 1945. It is one of three texts that Lacan planned to publish in a book under the title of *Collective Logic*[14] (the other two called "The Number Thirteen and the Logical Form of Suspicion,"[15] the other called "British Psychiatry and the War"[16]). At this time, he had already written his famous doctoral thesis on paranoia, he had written the mirror stage, and he was very much involved in the Hegelian humanist movement in France, under Kojève and his reading of Hegel, which is very much defined by its interpretation of Hegel's *Phenomenology* as the key to a kind of new anthropology. Kojève had a strong footing in Heideggerian philosophy as well and was of the existential disposition: for him, concrete dilemmas evoke ontological problems. In addition, Kojève was also a Marxist. He finds in the dialectic of master and slave a way to reconcile the two—a sort of dramatization of the constitution of desire connected with the problem of domination, also of an autogenetic nature. This account was Heideggerian since, in desire's coming into being, there is a passage from negativity into a more situated incorporation of desire in Dasein. At the same time, this account emphasizes the intersubjective relation, giving the Other an upper hand. This makes Hegel a particularly interesting thinker, as an anthropologist of desire whose concept of negativity involves something of intersubjective nature. On the other hand, Kojève connected the dialectic of master and slave to class struggle, superimposing the problem of domination to the problem of the end of history.

Lacan struggled with all this, as he liked some of its aspects (in particular, the fusion of different orientations) but not others. For instance, he did not care much for the Marxist tone of Kojève's teachings, but he was really intrigued by the idea of autogenesis through which the constitution of the desire of human being is connected to an assumption of negativity through the other, an account compatible with Lacan's mirror stage.

Coming from such involvements immediately prior to the war, "Logical Time" reveals yet a third aspect of Lacan's original thoughts, which happened to shape the very way in which his development evolves: the fact that Lacan was not simply interested in the dialectical view of desire; he was also interested in a logical and formal treatment. Although in "Logical Time," you find the problem of the constitution of identity, the riddle whose solution involves the other, the dialectic is not a prominent theme within its presentation, and intersubjectivity is not so much ontological as it is logical. In this way, it stands as a parallel universe to the mirror stage as a genesis of subjectivity.

Lacan, who during the 1930s was interested in surrealism, the poetry of the time, the general ambiance of the Spinozist humanist theory of desire, the fold of nature, Kojèvean intersubjectivity, and so on, began to write differently after World War II as if he was trying to separate himself from his earlier humanism. Lacan's collective logic consisted of a newfound formalism, a framework through which he struggles with the problem of dialectics during the 1950s. Through the theme of collective logic, Lacan moves closer to the problem of cybernetics, which will culminate in "Seminar on 'the Purloined Letter'," where he explicitly mentions it.[17] His second seminar is also filled with references to cybernetics. This new framework is contemporary to the first general model of cybernetics by Norbert Wiener.[18] However, this wasn't a sudden shift, as Lacan was always interested in the intersection of surrealism and mathematics from the 1930s and 1940s. But the shift in his work becomes markedly visible only after the war.

The topic of collective logic was also a topic that had been developed in France at the time. Guilbaud wrote a book called *Theories of General Interest and Logical Problem of Aggregation* and used problems by Pascal and Condorcet to discuss how the aggregation of many people and the problem of rational choice overdetermines the space of possibility—how connectivity carries over or creates constraints for the behavior of individuals, constraints connected to the very being of the collective and not to the inner structure of its parts. It was very much the beginning of complex studies in science. So, from Guilbaud, who studied elections and came from a background of economics and social theory (and who was a personal friend of Lacan) to Wiener, who shortly laid the foundations of cybernetics and introduced theories of information that allow us to study any system from the cybernetic point of view—Lacan was actively involved in the development of logical and formal theories of collectives and networks.

Another significant figure in this regard is Le Lionnais, who famously published a collection of important mathematical papers in France around this time, and was the one who evidently told Lacan about the prisoners' dilemma and the prisoner's sophism at a dinner party.[19] At that time, Lacan was looking for a formal field that would allow him to theorize the Kojèvean universe without being bound by its anthropological and humanist references. He was already trying to see what survives of the intersubjective perspective if we don't essentialize anything about human beings, and if, instead of anthropological or even ontological origins, we could depart from logic alone, which will eventually culminate in the linguistic orientation. Therefore, Lacan moves from collective logic to cybernetics to linguistics and structuralism, a progression toward a formal and logical framework that will finally allow him to think about the problem in a rigorous way. "Logical Time" is, therefore, a pre-linguistic, pre-Saussurean, and pre-Levi-Straussian text, though it clearly struggles with the same objectives: basically, how to have a logical theory of the subject without any substantialist commitment.

To understand the relevance of this text and the importance it had for Lacan—to understand this sophism—we have to understand how difficult it was in France to come up with a combination of logic and subject in the face of dominant intellectual currents of the time. You had to either take a humanist perspective on the side of Sartre and existentialism, where you could keep the category of subject, or you had to side with science and structuralism and the epistemological theory of Bachelard and lose the subject. Is there a middle ground between the philosophy of science developing in France, related to Spinozism and connected to thinkers like Cavaillès and Lautman, who published extensively on the rational basis of logic on the one hand, and on the other, the humanist tradition of subjectivity with no logic and mathematics? The important point here is this: the birth of cognitive science served as the initial mediator to bind together a deterministic theory of parts with a non-deterministic theory of the network of interactions in such a way that the resulting theory did not require any essentialist presuppositions regarding the constituents of the system, for example, whether they are essentially free or not, yet at the same time, was able to show how, in a closed system, the interactions of parts could result in constraints that appear to come from nowhere, that is, irreducible to the parts of the system and external determinations. For instance, we could have a set of individuals that, in their interactions, produce certain constraints of which they are totally unconscious, in the sense that the participants do not perceive these constraints or laws as coming from them willingly, but at the same time have no external causes either—the constraints are produced immanently by the dynamics of the particular network of communication. For example (and closer to Lacan's time), the study of elections and how electors behave, or the study of financial speculations developed by Keynes and other economists, showed that such behavioral constraints emerge in a given network of interactions. Another important discovery, mainly attributed to Von Neumann, was the study of the economy through a new branch called Game Theory that investigates logical constraints for the behavior of rational (self-interested) individuals in a given domain. This theory, which is based on the relationship between "the ego" and the "other," can be read as an attempt to unite the economic branches mainly based on confrontation (Hegelian) and coexistence (Cartesian).[20]

It is quite obvious that all of this interested Lacan immensely. These developments allowed him to bind together the two poles of French philosophy. On the one hand, the philosophy of reason and science, and, on the other, existentialism and the Freudian doctrine. In developing a logic of collectivity, Lacan was not pursuing political interests, nor any interest other than what mattered to psychoanalysis. His problem was how to let go of Freudian anthropology and other essentialist theories of the human, and at the same time, keep true to the constraints on individuals that the Freudian doctrine seemed to impose.

The dinner party discussion on the problem of sophism was just the spark that Lacan needed. This is a story that begins with a riddle about self-knowledge. Self-knowledge here comes from enriching my knowledge of the other, the fact that the other plays a role as the mediator in the constitution of my knowledge about myself. More crucially, the other plays this role in a very particular way, one not based on direct communication. Rather, this knowledge is based on what doesn't work—that is, what doesn't work constitutes the information within the process. There is an eruption of the cause, associated with the event of haste and the anticipation of certainty, that plays a part in the prisoner's capacity to decide and to know, which doesn't come from outside: there is no external information affecting the decision. Rather, something remains to be decided, and one must act upon this lack of decision. We think this is what probably interested Lacan since it resounds on the epistemological level with a formal treatment of social interaction without substantialist presupposition. Also, this sophism has a role for negativity, an indeterminism, that harkens back to the existential tradition because the act that Lacan seems to formalize is not the act of assuming what the other knows about me, but a sort of decision and movement that happens precisely when this assumption is impossible. This brings several interesting elements together for Lacan to begin a formal treatment of his Kojèvean existential problem.

These remarks harken back to the "Early Lacan" section of this chapter. As we tried to show then, there exists a separation between the concepts of model and theory. While Lacan's model has definitely evolved through various stages of his intellectual development, the core of his theoretical interests remained relatively stable. One way of summarizing Badiou's critique to Miller in the previous chapter is that Miller's famous texts in the 1960s attempted to suture the theory to a specific model—the signifier: a movement to which Lacan was not so much committed himself. This brief historical exegesis illustrates the separation of the two concerns clearly. Lacan's theoretical interest, his main theoretical *aporia*, stayed focused on an account of the indeterminism of the subject, as required by the theory of the unconscious, without committing to a substantialist view of the subject, as existentialism and Freud's own doctrines tended to do.

We probably find the purest form of Lacan's theoretical concerns presented in the text in a logical form—the form of a sophism. This is why Lacan himself repeatedly refers to this text in his middle and late periods.[21] The original text, written in 1945, was edited for print by Lacan himself in 1965, to be placed "between the before and after."[22] But Lacan repeatedly referred to it in his seminars in the 1970s. This, perhaps, is the only text to which Lacan refers on multiple occasions in his teachings. Among these, the reference in *Seminar XX* is worth quoting:

> If there is something in my *Écrits* that shows that my fine orientation, since it is of that fine orientation that I try to convince you, is not such a recent development, it is the fact that right after a war, where nothing obviously seemed to promise a pretty future, I wrote "Logical Time and the Assertion of Anticipated Certainty." One can quite easily read therein—if one writes and not only if one has a good ear—that it is already little *a* that thetisizes the function of haste.

> In that article, I highlighted the fact that something like intersubjectivity can lead to a salutary solution (*issue*). But what warrants a closer look is what each of the subjects sustains (*supporte*), not insofar as he is one among others, but insofar as he is, in relation to the two others, what is at stake in their thinking. Each intervenes in this ternary only as the object *a* that he is in the gaze of the others.[23]

"Logical Time": A Sophism

"Logical Time" is built around a thought experiment. A prison warden summons three prisoners and announces that, for reasons that don't matter to them, he shall free one of the prisoners. In order to decide which prisoner will be set free, he will rely on the outcome of a test. The three prisoners are to carry on their backs a disk with a certain color. The prisoners cannot see the color of their own disk, but only that of the other two. The warden will select the disks from a collection of five disks, two black and three white. They can look at the color of the disk on their companions' backs but they cannot, of course, communicate the color to each other in any way, which obviously would not be something the prisoners want to do anyway. The first prisoner who is able to deduce his own color, and describe his deduction based on logic, not probability, will be the one who shall be discharged.[24] In the end, the examination of the answers will be done on an individual basis. The prisoners acknowledge that they understand the terms. Then the warden adorns each prisoner with a white disk. The three prisoners contemplate for "a certain time," and then they take a few steps together, passing side by side through the doorway. They separately provide a similar response, something as follows (the so-called perfect solution):

> I am a white, and here is how I know it. Since my companions were whites, I thought that, had I been a black, each of them would have been able to infer the following: "If I too were a black, the other would have necessarily realized straight away that he was a white and would have left immediately; therefore I am not a black." And both would have left together, convinced they were whites. As they did nothing of the kind, I must be a white like them. At that, I made for the door to make my conclusion known.[25]

All three thus exit simultaneously, having the same reason for concluding.

To Lacan, this problem only has a "logical value"—but a very particular logic imbued by a remarkable sophism. Let A, B, and C represent the three prisoners and let the analysis proceed from the perspective of prisoner A. Let's assume A thinks he is black,[26] and then B would think: if I am black, then C would have left immediately. C would think the same. Now that B is waiting for C to leave and C is waiting for B to leave, to prove to themselves that they are each black, we have a deadlock. Since neither of them has left, which allows each to believe she is white. Their hesitations for but one instant would clearly suffice to reconvince each of them beyond the shadow of a doubt that she is white. For hesitation is logically excluded for whoever sees two blacks.[27] In other words, the very existence of the so-called deadlock is proof that neither B nor C is black.

Let's now assume that A has thus concluded that he is white (had he been black, B and C would have realized that they are white and would have left long before) and starts to leave; he must abandon his decision once he sees B and C are setting off with him. But, in this case, the dilemma is this: If A, seeing B and C set off with him, wonders again whether they have not seen that he is black, it suffices for him to stop and newly pose the question in order to answer it. For he sees that they, too, stop: since each of them is really in the same situation as him.[28] If A were black, then his hesitation would not have mattered to B and C. They would not have hesitated as a result of his hesitation. If they hesitate, it means that A's hesitation does matter, and thus A is not black.

Lacan's analysis shows that there are two moments of scansion: the moment of scansion between B and C and the moment of scansion among all of them. These moments have two different logical values: a single hesitation suffices for B and C to know that neither of them is black. The second scansion evinces different information: that A is not black.

Prima facie, there are two types of information the prisoners rely on: the color and number of disks, and the signals that their reactions send to each other. Lacan calls these the special information.[29] Lacan argues that only relying on spatial information will amount to a restricted version of the logic. In contrast, in order to properly capture sophism, Lacan proposes an alternate temporal logic. In the spatial logic, it is what prisoners do, whereas in the temporal logic, it is what prisoners don't do. It is only within a temporal frame that hesitations are noticed. If the logic only looks at the action after the pause and pays no attention to what the pause signifies, then no certitude could be attained. Only temporal logic is sensitive to the pause itself, as an absence of action. Without the scansions or the suspended motions, no certitude is possible. In this case, for Lacan, the suspended motions are not experiential data, but part of the logical process. In addition, scansions or pauses are seen as such if only we were to place them in a temporal sequence. In this case, the temporal sequence comprises three temporal instances. Laying out the three instances in chronological succession will amount once again to an alignment of signs, and to the spatial logic. What we need is a mode in each of these moments such that they preserve their hierarchy, to reveal in them a tonal discontinuity.

In the first modulation, Lacan presents the possibility of being opposite to two blacks, according to which one would know, immediately, that one is white. In this case, there is a logical exclusion, which is anterior to the subject's movement. In other words, the subjectivization, in this case, creates an impersonal subject: "One knows that . . ."—*protasis* leads to *apodosis* regardless of anyone who is involved in the logical process, and at the instance of the glance. Let's call this time t0. In this case, there are no black disks, so the moment of glance will not create an impersonal, "noetic" subject.

The second modulation starts with an assumption, that I am black, and a conclusion that if this is true, the two whites that I see would waste no time realizing that they are whites.[30] In this case, from the instance of glance (t0), we come to the process of comprehending. First, in the logical hypostasis, A assumes that he is black. Seeing B and C pause, he then concludes that B and C know that they are white since, otherwise, one or both would have bolted out right away.

At this time, if B and C—assuming A is black—get up and walk, then A concludes that he is black and will move with them too. Let's call this time t1. It is obvious that there is a pause between t0 and t1, and thus t1 > t0. So, the decisive factor here is the first pause. Here Lacan remarks again that this pause is objective—similar to the instant of the glance. Although the interval of the pause may vacillate, the meaning emitting from it is the same.[31]

So now A is at time t1: if I am black, then the first pause must show to B and C that they are white, and thus they should get up and go. Thus, t1 is the instant that a judgment is made based on the first branch of the original decision tree.

Now things get interesting (more subjective and less objective). The time t1 is the amount of time for A to conclude the first judgment. He then expects B and C to get up and walk, assuming that he is black. But they don't. There is one more pause. Now we reach another conclusion: I'm white. If I were black, then at t1, B and C must have left. They didn't leave—they are still paused—so I must be white, so I better rush to declare myself before they realize they are also white. Let's call this moment t2 > t1. The main supposition of A to differentiate is this: if I have reached time t2 (A's moment of conclusion), then it must be the case that no conclusion was possible at t1 (A's moment of comprehending) or t0 (the moment of glance).

This leads to what Lacan calls the "ontological affection of anxiety," the emergence of which is a signal for the third moment, which he calls the "subjective assertion": the personal form of the knowing subject which can only express itself as "I." Said in another form, the logical process that we have seen here as sophism "can only be borne by a subject who has formulated the assertion about himself" and not by anyone else. Tied to the personal affection of anxiety, this conclusion is also very personal—and hence is not reciprocal, reversible or assignable to anyone else.[32]

Understanding the type of subject correlated to each of these moments will highlight the difference between this "ontologically grounded subject," and the subject of the assertion. Associated to t0 provides only the general form of the noetic subject: "he can as easily be god, table, or washbasin." In contrast to this "noetic" subject, after t1, expressed in "the two whites" who must recognize "one another," introduces reciprocity since the one can only recognize himself in the other and only discover his own attribute in the equivalence of their characteristic time. This subject is still determined through the other and is not yet ontologically grounded. After t2, the "I," the subject of the conclusive assertion, is isolated from the other—that is, from the relation of reciprocity—by a logical "beat." Three moments and three subjects: the first two subjects are fully determined, either by the *hypothesis* (apodictic conclusion from the given premises) or the *apodosis* (inferences made from the assertions made by the other), but the third subject is not grounded deterministically—that is, its moment of subjectivization is free from the determination of what has transpired either hypothetically or apodictically. Lacan compares the third moment to the Heideggerian phenomenon of anxiety through which Dasein comes to encounter the possibility of its own being—dislodged from the home within which it usually experiences itself. The "I" that is instantiated in this way is the outcome of "a decanting of its own logical time," which "largely parallels its psychological birth."[33] This moment of anxiety is a specific moment for Lacan, whose experience in the logical time is coupled with a sense of competition that the subject experiences with the other. The subject's anxiety regarding its own identity is because of the uncertainty of the conclusion.

If, before reaching t2, prisoner A sees everyone else is leaving, A will leave with them nonetheless (at the time, since he hasn't reached t2 and is still at t1, he assumes he is black). The uncertainty of what has transpired is the most obvious from the point of view of a spectator, who observes that, in any case, they all move together: Is this move based on a correct computation, or is it a reaction to the movement of the other? The point is, the moment of subjectivization for Lacan remains the moment t2—the moment at which the subject is not externally determined. Regardless of whether all prisoners manifest their judgment through an act, the prisoner who is subjectivized according to the moment of t2, will act solely in accordance with the reasoning that gained him this moment of "internal" determination. The difference between this and the moments prior to t2 is that, for the prisoner passing t2, there is an anticipated certainty. And the anticipated certainty is what differentiates the subjective assertion from the other types of subjects (which are determined at moments t0 and t1).

"Logical Time": Dividing Lacan

Badiou, totally fascinated with the text of "Logical Time," uses this text to draw out two processes within which the subjective is determined and maintained. Though he is mostly in agreement with Lacan's analysis, Badiou questions its presuppositions. As we shall see, these un-presented elements are familiar to us: they concern the same impasse of the algebraic grammar of the subject as the one which we found in Lacan's fourth paradigm[34]—the overdetermination of the topological by the algebraic, leading Lacan to skip over the possibility that the outplace could found a new law, rather than merely interrupt a process—or, that the real could found a new consistency, rather than causing the reconstitution of the structure.

Badiou states that Lacan's presentation has a flaw and it cannot hit the mark at the end, which consists in presupposing that there exists absolute reciprocity, a strict logical identity between the three prisoners.[35] Lacan runs everything, leading to the conclusion, as three identical logical machines. If what Lacan insinuates was the case, then we can summarize the deduction on the side of each machine—the prisoners—as three deductions:

- R1: The moment of glance, personal and non-subjective. A sees two black disks therefore, A is white—this corresponds to t0 in the presentation we formerly saw.
- R2: The moment of the first inference: A sees one black and one white. If A were black, then one would walk according to R1. This means R1 is part of the reasoning R2—this corresponds to t1 in the previous analysis.
- R3: The moment of the second inference: A sees two whites. If A were black, the others would see one black and one white. So, they would reason according to R2. Thus, R2 is part of the reasoning R3. This corresponds to t2 in the former analysis.

This gives us the following inclusion: $R1 \subset R2 \subset R3$. This is obvious if we recall: $t0 < t1 < t2$.

This is the case only if the prisoners are identical logical machines. But if this is the case, then there is no reason for doubt for scansion in R3, as the time for both R1 and R2 is over. That R3 exists means that R1 and R2 were inconclusive. This is apodictic. In fact, the identical logical machines will all move at the same time with surplus certainty and not suspended doubt.

The problem is, therefore, that the scansion should no longer happen, since everything is apodictic. There is no subjectivization since there is no retroactive confirmation of the cause. There is neither the first haste (the result of anxiety or courage) nor the second haste (the result of superego or justice).

Regardless, in Lacan's own presentation, a periodization (the qualitative jump from algebraic to topological) does take place; both processes seem to be present. If algebra, however, could only produce the apodictic or objective logic, then where does the subjective logic come from? "There must be something *that Lacan does not say*."[36]

There is a crossing, observable in the real, from objective to subjective logic. We proved that algebra couldn't account for this crossing. The law of the splace couldn't account for this crossing. Either the logic follows the algebraic predication, with no hesitation, or there is hesitation and haste, the non-algebraic subjective logic, which can only be the case if our original presupposition is untrue (the three prisoners are non-identical. The experiential fact of scansion is only possible on the basis of the non-identity of the other: the fact that there are different "speeds" of reasoning. The fact of non-identity adds something that is beyond just the measurable speed of reasoning: it pertains to qualitative elements in every subjective determination.

Once the qualitative difference, as the only ground for explaining the experiential fact of scansion, enters the overall assessment, a whole slew of other possibilities will enter into the assessment that was simply absent from a purely algebraic logic through which one could evaluate the other: speed, haste, stupidity, and so on. Badiou calls this the logic of neighborhoods. The logic of neighborhoods extracts from topology not the "global" structures, but rather from the "local" relationship between determination and indetermination, which characterizes a point and its vicinities. Here, a place is not so much a determinate difference but a certain "open set" of differences: given a certain metric, all the points *p* in a certain region of space will be said to occupy the same place—but a different metric could distinguish them, producing a new consistency out of the interiority of the place. We could even say that the topological grammar was the depositary of a profound philosophical idea—to be later expanded in *Being and Event*—that infinity can be thought as a density in place (as an infinity of points constituting, for a metric, a discrete difference). So, what Lacan calls "competition" not only has an algebraic, that is, objective, element but also a qualitative element. This makes the subject the result of a purification of force in the tension to resolve the algorithm.

> By supposing that the other stands in a position of a possible difference of force, I subordinate my experience to the notion that every occupation of a place restores the law on the basis of the outplace.[37]

Every possible placement is, at the same time, an enablement of force. In order for the differential to play a role in my escape from prison, the entire field should be structured according to this differential from the get-go, as the necessary concept of the subject. In this case, the differential of force is subjective in the making—there is no subjective without such a qualitative difference. So, if the subjective is founded on the differential of forces and if certainty is dependent on an earlier moment of subjectivized anticipation, and thereby if consistency is always a confirmation of an earlier moment of cause, then it is only on the basis of "the disturbance in the force" (as Yoda would have put it!) that we may ever hope for a moment of consistency. This is what is expressed in the famous quote from Mao: "Disturbances are an excellent thing."

But the differential in the force is of no use to the subject during the calculation, and it is for that reason that the subject has no idea about it. That is why it is only appropriate that the subject takes off right after her calculation ends. The subjective/qualitative/neighborhood logic only comes in when the other steps forward and renders the outcome of the situation suddenly undecidable. But perhaps the calculation is of no use to the subject when the other steps forward, as what matters at the end is who gets to the warden. So "I hurry for the simple reason that being the first to exit is the only real that matters. The act takes precedence over the reasoning."[38] But that is not quite true. For there are two things that matter: being the first and being right. For if the subject has only guessed that she is white, that she doesn't have a convincing deduction, what's the use to be the first? So, she won't rush it. The real of the situation is not only the act but also the reasoning. The subject can only rush if she has finished R3, not when she is at R2 or R1. But this is where the moment of subjectivization comes into the picture, which during the process of calculation, forces each prisoner to abandon the calculation due to the subjectivizing effect of anxiety caused by the others' movement. In other words, it is the interruption of the algorithm, and not its execution, that has the subjectivizing effect. The force breaks with the law. Accordingly, also, the subjective process is to recompose the consequences of this interruption.[39]

This is exactly where Badiou's intervention comes into the picture. Badiou divides the moment of anxiety into two possible paths toward a solution. First is a solution based on anxiety: the moment I started to walk, due to B and C moving while I am still at R2 (or earlier), I halt since my reasoning was incomplete, leading me to work through R3 in the same place as B and C (the three of us have halted). Second is a solution based on courage: I finish R2, the others move and, seeing me move too, they halt—but I don't; I extract the information concerning R3 from their halt, without myself stopping; the courage lies in the fact that I trusted that I could reason according to R3 without actually being in the same place as the others. The point here is that we should "think on the go": trust the difference internal to the place (B and C "are" the same as A, algebraically. Hence we can conclude "through" them, but we remain different from them since we did not stop) and continue to move. Only by focusing on this second solution—which alone would truly satisfy the warden's request that only one should leave—that we show that Lacan is divisible (and not simply "already divided" into an "unconscious Lacan"), and though the solution based on anxiety and the superego injunction (Badiou's version of it could perhaps be stated: "turn every open set into a determinate point!") requires the topological to already be "in place," it is only the second solution which truly mobilizes the real as consistency.

We should make a point clear because it ties together the imbrication of SD and HD in our theoretical framework with the expansion in the field of action—that is, the emergence of some "maneuvering ground" within the system of placement out of which a new metric could be constituted. This move, from the algebraic transcendental to the topological immanence, is not only a generalization of our theory of structure—to show that the subjective was "always already there"—but also the proof that a discipline is possible. We must emphasize this, not only because it is the first concrete example provided by Badiou of a positive discipline, conditioned by a theory of regional structures, but because this "division of haste" will, later on, be the basis for the relation between evental site (a place which contains un-presentable elements relative to the placing system) and the generic (a subset of the placing system which counts what is indiscernible in the site) via the event (the perspective shift which thinks what is "indifferent" in the situation not algebraically but topologically).

Now let's recap the salient points. Lacan's 1945 text is a unique text as it captures the issue of subjectivization and subjective-process in a purely logical form. This is important first because subjectivization and subjective process are at the heart of the psychoanalytical cure—to some degree, they are two pillars of the psychoanalytical process: to evoke the subject and to maintain the relation with the evoked subject. This reminds us of the use Lacan made of the allegory of the cave: that you bring the subject to the opening of the cave and then you keep it there.[40] Second, this text shows both of these processes in a purely logical form, without any recourse to the existential or substantial presuppositions about the subjects. We saw this to some degree in how the prisoners initially function as little logical processors within an overall structure organized by a set of given premises. What are the lessons we could draw from this "little anecdote"? Badiou and Lacan both agree with the perfect solution, at least initially anyway: the prisoners run through a sort of deduction, and the moment they reach a conclusion, they dash to the door. But when they see their fellow mates also doing the same, they experience a moment of doubt. They all stop, and this stop gives them the certainty that their original conclusion was correct, and then they dash to the door again.

So, two significant moments: the idle time thinking that leads to the conclusion and the moment of doubt that leads to certainty: which Badiou terms subjectivization and subjective process, respectively. Subjectivization pertains to the cause with the anticipation of certainty (resulting in the first haste) and subjective process to consistency with the confirmation of the cause (resulting in the second haste). Consequently, cause and consistency are doubly bound to each other: if there were no cause, there would not have been any consistency, and without consistency, the cause is retroactively undone. Badiou calls this process periodization. The earlier subjectivization succeeds with a subjective process; the cause is supplemented by certainty. Clearly, Badiou is agreeing with Lacan up to this point. But then Badiou starts to take the matter a little bit deeper.

According to Badiou, Lacan's presentation of the process is lacking something. The way Lacan presents this process only accounts for the element of initial reasoning, which is ubiquitous among all three prisoners. This perceives the prisoners as identical logical processors. But in that case, the moment of doubt becomes unjustified. Inasmuch as Lacan tries to differentiate the objective from the subjective logic—as a purely algorithmic process of reasoning which remains veritably objective and purely progressive—the subjective cause becomes untenable: the subject will not be caused. Hence, Badiou's critique of Lacan is that purely algebraic reasoning will not cause the subject. It is clear that Lacan's goal in this text is to show not the objective but the sophist logic: the logic that goes beyond the automatic unraveling of conclusions from earlier premises; he aims for a logic that subjectivizes "the human animal." So, what went wrong; what is missing? "There is something that Lacan is not telling us!" Badiou proclaimed.

What is missing is that the field of placement is structured not only according to the laws of the game—such as the number and color of disks, the color of disks at each prisoner's back, the number and color of disks visible to each prisoner, the stipulated emancipatory conditions, and so on—but also what does not make it to the law: the qualitative differences—such as the speed of reasoning, the speed of movement, and so on—and it is precise because of what is not included in the law that we have the opportunity for the "human animal" to turn into the subject. Badiou calls algebra the field of differences captured by the law, and topology the field of qualitative differences that are "indiscernible" to the law (or it is the real of the law); the former is the logic of algorithms and the latter the logic of neighborhoods.

The latter logic precisely pertains to the qualitative differences in the structure that turns the structure into an unequal locus of struggle among the forces that it harbors, and this is precisely why we can have hope for the subjective: the subjective is only possible if "the splace is always *already* the locus of the subjective."[41] This takes us somewhat back to the beginning of this inquiry into Hegel and to what Badiou recognizes within the interplays of forces in Hegel's *Logic* in order to show the qualitative pass of the real beyond the symbolic—beyond the law of interaction of whole and its parts, and beyond the structural dialectic. First, we must realize that splace is always already subjective—because splace is never free from outplace. Subjectivization is not even possible if this were not the case, because the qualitative real is silently operational underneath the visible surface of the law. In the moment of subjectivization, the latent invisible outplace becomes visible and active—it passes to act. That is why the mere succumbing to the logic of signs cannot account for the emergence of the subjective. Furthermore, only through the periodization of this moment within the subjective process that an affirmation of the subjectivization is sealed: it is only through consistency that the cause is retroactively named.

The point is this: the process of periodization marks a qualitative shift, one that is not substantiated, but logically grounded. What Hegel and Lacan have succeeded in doing is showing how, within the pure process of logic, we can theorize qualitative jumps—that the very constitution of a structure involves facets that make such jumps legitimate. We saw this in two different ways. In Hegel, we saw that the truth of the relation of the whole to its parts, where parts remain ambivalent to the whole within which they are inscribed, is to see them as the relation of forces. We also saw this in the fifth Lacanian paradigm within a more mature and succinct form: the ultimate model according to which we may see the relation of whole and parts according to the second law of thermodynamics: the structured field of signifier already assumed an unstructured excess—the structured field of the signifier can never exhaust all of the regions that it is trying to render to logos. But we saw how this was what Lacan precisely meant to do. Lacan meant to achieve a framework within which he could meld the effects of the structure with the indeterminism that resides in the core of subjectivity. Furthermore, Lacan's solution is spot on: he does clearly call out the moments of subjectivization and subjective process, but he does not have a theory to account for them.

With this, we finally come to the moment of division: Haste is divisible.

> I can exist, without giving any thought to the qualitative difference of the other, because the real subjects me to an intolerable pressure and because the space of topological master is null. My passion to be free leads me to trust only the shortest algorithm, without tolerating any interruption.[42]

Here is the situation: to be free, the first person who arrives at the door must have the correct answer and be able to account for it. Either the first person is correct and convincing and is set free, or no one will be free. Under such circumstances, even if I am correct, it won't matter if I am not the first prisoner at the door. Because of this, "the real overwhelms me, in the guise of the vital importance of the stakes involved, I exclusively call upon the law, which fails to sustain me to the end."[43] Here is the situation: to be free, the first person that arrives at the door must have the correct answer and be able to account for it. Either the first person is correct and convincing and is set free, or no one will be free. Under such circumstances, even if I am correct, it won't matter if I am not the first prisoner at the door. Because of this, "the real overwhelms me, in the guise of the vital importance of the stakes involved, I exclusively call upon the law, which fails to sustain me to the end." So even if I am not finished with my reasoning, the moment I see one of the other prisoners rushing to the door, I will get up and run to the door too. We saw how this reciprocal moment of subjectivity centered on the "not yet" leads to subjective assertion.

Lacan chooses an example that challenges his own theory in terms of algebraic presentation: his theory cannot account for the second halt. Once we realize that there is a different and broader logic at play, which in essence, is what *Theory of the Subject* attempts to disclose, we are able to see its trace in the gap between moments of conclusion and certainty. For anxiety to come into play, something qualitative must be present. But Badiou then says more. He says this qualitative can be thought of under two modes: it can be thought of as a weak force, which is the real as cause, which can be recuperated under the dominance of the symbolic or can be divided so that a second halt on the side of the subjectivizing subject becomes unnecessary. So, haste is divisible under two modes. One mode gives us the necessary halt, and the second is the one that skips the halt (according to the operator of justice).

In this way, we can appreciate what a major feat it was for Badiou to go back to a rather obscure early text of Lacan to bring such a crucial point of difference: to tie it back to a framework whose genesis goes all the way back to Hegel, as we have seen in the foregoing. In addition, Badiou's treatment of Lacan's original solution manifests another trait, one that is of his trademark, which we also witnessed to be at play as early as "Mark and Lack," and which becomes more evident in his mature work: a trait of not dismantling the opponent's view but first to regionalize it and then to enrich it. Badiou does not dismiss Lacan's solution at all. Rather, as we saw, he first locates certain presuppositions with which Lacan's solution is laden, and then he provides a general solution that treats these presuppositions as circumscribing a regional alternative that fits within that general solution. This, in our view, is an early sign of a "Cohen-like" move toward Lacan: he does not prove Lacan wrong (in the same way that Cohen didn't of Gödel), so much as to prove that an alternative solution can be consistently derived from the premises when the solution is not encumbered with certain presuppositions—by us having "confidence in confidence" and being able to reason topologically.

As we've seen in the previous chapter, Lacan formulated the concept of both subjectivization and subjective process, but he could not bind the two: this is what Badiou meant by the subject being the dashes that bind subjectivization to subjective process. Lacan gives precedence to structural determination over historical. Lacan cannot imagine a situation in which the historical correlation of forces gives rise to a new system of placement; he only can think of this inscription to recoil into the old system. This is why Maoism was interesting for Badiou: the core of Maoism involves a subject in an uninterrupted sophism. The thought process imbued by structural determination would prescribe to the Chinese, after the revolution, to look at the Soviets and to follow their suit. Until the Chinese revolution, the idea was that once you take power, the logic of power is going to re-inscribe the political process that preceded it back into the dialectic of domination: a group will rule, and some others will be ruled. What was interesting about the Cultural Revolution was that, once the Communist party took power, they themselves suddenly inaugurated an unprecedented process by putting the power in a completely novel way in the hands of peasants and by stating that the party had to be completely rethought: in view of the peasants and not the opposite.

The "event" of the Chinese Cultural Revolution provided a crucial caesura that finally gave Badiou the model within which the HD and the dialectic of forces could be seen in operation—a model that could be deployed by the new political theory in ways that psychoanalytic models based on the clinic were unable to sustain. A similar move also supported Badiou's retort to Miller. To Miller's claim that every system of thought must be sutured to inconsistency, Badiou provided a different construction from the very field within which Miller attempted to show the traces of inconsistency: the field of mathematical logic. Badiou provided a model via mathematics to reveal and criticize the metaphysical presuppositions of Miller. We witness a comparable move this time in relation to Lacan, where we have the presupposition that every historical contradiction; that is, every contradiction at the level of force, ultimately, must be re-inscribed back into the structure. To this, Badiou again poses an alternate model, but this time from politics (although it could come from anywhere else): the Chinese Cultural Revolution. This situation consisted of a historical contradiction where the correlation of forces was singularly different, and it was possible to wager that the whole political system could be reconstructed from the standpoint of the peasants, according to which every political cadre was suspended from their political function and had to spend six months toiling in the rice fields before they could resume their jobs. This time, however, Badiou's intervention goes a little bit further than before: the matter here is not simply criticizing the theory of suture but showing where something else could appear. It is not simply a matter of saying that the subject does not need to be thought of as the gap in the structure. This is now going a bit further. Badiou wants to show we can find in Lacan's sophism a totally different logic at work, in addition to the familiar algebraic logic. Therefore, the critique here is not only to reveal the metaphysical nature of the claims embedded in Miller's "Action of the Structure" and "Suture" and so on, to empty out their presuppositions, but rather to find a different thesis that could enrich it. This is veritably where we go further than "Mark and Lack."

That is why *Theory of the Subject* delineates a turning point in Badiou's project (which, as we shall see shortly, is also a turning point in our project toward the axiom of Infinity). He takes one specific example of Lacan and shows it as incoherent unless we remove certain presuppositions that Lacan had to make based on a model that shaped his practice. But Badiou had already encountered this model in politics, the Chinese revolution, and accordingly, he was able to recognize the theoretical potential of Lacan's framework and to inspect the presuppositions that encumbered it. Badiou argues if we started from the standpoint of this model that Lacan didn't have, we could recognize the dialectic of force in play. In addition, if we recognize the dialectic of force, bound together with the problem of structure, giving us a more general theory, we will end up having more than one operation, and a bigger space for action, without stepping outside of the general theory of subjectivity. Therefore, Badiou accepts all the properties Lacan had regarding the subject and subjectivity (it is not an individual, it is not consciousness, it is not predictable, etc.), but he adds more tools and more operations.

Putting all this in context, in relation to the problem in "Logical Time," we can conclude that after R3, there is an ideological overdetermination about the color of my disk. There is an ambiguity, for sure, in what this color could be, and in resolving this ambiguity, I can continue to stick to the same line of reasoning that has gotten me to this point (cause—the moment of subjectivization), without giving into the action of the other by pausing (certainty: the subjective-process). It is, therefore, exactly the disciplinary action that amounts to doing more of what I did before: I continue to produce more reasoning, which allows me to cleanse the ideological ambiguity. This evokes what we said at the end of the first chapter. This is where the subject is no longer determined by the overall situation, itself an ideological disposition. Instead, by sticking to the operations that have sustained the subject thus far, the subject continues, in fidelity to thinking, in terms that will determine the other. Be it in every condition of the truth; the subject can determine the other only by thinking under the terms of the dialectic of force.

> *To put into consistency the haste of the cause*: therein lies the whole enigma of the subject.[44]

NOTES

1. Badiou, *Theory of the Subject*, 243.
2. "Subjectivization raises the question of its cause, to which Lacan responds with the real as object; subjective process, that of its consistency, to which Lacan

henceforth still responds with the real, but in what guise? This is where he leaves us behind, because the Borromean knot such as he conceives of it remains haunted by the logic of loss and of dispersion" (Badiou, *Theory of the Subject*, 244).

3. Badiou, *Theory of the Subject*, 244.

4. "There's no such thing as a prediscursive reality. Every reality is founded and defined by a discourse" (Jacques Lacan, *On Feminine Sexuality, The Limits of Love and Knowledge, Encore, The Seminar of Jacques Lacan Book XX, 1972 – 1973*, New York: W. W. Norton & Company, Inc. 1999, 32).

5. "There is in effect something radically unassimilable to the signifier. It's quite simply the subject's singular existence" (Jacques Lacan, *The Psychoses 1955-1956, The Seminars of Jacques Lacan Book III*, New York: W. W. Norton & Company, Inc. 1997, 179).

6. "There is in effect something radically unassimilable to the signifier. It's quite simply the subject's singular existence" (Lacan, *Seminar III*, 179).

7. "He is, indeed, insofar as he is drawn to the four corners of the schema: namely, *S*, his ineffable and stupid existence; a, his objects; *a*', his ego, that is, his form as reflected in his objects; and *A*, the locus from which the question of his existence may arise for him" (Jacques Lacan, "On a Question Prior to Any Possible Treatment of Psychosis" in *Écrits, the First Complete Edition in English*, 445–488. New York: W. W. Norton & Company, Inc. 1999, 459).

8. "It is this ex-sistence (*Entstellung*) of desire in the dream that explains how the dream's signifierness masks its desire, whereas its motive vanishes as being simply problematic" (Jacques Lacan, "The Direction of the Treatment and the Principles of Its Power" in *Écrits, the First Complete Edition in English*, 489–542. New York: W. W. Norton & Company, Inc. 1999, 525).

9. Badiou, *Theory of the Subject*, 245.

10. Badiou, *Theory of the Subject*, 245.

11. Badiou, *Theory of the Subject*, 246.

12. Badiou, *Theory of the Subject*, 246.

13. Badiou, *Theory of the Subject*, 248.

14. Fink in "Translator's Endnote" indicates: "In the original version of the paper, Lacan indicated here that this paper was a '*fragment of an Essai d'une logique collective*' (An Attempt at a Collective Logic)" (Lacan, *Écrits*, 783).

15. "The Number Thirteen and the Logical Form of Suspicion," *Autres Écrits*. Paris: Seuil 2001.

16. "British Psychiatry and the War" in *Psychoanalytical Notebooks of the London Circle: Vol. 4, Psychiatry and Psychoanalysis: Spring 2000*, 9–34.

17. Jacques Lacan, "Seminar on 'The Purloined Letter'" in *Écrits, the First Complete Edition in English*, 6–48. New York: W. W. Norton & Company, Inc. 1999.

18. *Cybernetics: Or Control and Communication in the Animal and the Machine* was written by Norbert Wiener and published in 1948. It is the first public usage of the term "cybernetics" to refer to self-regulating mechanisms. The book laid the theoretical foundation for servomechanisms (whether electrical, mechanical, or hydraulic), automatic navigation, analog computing, artificial intelligence, neuroscience, and reliable communications. A second edition with minor changes and two additional

chapters was published in 1961 (Cybernetics: Or Control and Communication in the Animal and the Machine (2023, February 3). In *Wikipedia.* https://en.wikipedia.org/wiki/Cybernetics:_Or_Control_and_Communication_in_the_Animal_and_the_Machine).

19. F.L. Lionnais, *Great Currents of Mathematical Thought, Vol. I.: Mathematics: Concepts and Development*, edited by F.L. Lionnais. New York: Dover Publications, 2004.

20. See Favert-2015.

21. We have seen references to "Logical Time" in Lacan's seminars IX, XI, XII, XX, and XXI.

22. Lacan, "Logical Time," 161.

23. Lacan, *Seminar XX*, 49.

24. Lacan, "Logical Time," 161.

25. Lacan, "Logical Time," 162.

26. At the first sight the following combinations are possible: two blacks, one white (trivial), one black, two whites, and three whites.

27. Lacan, "Logical Time," 163–164.

28. Lacan, "Logical Time," 164.

29. Lacan, "Logical Time," 166.

30. In his analysis of Lacan's "Logical Time," Fink asks why Lacan assumes that A first hypothesizes that he is black. Obviously, A could have gone to a different direction and assumed that he is white, but in that case the temporal analysis would have broken down. Furthermore, it is not clear whether that side of the decision tree is conclusive at all. Fink thinks that underlying this choice may be Lacan's early notion of paranoia at the root of personality: that everyone fundamentally thinks that they are different than others, so if the other two are white I must be black. This supposition is an automatic supposition of a paranoiac character: to put oneself in the minority status, and in Lacan's presentation this supposition seems to be an underlying silent motor that drives the logical process (See Bruce Fink, "Logical Time and Precipitation of Subjectivity," in *Reading Seminars I and II Lacan's Return to Freud*, 356–386. Albany: State University of New York Press 1996, 356–386).

31. Lacan, "Logical Time," 168.

32. Lacan, "Logical Time," 169–170.

33. Lacan, "Logical Time," 169–170.

34. See section "Middle Lacan" in chapter 5 of the present work.

35. Badiou, *Theory of the Subject*, 251.

36. Badiou, *Theory of the Subject*, 253.

37. Badiou, *Theory of the Subject*, 255.

38. Badiou, *Theory of the Subject*, 257.

39. Badiou, *Theory of the Subject*, 257.

40. "The place in question is the entrance to the cave, towards the exit of which Plato guides us, while one imagines seeing the psychoanalyst entering there. But things are not that easy, as it is an entrance one can only reach just as it closes (the place will never be popular with tourists), and the only way for it to open up a bit is by calling from the inside" (Jacques Lacan, "Position of the Unconscious" in *Écrits, the*

First Complete Edition in English, 703–721. New York: W. W. Norton & Company, Inc. 1999, 711).

41. Badiou, *Theory of the Subject*, 255.
42. Badiou, *Theory of the Subject*, 257.
43. Badiou, *Theory of the Subject*, 257.
44. Badiou, *Theory of the Subject*, 251.

Chapter 5

Toward a General Theory of Structure

We now attempt to locate the developments in part II in light of the work done in part I and the overall theme of this book. The result of part I was the (meta-critical) concept of *discipline*, which we recognized in Badiou's reproach to Miller: while Miller proposed a direct ontologization of the logic of the signifier as the "logic of all logics," Badiou criticized this move as illegitimate, since the critique of science already occurs within science itself, what we have termed the *interiority* of critique. In opposition to the epistemological critique proposed by Miller, which sought to locate the subject from the standpoint of a fundamental theoretical position, the logic of signifier, we propose that discipline is what pushes thinking beyond knowledge.

We began part II by stating that *Theory of the Subject* is not in rupture with this initial position by Badiou, but rather develops the mostly negative critique of Miller into a positive theory. We located this theory as taking place "after" metaphysics (as we criticized Miller precisely for such a metaphysical position) but also "before" a theory of structure (as that had made us incapable of thinking about the "other" of structure). In this regard, most of part II will deal with the conditions of the possibility of speaking about structures without structuralist commitments—if we had such commitments, our only possible critical standpoint would be the Millerian one, that of the "structure of structures," a metaphysical reading of the logic of the signifier.

This is how we read Badiou's choice of Hegel: Hegel as the thinker of interiority, who, with some minor "tweaking," can be separated from his own metaphysical presuppositions (the first chapters of *Logic*, leading up to the dialectics of Something), and still provides us with a more general framework than structuralism. With the dialectics of something and otherness, we are "before" the theory of structure and can still locate broader conceptual tools—identity, otherness, determination, limit, and so on—which only configure a structure if we added some extra constraints to them. At this point, the two "matrices" of Hegel emerge the circular process, which sticks to the metaphysical commitments of *Logic* and leads to the Structural Dialectics of place, and the periodizing process, which emerges in the underside of the first and leads to the Historical Dialectics of force. Therefore, our strategy in chapter 4 was to locate, with Hegel and through him, a theoretical standpoint which could accept the benefits of a theory of structure without accepting the ideological commitment that we are either within "a" structure or within "the structure of structures" (Miller's logic of signifier). The division of Hegel into SD and HD gave us precisely that: a way to approach structural dialectics as a constrained theory.

Thus, when we returned to Lacan (and to Miller, understood now as a rigorous heir of Lacan's SD period), we were in possession of positive theoretical resources to reframe Badiou's critique from "Mark and Lack": Badiou's negative position in that text can now be recognized as a positive espousing of HD as the only materialist dialectical position. Rather than merely criticizing the SD strain in Lacan—as we did through Miller in chapter 1—we are now able to find the resources in Lacan himself to at least situate Lacanian SD as a region of a broader theory, which Lacan did not directly develop. So, on the one hand, we criticize the improper generalization of SD (which leads to discursive idealism), and on the other, we accept that Lacan's theory remains a touchstone for any future HD—it remains a region of this theory/dialectics, such that any general framework will have to remain compossible with the discursive materialism of Lacan.

So, in short, this movement from Millerian Lacan to Hegel and then back to Lacan resulted in the creation of a theoretical standpoint which has the space for SD and HD as well as supplementing the theory of structure (SD) with the resources for a concept of discipline (HD). To have restricted the reach of SD means a first step towards a theory of situations, and to have promoted HD means a first step towards a theory of the generic supplement of situations.

The issue that is central to this development is the question of the infrastructure of the real. This question, as a central theme in *Theory of the Subject*, serves two purposes. The first is regional to *Theory of the Subject* itself: a concept of discipline for emancipatory politics. The other purpose is more global and methodological, which in our view, watches over the entire development of Badiou's thought: that of splitting the concept of the real into two—cause and consistency—which is not to do away with the inaccessibility of the real, but rather to provide two positions in which this inaccessibility can appear. As the cause, the real emerges as the impossibility for the subject to locate itself outside of the structure, which determines its place, and as consistency, it appears as the structure's own impossibility of determining the inexistence of another structure (another model) for the subject's determination. Although this debate does center on the real as the limit, we propose to frame it as a debate around the "infrastructure" of the real: what more can we think about something that remains inaccessible? Badiou's first response: we can think about its relative status—the fact that it is inaccessible *in* a given situation. His second response: once the real becomes a relative category (hence given a minimal structure, as it relates to a structure in its inaccessibility), we can think how the structure cannot legislate over another law, which means that we can locally produce another law which is not overdetermined by the first law of that structure. The relativity of the real brings into view the relativity of the laws of structure as well. This connection between "a" real and "a" structure is something that is promoted by the real (in the real) and not through the symbolic—hence it attests to the consistency of the real—we can trust the real to regionalize the law. This reading further reinforces our argument that the new operator of materialism—after inversion and decenterment—is regionalization!

This brings out a caveat—which is the first point of passage to part III and to the overall theme of this book—that, according to the theory we have developed up to this point, the regionalization of structured situations was only possible because of the concept of history. To historicize is to situate. But this is problematic, at least philosophically, because history can only operate in this way (and it didn't for Hegel) because of a suture between philosophy and politics, which bears the name "historical materialism." It makes sense why Badiou would adopt this position in order to critique Lacan, as it renders the real of a situation relative to that structure. The limits of knowledge are not absolute, but relative to a given historical moment. In a certain sense, however, this is philosophically inconsistent, the same as Heidegger's *aporia* (and thus, no wonder that Badiou begins *Being and Event* with that theme). If all structures are situated, can there be a general theory of structure? In the parlance of *Being and Event*: is ontology a situation?

The solution in *Theory of the Subject* is only to think of HD from the standpoint of an engaged discipline, politics. But Badiou himself crossed that line when he used politics to produce a critique of psychoanalytic ideology. Theoretically, the impasse centers on the division between the two grammars of algebra and topology. It is true that we are already one step ahead of Lacan here by having found an intra-mathematical distinction between ideological and disciplinary standpoints, a distinction that allows us to move between the theory of structures and their "sites" on the one hand and the genericity of disciplines, which extract their consistency from what structures cannot place, on the other. But this remains an arbitrary division: it is not a distinction between "quantitative" and "qualitative" in any true mathematical sense—the meaning of such a division remains extrinsically given by the fact that it serves as a model for a process which, in fact, functions as the theoretical basis for all of this: the post-Maoist political discipline with which Badiou is already engaged. This engagement is itself not included in the theory, even though it was strategically essential for his investigation: it was at stake in Badiou's decision of where to begin in Hegel's *Logic*, and even though it was uncovered as already being implicitly at stake in Lacan's treatment of "Logical Time," it was only through Badiou's political treatment of the text that such commitment came to light. Even though the bulk of *Theory of the Subject* is dedicated to spelling out that one can count on the real to produce a "law out of the non-law," nothing is said of the shift which allows us to pass from SD to HD, to move from the exteriority of a place to the interiority of a neighborhood, from the algebra of finitude to the infinite swarm of topological points. This missing piece will become the theme of the "Generic," which allows mathematics to emerge as a general theory of situations.

Hence, the un-suturing of his own theory of situations from politics is precisely what will lead Badiou from Marx to Plato: the meta-ontological decision to think of the count-as-one as set-hood allows Badiou to think of situations independently of politics. This return to Plato also offers a new critique of the Heideggerian paradigm, itself implicitly criticized in Badiou's critique of Lacan. The theory of consistency as an effect of *Logos* (therefore antithetical to Being) is criticized in *Theory of the Subject* with the notion of real as consistency, but this is done by moving from ontology to politics and the political concept of history. In *Being and Event,* this critique will remain within ontology itself—by proving that there is an ontology which is not Heideggerian and in which consistency is not antithetical to being, but what allows us to think the being of situations in the first place.

But more than this, the move from *Theory of the Subject* to *Being and Event* will allow Badiou to complete the movement we have been tracking thus far: the movement from "Mark and Lack" to *Theory of the Subject,* which moved the negative critique of SD to a positive theory that regionalized the reach of the logic of the signifier. The operator of passage in this movement was the idea of discipline as an immanent process of rupture with the limits of a given structured field of knowledge. With the move to *Being and Event*, it is the very *infra*-structure of a discipline that is under consideration: Badiou drops the algebraic/topological distinction of his dialectical "grammar," and instead, we get a unified theoretical language, that of set theory, which writes both the properties of the state (SD) and the generic (HD) instead of two competing forms of thinking. This seems to consolidate the idea that "regionalization" is the materialist operator in this movement: Badiou's later ontology fully admits that the articulation between a situation and its generic supplement is not one of inversion (the truth is not hidden below, it's not the site as the bottom limit) nor decenterment (the site is not the remainder), but of local supplementation (the site as the possibility of a generic supplement). And it is able to think of this materialist operator without falling into the impasse of having to think about its possibility from within a given particular discipline—we can now separate the being of truth from the thinking of Truth in general.

Furthermore, this finally leads us to the problem of infinities. For us, the marker for this move is Badiou's statement in the introduction to *Being and Event* that, for Lacan, the real remained a category of the subject. The critique of the real (only) as the cause in *Theory of the Subject* is the clue, we propose, to understand Badiou's critique of Lacan's (appeal to) "intuitionism" and his rejection of the axiom of Infinity: because Lacan mistook SD for a general ontology, he generalized the inaccessibility of the real for a situation to mean its absolute inaccessibility. In this way, he sutured the infinite (as the thinkable condition for possible generic supplements) under the countably inaccessible (the constructivist world of SD, we could say). This is why the title we chose for this book, *The Infinite and the Subject*, locates, in our view, precisely our ongoing problem in the new framework: it's a critique of the limits of Lacan's theory of structure and subjectivity—so, a continuation of "Mark and Lack" and *Theory of the Subject*—but using the thinking of the infinite as our critical grammar.

So, as we move to part III, we would like to summarize what we have achieved thus far finally:

1. *Theory of the Subject* provides us with positive content for the negative critique of "Mark and Lack."
2. This movement was possible by shifting the perspective from a purely scientific one to a political one. The place from which the possibility of a new framework could be seen is historicism.
3. This positive theory brings to the fore the operation of regionalization, an essential materialist concept.
4. This afforded the possibility of thinking of consistency as condoning novelty that is inaccessible to structural causality.
5. The theory of situations, which now has space for disciplines, is still sutured to a particular discipline—politics—and this compromises its rigor.
6. Finally, to overcome this impasse, we need to find a different ground for the structure that is not history—this is what infinity will provide.

In light of these conclusions, we are now ready to turn the chapter to the main idea determining our entire inquiry: axiomatic thought and, in particular, the axiom of Infinity.

Part III

SUBJECT AND INFINITY

INTRODUCTION

Badiou, in *Theory of the Subject*, began with the category of Something in Hegel's *Logic*. This was a strategic decision for several important reasons, which we reviewed in part II. First of all, a return to Hegel was strategic on its own insofar as Hegel was the philosopher of interiority, a constitutive aspect of the concept of discipline. But why Hegel and not Spinoza? We will soon discuss this question in detail, but in summary, Spinoza would have seemed a more "natural" choice given Badiou's Althusserian pedigree—Spinoza's flavor of immanence informed Althusser's structural causality, and conversely, Althusser was not particularly a fan of Hegelian dialectic—but Badiou chose Hegel as an ally in the effort to build a concept of discipline. The reason is that a theory of interiority could not be founded on causality alone, especially causality that is rooted in the action of the structure. Why? Because this causality cannot account for novelty, and without the possibility of novelty, there is no room for periodization—which we've seen is the operator that shifts an idealist theory of discourse to a materialist concept of discipline: there is nothing in the structure that was not always already determined, either implicitly or explicitly. The centrality of epistemology in French structuralist and post-structuralist thinking derived precisely from this impasse and its Spinozist solution: if there is nothing ontologically outside of the structure, novelty is only concerned with knowledge of structures and their underlying necessity—hence the "suture" of political epistemology to science, which Badiou took as an obstacle to a theory of mathematical discipline, for example.

For the concept of discipline that Badiou was after, novelty—a novelty that lies beyond the grasp of structure—was a necessary requirement, for which Badiou needed a theory of indetermination, of why and how novelty emerges. This requirement for a theory of indetermination is, for us, the principal reason Badiou's path diverged from Spinoza, a conjecture that we will revisit in more detail—and for the same reason, the requirement for a rational theory of indetermination was one of the main drivers toward Hegel.

Secondly, Badiou's rather abruptly selective ingression of the category of Something in *Theory of the Subject* also proved to be an important move. The category of Something possessed at least two advantages. First, Something encompasses the entire apparatus of the dialectic of determinate/indeterminate and thus was already etched with the scission that would allow Badiou to bootstrap his splitting of Hegel into two dialectical matrixes of SD and HD—which we covered in detail in chapter 4. Second, Something is a reasonably supposition-free construct, allowing for safe adoption without being burdened by the question of "beginning"—the question that, as we saw in chapter 4, was of significant challenge to Hegel in *Logic*. For Badiou, the question of the beginning was shrouded in metaphysics. To find what we metaphorically referred to as "one foothold before" structure, Badiou did not see an engagement with the issue of beginning a required step; it was out-of-scope for the kind of theorization he was about to embark on. As a result, the decision to start from Something came with no "founding theory" in *Theory of the Subject*.

Why is this observation important for us now? For one, Badiou himself avowed his "turn" (or "return," if our previous conjecture is true that he actually considered the question of beginning and then turned away—as we developed in chapter 4, where we read *Logic* along with *Rational Kernel*, which was published before *Theory of the Subject*) to the question of beginning in *Being and Event*. As important as this announcement is for us, it would be of particular value to our project if we could position this turn (or return) within the theoretical narrative that we have developed so far. This positioning allows us to see not only a continuity between *Theory of the Subject* and *Being and Event*, but a continuity whose contours can be sparsely sketched based on Badiou's own commentary on the matter (or the secondary literature that is particularly impoverished in this regard), but also allows us to position *Being and Event* as answering to inquiries catechized as early as "Mark and Lack" (if not earlier). In other words, we have a vested interest in the question of the beginning, and we think through the investigation of this question by the theoretical framework that we have established so far, we could also shed significant light on the project of *Being and Event*, whose apogee we will capture through three ruminations: axiom, infinity, and subject. To us, these three thoughts precisely fulfill the matter of beginning, as well as progression, for Badiou, and it allowed him, in keeping with the original Hegelian considerations to which he remained faithful, to finally break away from Hegel—a breakage that in our view heralds a new epoch for thought.

So how is this turn justified by what we have seen so far? Put differently, how does Badiou's splitting of Lacan, which finally established two distinct grammars of SD and HD, lead to the question of the beginning? How does the problem of the real beyond its original algebraic and epistemological grounds require stepping back from Something to face this question?

We have positioned Badiou's philosophical development as the gradual construction of the concept of discipline. We saw that he first sketched the contours of this concept in a negative way by debunking the theory of discourse and the logic of signifier in relation to science in general, and mathematical logic in particular, but when it came to politics, Badiou had no choice but to construct a positive concept of discipline to show how politics, like science, can also think its own proper content. This disciplinary interiority is only possible if political thought is not just a discursive narrative of the political domain, but is itself constructing that domain (political thought is of the character of ontology and not epistemology—Hegel's equation of ontology and logic). In addition, political construction does not exhaust the domain (there is always a remainder that escapes political determination—first scission of Hegel, the Dialectic of Two), and the law that structures the domain does not operate only according to the weak reactive relation of the whole with its parts, but also strong active relation of forces (parts/identities interactions that do not always repeat the whole/place, but periodize the place and reformat the whole—the second scission of Hegel, the Dialectic of Four).

It is precisely the success in establishing another region of thought as a discipline—politics as a region separate from science—that now calls for a more general theory capable of describing the multiplicity of disciplines. If both science and politics are disciplines, then what is the general theory according to which we can justify the existence of more than one discipline? How can we theorize multiple regions of thought that can be simultaneously and independently designated with the status of a discipline? This requires a theory that, while treating regions of thought as properly self-governing, can also account for what constitutes their sovereignty—the fact that they are exterior to each other, the fact that they are "counted as one" region—considerations missing from the concept of the discipline itself. This theory is meta-disciplinary, establishing conditions for a multiplicity of disciplines, without specifying the content, methods, or laws of those disciplines. This is a theory of situated disciplines: it speaks of disciplines as regional spaces of thought with laws that govern their independence and sovereignty.

The first question about this theory of situation is this: From where does this theory speak? What is the status of this "space" where multiple of disciplines are situated? This is clearly a question pertaining to the beginning: the question that was under-treated in *Theory of the Subject*, with a pseudo-axiomatic gesture by Badiou to start from the category of Something. In our view, the new discourse, the meta-disciplinary discourse, is the discourse of ontology, and furthermore, as we shall see in the chapters that ensue, this discourse also satisfies the requirements of a discipline—ontology itself is a discipline.

Before discussing this claim in detail, we should clarify one crucial point. We evidently observe a pattern emerging between part II, where we moved from a localized and negative concept of discipline, focused only on scientific practices, to a generalized and affirmative theory, inclusive of both politics and science—and any region of thought that manifests the characteristics of SD and HD grammars[1]—and part III, where we move to a yet more general theory to distinguish the multiplicity of regions of the discipline. In short, both parts evoke a movement of generalization. Should we then expect a more generalized theory for that theory, and so on and so forth? The intriguing answer here is no: the next theory we are about to engage with does not have a more generalized theory to ground it. The "magic" that stops this seemingly infinite sequence is the thought of axioms. Yet another theory does not circumscribe the next theory, but by a set of axioms and by axiomatic thought in general.

For the later Badiou, the issue of beginning became far more important than it was in *Theory of the Subject*, where the concept of discipline involves primarily a theory of progression, not beginning. What became evident was that a materialist theory of progression cannot be ambivalent, as *Theory of the Subject* might have been, to the question of beginning. In fact, it is only through a proper theorization of the beginning that a complete theory of progression is possible. In that sense, Badiou calls his *Theory of the Subject* still a logicist project.[2]

It was not only within the analytical philosophy developed by the Anglo-American tradition that formalism, in the manner that Badiou speaks about, became a dominant orientation in thought. We also showed that the French intellectual milieu of the 1950s and 1960s, like the young philosophers gathered around *Cahiers*, was also driven by a formalist and logicist agenda based on structuralism and epistemology. But in our view, Badiou is correct to mention that *Theory of the Subject* was "caught in the grip of logicist thesis" insofar as its attention was mainly toward a theory of progression rather than beginning.

We saw from Hegel's *Logic* that the question of the beginning was the dominant question. In this regard, dialectics is not just a method for thinking, but a principle of being. In that sense, for Hegel, logic is the same as ontology. Dialectics for Hegel is the consequence of the thought of One, which itself is an ontological principle. This ontological principle, as we developed in chapter 4, was the principle that informed the entire progression of *Logic*. Badiou, in *Rational Kernel*, a text that he wrote before *Theory of the Subject*, engaged with this question of beginning to demonstrate that this principle of One and the ensuing dialectical progression are not coherent, that Hegel is unconsciously committing to Two rather than One as the founding principle, that the beginning is not the splitting of One to arrive at Two, but uniting Two to derive One. In that sense, Hegelian ontology begins with two separate thoughts of determination (Being) and indetermination (Nothing), and a method of bringing them together in one thought, Becoming—whose stable mature version will be the thought of Determinate Being as Something. This, as we theorized, was the first splitting of Hegel: to the idealist Hegel (whose principle is One, and sees multiple only as the outcome of an operation) and materialist Hegel (whose principle is multiple, and sees One as the outcome of an operation). The thrust of Badiou's critique is this: what Hegel regards as the disappearance of Being and Nothing into each other—which yields the category of Becoming—is only possible if they were the same in the first place. Nothing is the same as Being, and this is why they disappear into each other. At the beginning of the *Logic*, Hegel introduces the thought of pure being, and we simply select two names for this inaugural thought: Being and Nothing, and then we iterate between two names for this same thought, and we name this iteration something else: Becoming. This is what Badiou calls Hegel's "sleight of hand." The problem, again, as we saw in chapter 4, is not with what Hegel wants to achieve: by the time the *Logic* reaches Determinate Being, Hegel is on the right track, in Badiou's estimation. The problem is with Hegel's principle. He cannot derive two from one. Nothing is not the Other of Being, for there is no Other: this is the principle of One. Nothing is simply another name for Being: Hegel cannot "fool" us!

By Determinate Being, we had a proper architecture for the coexistence of determinate and indeterminate, but this architecture is only supported by heavy secondary comments through which Hegel is "appealing" for the support of the opposition between Being and Nothing, and not by a demonstrative dialectical progression, as Hegel wants us to believe. Hegel cannot maintain the opposition of Being and Nothing while deriving a continuity from Being to Becoming. This meant that this opposition was a presupposition prior to the beginning of logic, which is meant to be supposition-free. By the time we arrive at the category of Becoming, as the sublation of Being and Nothing, we all of the sudden witness Hegel "forcing," as Badiou called it, this opposition. In simple terms, Hegel wanted "to have his cake and eat it too": he wanted to have his One, and he wanted the "real" beginning of logic, which is dubbed Determinate Being, to have Two, and the linchpin of this "sleight of hand" is the category of Becoming in the *Logic*.

The foregoing critique is already laid out in *Rational Kernel* prior to *Theory of the Subject*. The point is that, in *Theory of the Subject*, Badiou does not revisit any of this, as if he had made a decision to start fresh from an architecturally robust concept in Hegel, the category of Something, and not get bogged down in the murky swamps of the beginning of logic and ontology. We propose that as sound as this decision may have been for getting the project of *Theory of the Subject* off the ground, by the time we reach its conclusions, it becomes evident that a (re)turn to the premise is indispensable: we need a theory that can account for the multiplicity of disciplines.

But what does the latter question have to do with the beginning of Hegel's *Logic*? The key is the notion of the "infra-structure" of the real, a concept we introduced previously as one of the main consequences of reevaluating the conditions of the real "between" structures themselves, and not as the consequence of simply having a structure (and hence our choice of the hyphenated term "infra-structure"). For structuralism of all forms, but in particular, the one theorized by Miller and the middle Lacan, the concept of the real is an absolute concept: there is only "one" real—there is no way to differentiate "multiple reals" from each other as differentiation is a symbolic operation, and the real is the mere collapse of all symbolic operations. Thus, by definition, the real is indifferentiable. This absolute otherness of the real, properly manifested as the emblem of Lacan's fourth paradigm, reduces the traces of the real to epistemological gaps in the symbolic—gaps that, properly speaking, should not be even traceable within the symbolic. We saw the late Lacan depart from such an algebraic interpretation. For the late Lacan, radical otherness is a retroactive placement to account for what cannot be symbolized in the normal experience of the subject: jouissance is not in the Other. It is the remainder of what exceeds symbolization. It is a symbolic gesture to ascribe this remainder to the Other in order to fill the gap in meaning. What is important in Lacan's paradigm five was the move toward the ontologization of the real of jouissance: the real is not defined as the limit of the symbolic in terms of its epistemological access, which turned to a positive definition (middle Lacan); it is very much part of the material condition of the situation of bodies and languages—it is part of the same living experience—an embedded living experience that, under the influence of language, always creates an excess inexpressible in linguistic terms.

Badiou's formula of identity and asymptote changes this view of the real and the symbolic into one where exterior regions of experience are either imbricated or knotted in order to create an effect of the whole. These regions are seen under different principles and different logics: the principle of identity precipitates toward the algebraic logic to organize experience as a structured reality, with un-symbolizable residues captured in the imaginary and the real registers, whereas the principle of asymptote approaches toward the topological organization of experience that allows for qualitative neighborhoods, differences which would be lost if the swarm of points were approached algebraically. Revisited under this theory, the real is not within a different "register," which is "exterior" to the register of the symbolic—it is not the limit of structure. It is very much embedded within the same material reality, only analyzed topologically. This view completely relativizes the "real" relative to a region of being.

The consequence of the above for discipline is this: for multiple disciplines to exist as independent regions, they must be regulated not only by what their structural parts represent but also by what they don't represent: they must be of different "reals"—their "reals" act as an organizing principle that differentiates these regions—that is, while they are the "inaccessibles" within their respective regions, they are different from each other. This means that the differentiating principle that effectuates the multiplicity of principles must provide a theory that describes how the real of each region is only the real of that region, and not another region. Why does a theory of multiplicity of disciplines have to take into consideration multiple "reals"? Because the architecture imposed by the concept of discipline requires that the real of each discipline is embedded regionally. This theory is no longer merely a logical or a formal theory. It must consider not only what differentiates the discourse of a discipline but also the material a discipline is composed of. This is what we termed in the previous part as the operator of regionalization.

This is where the issue of the beginning appears. We need a theory of multiplicity, not just as an organizing theory, but also as a theory that treats the very being of the multiplicities and their regionalization. This theory is no longer merely a logical but an ontological endeavor, one that treats the being of the multiples as well. What this means is that the issue of beginning must address how multiplicities are possible, not simply that multiplicity is the foundation: the issue of the foundation is secondary. This important differentiation between the foundation and the beginning will be something that will become clear in the first chapter of this part. This is the theory of being qua regionalization—ontology.

The next question is why ontology could satisfy the requirements imposed by the concept of the discipline. There is obviously a point of comparison here with Hegel, and regardless of whether or not Badiou developed his program in *Being and Event* in a direct dialogue with Hegel, this comparison plays an important pedagogical value in our understanding of Badiou's program.

What are the main guidelines that we could extract from comparing the requirement for a disciplinary ontology with Hegel? We saw that one requirement pertains to the discourse of beginning. This requirement is formulated in what we have termed Dialectic of Two, vs. Dialectic of One: the beginning, insofar as it concerns ontology, cannot be One but multiple. It further tells us that One is the result of an operation of unification, or aggregation. Obviously, this operation is non-existent in Hegel, as One is a given for Hegel and not the result of an operation. The second requirement is a direct consequence of this first thesis pertaining to the beginning: once the beginning is not One, then the method of progression will no longer be dialectic: ontology based on multiple needs another method for progression. Therefore, the operator of periodization, which in *Theory of the Subject* captured the thought of disciplinary progression, can no longer be the operation of sublation, as this operator invokes the One, whereas periodization goes precisely beyond the One. Hence, the new ontology must ground this operator differently so that it invokes the principle of multiple as opposed to One. Once our principle is not One, this ontology is in rupture with all of its metaphysical past, and in particular with its Hegelian past. In fact, as we will see, this ontology is in rupture with all philosophy. This is what the formula ontology = mathematics is going to tell us: the rupture of ontology once and for all with philosophy. So, we have these three requirements: multiple instead of One, mathematics instead of metaphysics, and generic instead of dialectic.

In our view, however, the most important aspect of this rupture lies in the thought of "actual" infinities. This theory, from the outset, is based on the thought of infinity and infinities: the beginning deals with infinite multiplicities, the regions circumscribed by the theory are infinities, and infinity constitutes the substructure of the actualities of the real. In that sense, the most important rupture from metaphysics, phenomenology, and linguistics is that the thought of infinity is the "foundational" axiom of everything that the new theory stands for, so much so, that, in our opinion, we should call the new epoch in thought the era of infinity.

According to the foregoing remarks, our inquiry in this part of the book is organized around four important topics that go through the conditions of the new ontology in four chapters:

- Axiom is the title of the chapter that will capture the question of the beginning, the topic of One and multiple as the beginning of ontology, presentation vs. representation, the operator of regionalization, philosophy vs. ontology, the formula of ontology = mathematics, the axiomatic thought, which includes the axiom of Infinity—as the most crucial differentiator of all axioms. This will capture Badiou's first rupture with Hegel.
- Infinity is the title of the chapter that captures the most important idea that informs this entire book: the thought of actual infinities captured as an axiom of all ontology, the difference between this ontological (i.e., mathematical) thought of infinite and metaphysical infinite, especially in contrast to Spinoza, Hegel, and Lacan; this will capture Badiou's second rupture with Hegel—in particular Hegel's thought of infinity as the progression,

and realization of this progression, of finite, the thought of generic and truth, the operator of periodization in the regionalized thought of multiples—the operator of periodization that is stratified to operate within a region circumscribed by the operator of regionalization.

- "Subject" is the title of the chapter that contains the foundation of the new theory of the subject by recuperating what we have learned from Badiou's theory of discipline, the crossing of this theory with the philosophical account of cogito and the Lacanian notion of infinite jouissance.
- Thinking Discipline is the title of the conclusory chapter that discusses the consequences of the thought of discipline on general logic and, in particular dialectic, on Badiou's Platonism, on the new discourse of infinity and finite that in our view should gradually replace the dominant twentieth-century discourse of other.

NOTES

1. We are making a claim here that is neither fully justified in *Theory of the Subject* nor in our work. The claim is that the concept of discipline that we have developed using "Mark and Lack" and *Theory of the Subject* encompasses both politics and science symmetrically. We have shown that this theory is conceptually ambivalent to a specific region of thought engaged in the disciplinary evaluation but as we know, large parts of the material that is used to develop the theoretical apparatus for the theory are coming from the political region. Thereby a large part of the theory is sutured to politics. For example, the notion of history determined by the concept of periodization is developed completely within the political paradigm in *Theory of the Subject*. However, we do maintain a formalization effort that can transpose the key concepts such as SD, HD cause/consistency and periodization to the scientific region. For example, the Bachelardian notion of the epistemological rupture is a good candidate for a type of "historicization" in science, and thereby used to show the effect of the operator of periodization in scientific practice. In the absence of a properly formalized theory, we would like to caution against the equivocation we are making of the concept of discipline. But we have taken that risk hoping that the illustrative appropriation we are making of this equivocation will compensate for that risk.
2. Alain Badiou, *Being and Event*, London, New York: Continuum 2007, 5.

Chapter 6

Axiom

In our study, the concept of discipline emerged from a critique of epistemology from the perspective of the latter's treatment of the real: epistemology treated the real as an impasse to knowledge—the limits or gaps of our knowledge that are further disguised by the function of misrecognition into building a tertiary-imaginary constitution that fills the gaps of meaning. The concept of discipline turned the impasse of the real from *a* region beyond the epistemological border to *regions* that are internal to the disciplines themselves. This change allows the concept of discipline to do two things with the real:

- It regionalizes the real local to a given discipline—the real is not globalized to all disciplines; every discipline has its own real that is properly internal to the discipline; and reals of disciplines are different regions of being
- It allows for the concept of discipline to do something productive with its real—instead of the real to play exclusively a negative role, it constitutes a lacuna for novelty in the discipline; in that sense, the real acts both as cause and consistency for each regionalized discipline.

The real, thus defined as the impasse of the regions of disciplines, is then multiplied according to the multiplicity of the disciplines. Discipline is thus the name of that region of being whose being is conditioned by the very being of multiplicities. Therefore, the condition of possibility of disciplines is the condition of the possibility of multiplicities of regions of being with a sort of interiority that is required by a discipline. We call this the question of the beginning:

- The ontological question of the beginning: How is the being of multiplicities possible?

- The logical question of the beginning: How can multiplicities sustain an interior architecture without dependency on the outside?

Our purpose in this chapter is to show that axioms will precisely address both of the above senses of beginning. But this is a radical thesis that is in rupture with the prodigious background of this question. Where shall we begin from? That is, what sort of thinking should we adopt such that we are not submerged in commitments of which we are unaware? This is precisely the kind of orientation that the inquiry into the possibility of regionalized multiplicity—that is, multiplicities whose impasses are interior to their being, as opposed to being interior to the One impasse of all formalizations—would require, and this is precisely what Badiou's inquiry in *Being and Event* achieves.

But this is where the axiomatic thought enters this inquiry: since the first impasse that this proposal needs to tackle is how to start thinking without grounds. And even further: How is it possible to begin without a foundation and without presupposing the One?

THEORY OF REGIONALIZED MULTIPLICITIES: A THEORY WITH NO FOUNDATION

The axiomatic paradigm is instrumental to any theory that attempts to be supposition-free. It would be naïve to think, therefore, that the usage of axioms is only a stylistic side effect of equating ontology with mathematics. Axiomatic thinking is the very medium of a theory that has no foundation.

It should come as no surprise that the home for such an orientation is not philosophy. Almost all branches of science flourished by leaps and bounds, especially in the twentieth century, by abandoning the method of thinking that constantly harkens back to a set of foundational principles in their domain. The orientation of thinking that de-stratifies and encourages interdisciplinary liberation of concepts has been the touchstone of many advances, most notably in physics and mathematics. Among the leaders of this orientation in the twentieth century, Badiou praises, in particular, Albert Lautman, the French mathematician who Nazis executed in 1944.[1] Lautman's open debates with linguistic philosophy, the logical positivism of Wittgenstein and the *Vienna Circle,* and the logicist orientation in the mathematical philosophy of Whitehead and Russell in the late 1930s were among the high points of this style of thinking. For Lautman, beginning without presupposition was precisely how sciences achieve the liberty they need to advance—contrary to the logicist position. The mathematical domain does not progress based on the purity of the model but based on "mixes" and interactions, and so on.

Scientific practice does not develop based on generalization alone but rather primarily by dissociation, which invokes a different kind of generalization supported by axioms. Axioms provide a "pragmatic" method that architecturally sustains an open and dynamic dialectic, permeating many branches of science/mathematics, whose practices, unlike what logical foundationalism envisions, are not fully loyal to original logical concepts that underpin tautologous exchanges.

In an introduction written for Lautman's collection of work, Fernando Zalamea speaks of Lautman's concept of "effective mathematics."[2] Lautman's term refers to the structure of mathematical knowledge in relation to the concrete action of the mathematician to build its edifice gradually. The term "effective" denotes the practice of mathematics using mathematical discourse, theories, and methods. This progression, Zalamea adds, takes the concepts and examples and connects them through structural processes of liberation and saturation, resulting in mathematical creations like mixes between opposite polarities.[3] Focusing on the complex ways in which the advanced branches of mathematics develop (methods of liberation and saturation), Lautman rejects the reduction of mathematical theorems to tautologies. This effort in the early twentieth century has leveled the broken terrain of mathematics.

The emergence of mathematical structure in the early 1930s played an important role in the development of Lautman's ideas. This notion needed time to become mature and unravel in all its generality (which largely took place in the work of the Bourbaki group), yet mathematical structures were utilized across a wide network, in particular the German school (Hilbert, Emmy Noether, Emil Artin), which Lautman studied in *École Normale*. This helped Lautman to slowly see mathematics as a network of ideas, not a hierarchy of concepts.

Lautman researched the dialogue and interplay of many of the branches of mathematics with each other: for example, the methods of algebra in analysis, analysis subordinated to topology, the ubiquitous appearance of the geometric idea of a domain, and the agreement of the complex variable in arithmetic. This organization of mathematical branches under a structural network provided a synthetic view of mathematics that prescribed a new alternative to the dominant school of logicism, whose primary goal was the search for "founding" notions of mathematics, normally focused on elementary notions in arithmetic and set theory. The unity of mathematics is not derived from its ubiquitous founding principles and an analytical method to deduce the body of mathematical knowledge. Instead, mathematical unity lies in the living interaction of a network of mathematical ideas and theories that, while influencing and interacting with one other, are neither essentially nor pragmatically reducible to one other nor to any set of founding principles.

The real unity of effective mathematics disappears when the plurality of mathematical knowledge is reduced to a fictional reconstruction.[4] Mathematics is an open network, and far from having "Chinese walls" among them, their mutual interaction and influence are necessary, a process which Lautman calls "Mixes."

The richness of mathematics runs precisely opposite to the purification effort in the foundational arithmetic of Whitehead and Russell. The proof of this orientation in mathematics, as Zalamea points out, is proof of Fermat's last theorem in 1994.

The proof of Fermat's theorem is a modern-day manifestation of mixes. Mixes are the logical forms Lautman explicitly contrasted to Russell's hierarchical model that tends to establish the laws of logic *a priori*. Lautman connects "mixes" to the Platonic tradition (in references to *Sophist* and *Pilebus*[5]) by emphasizing their dialectical liberation and composition. In that sense, mathematics, on the one hand, divides the contents of a concept through definitions (syntax) and derivations (grammar), thereby releasing simple components, and how, on the other hand, it constructs intermediate relays through models (semantic) and transport theory (pragmatics), thus rekindling the existence of simple filaments and allowing for their reorganization into new concepts.[6]

Lautman's philosophy is driven by contrasting complementary concepts—global/local, whole/part, extrinsic/intrinsic, continuous/discrete, and so on—under the governance of terms that he has introduced: notion, as one of the poles of a conceptual tension, and an idea as a partial resolution of this polarity.[7]

In a presentation at the international conference of scientific philosophy in 1935, Lautman captures his disagreement with the program of the *Vienna Circle*, which states: "the formal study of scientific language should be the sole object of the philosophy of the sciences."[8] For Lautman, this preoccupation with what makes a logical statement meaningful fails to capture the fact that "[t]here is a physical real and the miracle that is to be explained is that there is a need for the most developed mathematical theories to interpret it."[9] Furthermore, there is the real of mathematics itself as a mere formal theory with no reference to the physical world. That is, there are mathematical domains which resist exploration, similar to physical real, until new theories are developed. This poses a huge challenge to the logical positivist theory of truth as correspondence. If, as Wittgenstein asserts, truth is correspondence, then unless there is an event that corresponds to a proposition, the proposition is not true. However, the same must be true with "meta-propositions" of the system, such as Carnap's propositions, whose objects are the propositions of the theory—which are the propositions with truth values. Accepting the latter set of propositions as true means that the whole set of theoretical propositions is exhausted (as the former propositions are about the theoretical propositions of the system), which in turn means that every event in the world has already been exhausted, which is an absurdity.[10] This noticeable absence of "realism" in logicism also marks the difference between Hilbert's (formalist) and Russell's (logicist) programs. The latter tries to establish a hierarchy of mathematical concepts, whereas the former tries to enrich independent fields of mathematics with axiomatic thinking so that new domains of practice can be identified.[11] In that sense, the logicist emphasis on foundations misses the independent development of mathematical domains such as group theory, the theory of ideals, systems of hyper-complex numbers, and so on, which are all axiomatic systems in their own right.[12] Nor does the question of rigor justify the search for a foundation because, for Lautman, a logical theory of the foundation of mathematics does not provide that rigor: first of all because even the best logical theories could only deal with elementary concepts in arithmetic, and at the time of Lautman's presentation, complex formalisms such as the Hilbert Space could not be derived by logical primitives. Second, rigor is not sustained through a hierarchy of concepts but through independently structured domains whose architectures are sustained by axioms local to each domain.

> The object studied is not the set of propositions derived from axioms, but the organized, structured, complete entities, having an anatomy and physiology of their own. As an example, the Hilbert space defined by axioms that give it a structure appropriate to the resolution of integral equations. The point of view that prevails here is that of the synthesis of necessary conditions and not that of the analysis of primitive notions.[13]

The point Lautman is making is clear and all too familiar for us, having gone through a similar effort in a different field that attempted to build a logic for all logic, the logic of the signifier: the advancement of the mathematical branches is by no means indebted to the formalism of "one" mathematical system that provides a hierarchy of concepts. On the contrary, the interplay of the branches is too ad hoc, too local, and too dynamic to be formalizable by an independent epistemological edifice that is developed outside or independent of the mathematical theories themselves. Mathematical knowledge is organized around sets of axioms that are localized to specific regions. This regionalized architecture allows for the dynamic interplay and compositions of these regions to create richer mixes of theories or may subdivide regions into specialized theories. In short, there is no meta-axiomatic: there is no axiomatic reciprocity between the One and the beginning—because different regional systems of axioms are not reducible in form or function to some general epistemology of what axioms should do or be—nor between the axiomatic beginnings and foundation—since axioms are not motivated by their relation to some primitive normativity.

The final question that we face is this: Given that the regions of mathematics are developed around a local set of axioms, how are these axioms defined and decided? What sort of theoretical or practical activity defines/decides "the" set of axioms for that region? First, let's realize what axioms do for a given system:

> Axioms are considered both, in comprehension, as a system of conceivable conditions independently of the mathematical entities that they realize, and, in extension, as defining the most extended class of entities likely to realize them.[14]

Axioms are both comprehension and extension. Comprehension produces the conditions that realizable entities satisfy. The extension defines the most extended class of entities likely to realize such conditions. But the catch is this: the function of the extension is so well aligned with the concept of classification into genus and species that this may create the semblance that axiomatization and generalization are the same and, therefore, that the de facto way to reach a set of axioms is to attempt to generalize objects by type. This is the sense in which the extension principle could be easily misconstrued as a generalization. There are examples where this method has been used to derive a set of axioms for a given region; for example, "Frechet envisages the axiomatic constitution of abstract spaces from the point of view of successive generalizations of the notion of space."[15] However, and this is where Lautman's genius strikes: we should distinguish between abstraction by *generalization* and abstraction by *generalities*. Generalization is understood as the familiar Aristotelian subsumption of particular under the general according to the hierarchy of genera. However, generality could be achieved in a different way and has nothing to do with generalization. In contrast, generalization involves disassociation comparable to what conditions the progress of physical sciences.[16]

The model for arriving at generalities, as opposed to generalization, follows what Lautman refers to as the process of "dissociation," which is often used in analyzing physical phenomena. The complexity of a phenomenon is dissected to a map of interacting and/or coexistent facts:

> In atomic physics, it is the dissociation of spectral lines into doublets or multiplets, the dissociation of molecules in ortho and para molecules, the discovery of the positive electron, de Broglie's hypothesis on the complementarity of two half-photons that constitute the photon. It is the same in the theory of relativity: the plurality of time or the duality of masses.[17]

In these examples and many others, a generality is an outcome of splitting an already compact phenomenon into the composition of different interacting facts, as if the generality, in this case, is reached through a process of particularization.

The point that Lautman is making is, of course, not the familiar assertion that the manifestation of physical reality influences the system of mathematical "representation" of this reality. That is not the point (notwithstanding that mathematics here is by no means a "representation" of the physical world). What Lautman points out is that, within the mathematical real, the emergence of generalities follows a similar process as what takes place normally in the physical world: we arrive at generalities through the dissociation of a complex mathematical phenomenon—as if the phenomenon is a trope composed of different abstractions, either due to the intrinsic nature of certain entities involved in the phenomenon or the action of entities unto each other. Generalities are not the reorganization of the encyclopedic knowledge of what we know about mathematical abstractions, but they are ruptures within the system of knowledge of every mathematical situation by the discovery of what was not there. These generalities are not derived from particulars but rather regulate and prescribe the particulars (comprehension principle) without determination (extension principle). Therefore, axioms should not be construed as generalizations of some pre-existing material.

> The passage from notions said to be "elementary" to abstract notions doesn't present itself as a subsumption of the particular under the general but as the division or analysis of a "mix" which tends to release the simple notions with which this mix participates. It is therefore not Aristotelian logic, that of genera and species, that plays a part here, but the Platonic method of division, as taught in the *Sophist* and the *Philebus*, for which the unity of Being is a unit of composition and a starting point for the search for principles that are united in the Ideas.[18]

The anatomy of a complex (mathematical) situation should then be treated as a composition of abstractions, as what is thinkable of that situation and not a hierarchy of types. Axiomatic definition characterizes not a "type" in extension but the possibilities of composition, structuration, and integration conceived in a dynamic and organizational way.[19] Lautman teaches us to move back from Aristotle to Plato.

> The distinction that is thereby established within a same notion between the intrinsic properties of an entity or notion and its possibilities of action seems to be similar to the Platonic distinction between the Same and the Other that is found in the unity of Being. The Same would be that by which a notion is intrinsic, the Other that by which it can enter in relation with other notions and act on them.[20]

But what do this excursion to mathematical logic and Lautman's conception of the axiomatic tell us about our own inquiry? The points can be summarized/recapped as follows:

- The theory of the possibility of multiple disciplines requires a theory of regionalized multiplicities in general.
- This theory will speak to what regionalized multiplicities are according to a set of conditions that themselves have no (or an accepted set of) presuppositions.
- Such a theory of regionalized multiplicities is not founded on a more general theory.
- Lautman philosophy of mathematics describes mathematics as a similar unfounded discipline, such that a more general theory does not justify its conditions—the foundation.
- Axioms are what architecturally support theories that are not founded on first principles.
- A Platonist conception of the theory of forms supports this conception of axiomatic.

To Lautman, a foundationalist stance does not reflect the way mathematical branches are structured and thereby cannot account for the osmosis of mathematical activity. Rather, mathematical branches are architected and erected on the basis of axioms. Axioms are generalities, not generalizations, and are defined through the process of dissociation, through which a complex mathematical entity is broken down into its constituent parts. The progress of mathematics depends on the participation of many generalities in creating new mathematical entities. These two basic operations, dissociation and participation, are two basic operations that are involved in the construction and interaction of mathematical entities. In this sense, Lautman sees the structure of mathematical reality as best represented by Platonism and the foundational operations of the One (Same) and the Dyad (Other). Furthermore, mathematical reality in this sense encompasses an entire lattice of ideal constructions that, although they exist "in-themselves," that is, independent of physical reality, they manifest themselves through the intellectual activities of mathematicians. They construct a Platonic space of "ideal reality." Lautman insists, however, that the Platonism that he subscribes to is different from the type of Platonism that manifests "superficial knowledge of Platonism, and that we do not believe ourselves to be referring to it." This superficial version designates under the name Platonism "any philosophy for which the existence of a mathematical entity is taken as assured, even though this entity could not be built in a finite number of steps."[21] Referring to modern commentators[22] in the conclusory part of his PhD thesis, Lautman offers a different conception of Platonism: rather than immobile and irreducible essences of an intelligible world, Platonist Ideas are related to each other according to the schemas of a superior dialectic that presides over their arrival. The works of these commentators, through a close reading of Plato's dialogues, reveal that mathematical objects/realities are organized "around a complex imbrication of levels, situated below a schematic echelon formed by ideas."[23] Examples that Lautman shows from the work of these commentators are the generation of Ideas-numbers through the operators of One and Dyad "by a successively repeated process of division of the Unit into two new units," the "schemas of division of Ideas in the *Sophist* are organized according to the same planes as the schemas of generation of numbers," "the constitution of bodies in the *Timaeus* assumes a matter which, before the existence of the world, has already been the receptacle of a geometric qualification."[24] The most amazing example presented by Lautman is from Julius Stenzel's work, which shows the degree of proximity between modern mathematics and the Platonic theory of ideas. This we quote in its entirety:

> The introduction of becoming within Ideas, in Stenzel's work, takes all its value from the genius of the text of the *Nicomachean Ethics* (EN 1.4, cited by Stenzel 1923, 118), where Aristotle says that Platonists did not admit the ideas of numbers because they did not admit as ideas things in which there exists the before and the after. Stenzel gives an explanation of this text the importance of which cannot be exaggerated: ideal-numbers being the principle of determination of essences according to the order of the before and the after, it is not possible that there are "numbers of numbers," that is, "a principle of the division of essences that is superior to this numerical division itself" (Stenzel 1923, 119). Metamathematics which is incarnated in the generation of ideas and numbers does not give rise in turn to a metametamathematics. The regression stops as soon as the mind has identified the schemas according to which the dialectic is constituted. We can thus see how well our reference to Platonism is justified, as regards the relations that exist between mathematical theories and the Ideas that govern them.[25]

Of course, it took two and a half millennia for Aristotle's assessment of the lack of an idea of ideas (i.e., ideas of numbers) to find its explicit statement in the Zermelo-Fraenkel (ZF) version of set theory axioms as the axiom of Separation. Later, we will use this connection to show how mathematics is the theory of historical situations (in the sense that Badiou defines these terms). In this sense, the discipline of mathematics as a region of thought has an interiority that is proper to it, but this interiority is structured in a way that its sub-regions, or branches, can remain concomitant to each other: algebra, geometry, calculus, set theory, complex number analysis, and others do not only coexist, they crossbreed. But politics, as a different region of thought with its own interiority, is not structured in the same way. Political sequences can only be understood in their own interiority, and there is no inherent or contingent interoperability among them. This conclusion, above anything else, may qualify us to say that Albert Lautman, for the disciplinary thinking of mathematics, had the same role that Sylavain Lazarus had for the discipline of politics. As Lazarus provided the interiority of politics in a singular form, Lautman thoughts bestow a unique and singular form of thought that is proper to mathematics.

The purpose of this brief overview is, of course, not to delve into the amazing work of Albert Lautman, which merits a dedicated work on its own, but to illuminate the itinerary that we have been trying to construct. This itinerary that started from part I motivated a need for a theory of regionalized multiplicities whose impasses are local to the being of those multiplicities. This theory cannot be a foundational "ground" of the understanding of multiplicities—which would mediate this understanding by a founding principle—something that is a non-starter for the present inquiry. We find ourselves seeking a theory that, while at the same time deals squarely with the question of the "beginning" of multiplicities, is not erected upon foundational principles. This is where Albert Lautman's ideas have helped. Axiomatic mathematics, based on the operation of dissection and participation—the Platonist versions of the One and the Dyad—is the paradigm of such a discourse for an "originary" theory. In our view, this discovery is, at the same time, the pathway through which Badiou transitioned from Marx to Plato, and that, indeed, must be part of the debt Badiou owes to Lautman.

We presume this encounter with Lautman was the outcome of Badiou's long engagement with mathematics, logic, and formal systems beginning as early as the 1960s. In fact, in his 1965 review of Althusser's *For Marx*, he anticipated his future engagement with mathematical set theory.

> There must exist a *previous* formal discipline, which I would be tempted to call the *theory of historical sets*, which contains *at least* the protocols of "donation" of the pure multiples onto which the structures are progressively constructed.
>
> This discipline, which is closely tied in its complete development to the mathematics of set theory, no doubt exceeds the simple donation of procedure of *belonging*, or of an inaugural system of empty differences.[26]

In the same vein, to introduce scientific formalism into structuralist epistemology, Badiou, in his essay "Infinitesimal Subversion,"[27] introduces the concept of "infinity-points" to demonstrate the interruptive, that is, the subversive power of scientific formalism on ideological continuity. This early text, which in its architecture and message, has a striking similarity to Badiou's mature work, posits the "infinity-point" as a point of "inflection" for a formalist intervention that changes the entire "routine" of progression of the formal system. The power of formalism, as Badiou suggests, creates precise points where such an intervention is possible and may even succeed. These "inflection" points, at the cusp of a system, are places where the system could be enriched with new axioms, yielding formalizations that are novel. Badiou used this strikingly refreshing approach of incorporating axioms and revolution to capture the "rational kernel" of the Lacanian real as a special type of an "infinity-point," where the consistency of the formal system is expressed: in that sense, the real is clearly the ground for the consistency of the formal system.

Around the same time, in his 1968 lecture course, *The Concept of Model*, Badiou articulates a difference in the way in which logic and mathematics treat axioms. Logical axioms perform formalization through unification, whereas mathematical axioms are always situated and thereby require models which satisfy the latter. An axiom is logical if it is valid for every structure and mathematical if it is valid only for certain structures.[28]

The point is this: Badiou was aware quite early that what defines a mathematical axiom is regionalized to a specific structure, such that there is no universal mathematical axiom. The Logicist treatment of axiom is still in the pre-modern paradigm, constituted of intuitive and self-evident postulates, like Frege's idea of identity as a primitive postulate of thinking, that tries to produce a general theory of structure.[29] The concept of model, as opposed to the notion of model, opens up this logico-mathematical differentiation. The classical notion of a model is an ideological construct that opposes the primitive structure of structures against reality. On the other hand, the mathematical concept of a model relates regions of mathematics to each other. Here we observe that model is the first name for what we have been referring to herein as discipline: therefore, Badiou's *The Concept of Model* is the first announcement of a conceptual apparatus that turns out to be one of our main theoretical pillars. In this gem of a text, Badiou criticizes certain ideological commitments that conflate formal and non-formal elements in the notion of the model and thereby orient thought toward a "system of structure as such," instead of the proper mathematical concept through which we obtain an intra-mathematical synthesis of structures.

The purpose of the forgoing remarks is to show that there has been a constant engagement for Badiou with the thought that mathematics allows for a plural thinking of axioms, for a theory of axioms which is not so much different from the physical real as it is indifferent to it. Axioms offer productive closure while at the same time resisting becoming the instrument for abstracting properties into definitions. They are irreducibly plural.

When we bind all of these themes together, we arrive at the idea that axioms really do provide a resolution to the overdetermination of the three terms of beginning, ground, and the One.[30]

ONTOLOGY: THE GENERAL THEORY OF REGIONALIZED MULTIPLICITIES

The foregoing gives us the following conditions for a theory of multiples:

- It must articulate the possibility and structure of generic multiplicities—the formal requirement.
- It must not regress to another more foundational one—the axiomatic requirement.

We know that the theory must be formal and axiomatic. Both of these conditions are satisfied, per Lautman, by modern mathematics. So, we have a strong clue that this theory will be mathematical in nature, and furthermore, since we are speaking about generic multiplicities and the rules of their formal and material conditions (causes), the general theory should fit the mathematical set theory. But why must this theory speak about Being in addition to being a formal theory of multiples? That this inquiry is mathematical (formal and axiomatic) is fairly conceivable, but why is it also ontological? Why the theory of general multiples is the theory of being qua being? The concept of discipline helped us to relativize and localize the real of each autonomous situation. It extracts its consistency, not from the structure of the situation but from that which exceeds that situation. This entails, first, that the real must be an intra-situational concept relative to the structure but also that it must have enough "infra-structure" to serve as the basis for a new law or principle of determination. A concept of the real which is *not* reducible to its emergence as an impasse for the structure—the concept of the unstructured real, is an ontological concept. The real becomes, in this way, *the name of being relative to an autonomous interiority*. A certain analogy using Lautman's dialectics of generality is helpful here: it is precisely by decomposing the real into (more primitive operators of) "the real as the cause" (of a symbolic structure's circularity) and "real as consistency" (of a structure's periodization) that Badiou opened up a renewed entry point to the question of being, as that underlying invariance is consolidated as the *cause* if certain variables are fixed "algebraically" or as *consistency* if certain conditions are met "topologically." In short, the grounding of disciplines in the interiority of their point of the real is *precisely* to open the problem of the real back onto the broader question of being—since it is now a matter of decomposing the real into its un-presented, that is, un-situated, concept while accepting that certain conditions must be "added" to this concept for it to emerge as the "real" of a situation. Thereby, the task that we take on in this section is to show that the generic and the genetic theory of multiple, which is a theory of the multiplicity of disciplines, is also a theory of being qua being. It follows from this that mathematical set theory is ontology. The latter task is what this section will show by following the first two meditations of *Being and Event*.

The first meditation of *Being and Event* centers on two questions: "What presents itself?" and "Is it one or multiple?" These are specific questions and are not free from presuppositions. Let's see if we can unearth these presuppositions.

The question of presentation is a very significant choice: it is essentially a question about the being of beings, as Heidegger would have said. What constitutes the being of beings? Heidegger answers that *presence* is the only access we have to the being of beings—anything beyond pure presence is a veiling of being, which is characteristic of knowledge. Beginning an inquiry into being with a question regarding presentation seems to show some Heideggerian presuppositions.

The second question, regarding one and multiple, is different as its roots go back to Parmenides and Plato. The question which central to Parmenides and Socrates' dialogue pertains to the oneness of being and no-being. Aristotle perhaps has a more pragmatic focus: his question regarding being was not so much about the *being* of beings but the being of *beings*. That is, the focus is on what makes *a* being a *being*. As we know, Aristotle alters his answer from *Physics* to *Metaphysics*. In the former, he argues that what makes *a* being is its matter (the material cause), not its form (the formal cause). In *Metaphysics*, he more or less reverses his position and concludes that it is the form (essence) of *a* being that constitutes its *being*. In either case, for Aristotle, a certain *oneness* allows us to differentiate beings and consider them as separate. This question of oneness is anchored in the dialectic of the same and different, as opposed to Plato's dialectic of being and nothing.

Scholastics maintain the Aristotelian sense of this thesis. Leibniz returns to this scholastic dictum, contra Descartes, that "being" and "one" are equivalent. He writes to Arnauld: "To be brief, I hold as axiomatic the identical proposition which varies only in emphasis: that what is not truly *one* being is not truly one *being* either" (April 30, 1687). For Leibniz, in order for something to count as a real being—a substance—it must be "truly one" or an entity endowed with genuine unity. And, as we saw above, in order for something to be a genuine unity, it must be a simple, indivisible entity. "Substantial unity," he writes, "requires a complete, indivisible and naturally indestructible entity" (to Arnauld, November 28, 1686). But the matter is extended, and thus, Leibniz believes, infinitely divisible. Hence, there is no such thing, for Leibniz, as material substance.[31] The sense in which Leibniz is referring to this thesis is again slightly different than the debate we shall review in this meditation. The question is not the *being* of beings but what makes *a* being a *being*. In contrast to Aristotle, who contemplates the matter or form, for Leibniz, it is self-identity: it is impossible for a thing to be not identical to itself, and thus it has to be *one* thing in order for it to be.

Badiou takes as an axiom that what is presented (presentation) is not one: there are presentations, not *one* presentation. Badiou is thus tying two questions together, both the Heideggerian question of being as a presentation and the Platonic question of oneness of being. This provides a nice way of progressing through a logical contradiction:

1) What is not one is multiple (from the logical relation of one/multiple).
2) Presentation is not one (Plato).
3) Presentation is multiple (from 1 and 2).
4) Being is one (or one is) (Plato).
5) Multiple is not one (from 1).
6) Multiple is not (from 4 and 5)
7) Presentation is not (from 3 and 6).
8) Being is the presentation (Heidegger).
9) Being is not (from 7 and 8).

We reach a logical absurdity. The culprit is statement 4, and in no unequivocal form, Badiou proposes to abandon it.[32]

Saying "one is not" is not the same as saying "there is no oneness." Badiou preserves oneness as an operation rather than a condition of existence. It is wrong to say that "*there is* one," because "there is" implies existence. Oneness is only the result of an operation.[33] The one that is yielded by the operation of count-as-one is a number. This is an important assertion. Recall that the One in Hegel is not yet a quantity. Rather, the unity at the beginning of *Logic* is quality. Therefore, we have no reason to believe that being qua being is number.

Badiou shares with phenomenology the idea that our only access to being is through its presentation: being is what presents itself. But the similarity ends there. Whereas, for phenomenology, presentation is always a presentation to someone, and therefore is an extrinsic affair, for Badiou, there is no external someone: presentation stays within the realm of being.[34] We can make no assertions regarding being qua being (being is neither one nor multiple) except as a presentation. We assert instead that presentation is always a multiplicity, and oneness is the operation over this presentation (and not over being qua being). We further define the *situation* as any presented multiplicity. Every situation has its own operator of the count. This is a structure in its most general term. As a result of the count-as-one, a situation functions like a structure. Membership in a structure is allowed insofar as every member is also subjected to the count-as-one operation. So, a situation is always already the retroactive effect of the count-as-one operation and, in that capacity, is always already functioning like a structure.

The operation implies that there must be something to operate upon, a domain that is anterior to the operation, and thereby also an outcome of the operation, that which is one. This sense of anteriority to the count implies that the domain of operation is not originally one. This further implies that the thought of multiple is a retroactive outcome of the thought of the count, or as Badiou calls this, the (pure) multiple is the inertia which can be retroactively discerned, a predicate of what is structured because of structuration.[35]

This gives rise to two types of multiplicities. There is one multiplicity that is the presentation itself: presentation is multiple. This multiplicity is derived backward from the count and is a pre-counted multiplicity. The only sense in which we know that presentation is a multiplicity is that it is not-One. The second multiplicity is the sense in which we compose a multiple of several ones, each counted by the action of the structure. So, there is a multiplicity linked to the inertia of presentation and another multiplicity to the composition of the One. We will call the first the inconsistent multiplicity and the second the consistent multiplicity.

> A situation (which means a structured presentation) is, relative to the same terms, their double multiplicity; inconsistent and consistent. This duality is established in the distribution of the count-as-one; inconsistency before and consistency afterwards. Structure is both what obliges us to consider, via retroaction, that presentation is a multiple (inconsistent) and what authorizes us, via anticipation, to compose the terms of the presentation as units of a multiple (consistent).[36]

For something that is not, One plays an important role in both obligating what the presentation is and authorizing what the situation can do. The double aspect of obligation and authorization is exactly what the Law provides. In that sense, we could say that One is the Law of the multiple. This Law of One legislates how to think about presentation as such before it is subjected to the count-as-one and also authorizes how the structured presentation behaves. Under this Law, the discourse of ontology requires that there exists nothing but situations. It immediately follows that ontology itself is a situation.

Badiou summarizes the opposite line of thought, in which the presentation of being is not multiple. This asserts not only the oneness of the presentation of being but of all presentation: the exteriority of being is revealed through its heterogeneity to all presentment and predication, "for presence is the exact contrary of presentation."[37] This is the mystical alterity which calls for the annihilation of all thought and experience of being through presentation: "this path consecrates itself to mystical annihilation; an annihilation in which, on the basis of an interruption of all presentative situations, and at the end of a negative spiritual exercise, a Presence is gained, a presence which is exactly that of the being of the One as non-being, thus the annulment of all functions of the count of One."[38] Badiou says this mystical orientation institutes another pathway to being, that constitutes being as One, the absolute otherness of non-being. This "negative" orientation to the experience of being could be hardly named ontology, for the simple reason that the part that has to do with *logos* is precisely what this orientation intends to abandon. The practice of expressing this experience of being, in artistic and especially poetic form, is a testament to its divergence from the discursive treatment of being.

In our view, Badiou believes that only such a "negative" orientation would not take ontology as a situation. Why is this important?

If ontology, as the access of thought to being, is not "the" access but only "an" access to being, then it is possible to resuscitate being as oneness, for example, through the oneness of non-being. Badiou mentions this to emphasize why *any* access to being must be discursive. Ontology is primarily a discourse which falls under *logos*. The negative experience of being denies any discursive access to being, any access through *logos*. In that sense, the negative experience of being cannot make a claim that ontology is or is not a situation because it can make no ontological claims at all. In *Theory of the Subject*, Badiou does mention this negative orientation, but he also mentions that neither materialism nor idealism follows such an orientation. At this time, he doesn't refute this negative orientation but only claims that it has no place in ontology in the first place. By that same token, the sense of oneness that it invokes is not a discursive oneness and, therefore, cannot be smuggled into metaphysics through a backdoor—any negative doctrine of oneness has no place in ontology either.

But unlike in *Theory of the Subject*, Badiou here takes upon himself a refutation of this negative orientation toward being. Perhaps we should connect this observation to our critique of *Theory of the Subject* at the end of Part II: Badiou himself confirms (in the introduction to *Being and Event)* that he was still "haunted" by the specter of Heidegger in *Theory of the Subject*—who is precisely a thinker of this "Great Temptation of Ontology." To critique this orientation would also be to endanger Badiou's own project, which also sought to inhabit a place prior to structure (to the transcendental of a world, for Heidegger) but after metaphysics (without a commitment to the One). For Heidegger, this place is only possible for a negative ontology, an ontology which criticizes the relation between being and logos, or the idea. And we saw that Badiou had trouble producing a general discourse on the interiority of disciplines before his engagement with axiomatic thinking. Furthermore, this negative orientation is the point on which the whole test of Badiou's apparatus hinges: if there are multiple presentations *except* the presentation of ontology, *then* we are still within the grammar of a negative architectonic foundation (Hegel: breaking the link between foundation and beginning, but not with the One) or the grammar of hermeneutic/linguistic/textual beginning

(Heidegger: breaking the link between One and beginning, but not with foundations). The former allows for ontology to be a thinking, but only if it is logically grounded on a metaphysical presupposition (One); the latter does not allow for ontology to exist because "nothing," which is not the One, can begin thought. It can only disrupt reason, as in a poetic perturbation of sense. If Badiou truly desires to set a new grammar for thinking, a way to radicalize the assertion of the regionality of structures (a veritable theory of relativity of Reason), then he must be able to conceive of regions that appear *amongst* other regions. Situating ontology is therefore essential for the consistency of this new approach which Badiou has been constructing.

Badiou stands by the assertion that ontology is a discourse; it is within the positive regime of predication and formalization; its experience will be of deductive invention, and its result is fully transmittable through knowledge: to the temptation of the presence we oppose the rigor of subtractive, in which being is said solely as that which cannot be supposed on the basis of any presence or experience. This is an important difference, one we also saw in the introduction. Being is subtractive, not withdrawn. Both Heidegger and Badiou instigate a kind of discursive limit on our access to being, but Heidegger sees this limit as a traumatic consequence of discursivity, whereas Badiou sees it under the materialist orientation as the relation of identity and asymptote, harkening back to *Theory of the Subject*. It is a similar limitation, with opposite reactions: the pessimistic despondence of Heidegger vs. the optimistic enthusiasm of Badiou.[39]

The above, then, is not so much a refutation as it is a re-appropriation of what ontology, as a discourse on being, really is. Ontology is informed by the materialist principles of identity and asymptote in the relationship of being and thought: being and thought are the same, but only asymptotically so. This means that, in their relation, there is always a remainder: the remainder of being where the subtractive trait of being inscribes itself. There is no privileged access to being through the backdoors of direct experience of presence, and even so, this experience cannot enter the domain of discursivity. Therefore, the first principle of ontology is that being always presents itself discursively to thought. The only way for the science of being to present being is by presenting its presentation: ontology is a presentation of presentation, as the only place that access to being is possible. What does it mean that presentation is itself presented? Recall that presentation is irreducibly multiple and inconsistent, whereas the situation is a consistent multiplicity. So, when we say presentation is presented, do we imply that the inconsistent multiplicity is presented in a situation called ontology?

This is certainly what Badiou asserts.[40] But if this is the case, we run the danger of a clear regression: we claimed that presentation, as the only way in which thought could access being, is pure multiplicity. Through an operation of the count, this pure multiplicity became the structured multiplicities that we call situations. The presentation had primacy over the situation: there are presentations which are worked on by the operation of the count to make situations. Now we are saying ontology, itself a situation, presents the presentation. Is this not stating that there exists a situation that has primacy over presentation? That cannot be the case due to the law of the primacy of multiples. So, there must be a presentation from which the situation of ontology is derived, and so on, ad infinitum! Another problem is that what is internal to the situation is also counted as one. Externally, the operation counts a situation of being as one, and internally, each member is also counted as one. The issue with the presentation as a member is that it cannot be counted as one because the presentation is supposed as prior to the count. You can count "out of" the presentation but cannot count the presentation itself.

These are two significant problems, both familiar to us: the first is the problem of regression we saw in Lautman's presentation of Plato's ideas and numbers (that numbers are not ideas). The second problem is the recurrence of the One. Both problems are in danger of falling into the foundationalist trap. This is precisely why the beginning for Badiou becomes crucial here: we should bar regression, as well as the emergence of the One. But this is precisely why axioms are called in: axioms prevent regression through a set of "well-established" and "accepted" generalities, and generalities are not foundational—that is, they are not generalizations. The latter is formed from including parts into the whole—particulars under the generals; as a result, the general must speak of the particular. Whereas the former does not speak about any particulars: generalities have no reference to any specific particularity. Nor do generalities speak of each other: they are mutually agnostic. We saw this as the property of the extensibility of axioms. Generalities are prescriptive, whereas generalizations are deterministic.[41] The latter only admit and approve of certain particulars, those that are species of their respective genera. This is the reason why axiomatic nature satisfies both requirements for ontology. As such, if there is such a thing as ontology, it must be an axiomatic discourse.[42]

Although Badiou is tight-lipped about the theoretical backup of his decision, what we have reviewed makes it quite clear that this decision could be right out of Lautman: "This axiomatic transformation into consistency avoids composition according to the one."[43] Why? Lautman answers: because axiomatic extensibility, unlike generalization, is prescriptive and non-deterministic. "Nonetheless, its obligation remains."[44] How? Lautman answers: because axioms are "in comprehension" as a system of conceivable conditions.

By taking ontology as an axiomatic domain (a situation), we can grasp that what it speaks of axiomatically—the presentation itself—does not count. What it does not count, it speaks of through the conditions that regulate it.

Therefore, by speaking of the presentation itself, ontology does not speak of consistent presentations (individual specific situations: stars, solar systems, stocks, etc.), and thereby, that which it does speak of can only "in-consist" in the situation of ontology, that is, it is prohibited by the generalities of which ontology consists. But, at the same time, what it speaks of, inconsistent presentation, consists, as they are in comprehension of the system of axioms that comprises ontology.

So, what do we know so far? We know that the theory of regionalized multiplicities is an axiomatic theory. We also know that ontology which, per Aristotle's definition, is the science of being qua being, is also an axiomatic science. What ties these two is the assertion that presentation, thought's only access to being, is multiple, and every presentation that presents itself is a regional configuration of presentation as pure multiple. Therefore, the theory of regionalized multiplicity is a special case of ontology. Regionalized multiplicities (situations) are the outcome of an operation: situations as consistent multiplicities—the only things that effectively present themselves—are the outcome of the operation of the count "out of" inconsistent multiplicities. Our access to inconsistent multiplicities is only an indirect, retroactive access—they never present themselves as such, as every presentation has already been counted and made into a situation. So, what licenses us to think that these presentations do not saturate being qua being? That is, what is Being besides beings? The answer is that beings are counted: they are structured as one. Therefore, there is a principle that underlies beings, and that is the principle of the count—the one. If beings were all there is, then we are obliged to agree with Parmenides that it is the one that is the principle of what there is—and what there is, is the outcome of the principle of Dyad, which creates multiplicities of what "there" is. This would posit an infrastructure that contradicts the requirements of our theory: the primacy of multiplicity. The theory of regionalized multiplicity does not require the being to be multiple—it only needs three things:

- A theory that establishes the possibility of independent regions of being, each of which has its own content—we shall call this the interiority requirement.
- That the theory has resources to think the impasse interior to them—we shall call this the novelty requirement.
- To the above two requirements from part II, here we just added the requirement of the axiomatic (that is non-foundational) nature of this theory—we shall call this the beginning requirement.

These conditions were not thought prior to *Being and Event* (in "Mark and Lack" and *Theory of the Subject*) and remained as outstanding tasks. As we saw, Badiou started with the assumption that there are multiplicities (starting with the category of Something in Hegel's *Logic*) —the first condition above—but by the time we reached the dialectic of HD and the operation of periodization, we witnessed that the theory can assume the second condition but cannot derive it—solely on the basis of its original assumptions.

But in *Being and Event*, Badiou takes on a broader requirement, something that goes even further than what the theory of situations (regionalized multiplicities) would require. The examination of the possibility of presented multiplicities or situations calls into question the primacy of one or multiple, and in that relation, the question is whether being is one or multiple: whether multiple is and, therefore, one is not (the Dialectic of Two of unconscious Hegel, mathematical set theory) or one is, and multiple is not (the Dialectic of One of conscious Hegel along with the rest of metaphysics).

It is in this sense that ontology becomes a general theory for situations or regionalized multiplicities. It resolves or satisfies all three conditions through a theory that regulates and prescribes the entire domain of presentation, and thereby speaks to being qua presentation (the closest we have to being qua being). This further tells us that ontology is a structure of an axiomatic science, the axiomatic science of multiplicities in general: with no determination of what the nature of its objects is. So, ontology cannot be any branch of natural sciences as they have the minimal deterministic requirement of their objects. The only branch of science that can satisfy the prescriptive and extendible aspects of axioms that speaks to multiplicities, in general, is mathematics. And thus, the equation of ontology = mathematics.

Axiomatic thinking, through resolving the ideological overdetermination of the question of the beginning, by de-suturing this question from the presupposition of the One (Spinoza, Hegel) and the requirement of the ground (Heidegger, logicism), increases the maximality of thinking: the function of axioms is precisely this, and it is through this maximality that thinking will have access to being (Lautman, Badiou). For Plato, the statement that being and thinking are the same implies an ethics for thinking that adds the least possible in order to arrive at being: reduce your presuppositions and remove the unnecessary determinations in order to arrive at a *thought* that is adequate to the *object*. The concept of discipline extends this original Platonic prescription to the maximization of thinking: when a discipline, which is an immanent practice of thinking, thinks "as much as it could," which is to exceed the structured discourse that defines the borders of meaning (the dictionary of knowledge in Badiou's terminology) and thereby proceeds to think "nothing" from the standpoint of the structure, it furnishes privileged access to being that is the only "ground" for the thought of "nothing." So rather than the Platonic paradigm of finding a thought adequate to the object—we have a thinking that is as far removed from the object as possible. That is, it is almost exclusively grounded on the subject. In other words, it is where the (objective) ground is absent that thinking and being touch each other.[45]

The lesson of thinking maximally comes from Lautman: thinking maximally is precisely thinking outside of, or ambivalent to, the borders created by dominant trends of thinking, such as analytical philosophy, theory of truth as correspondence, linguistics, propositional logic—all theories that have taught us to think only in very restrictive boundaries of evidence and statements. In contrast, the posture proposed by the concept of discipline is in order to access the truths, which is the same as rekindling the very potentiality of thinking, is to go beyond the perceived boundaries of the real to "nothing" itself, and bring this "nothing" into disciplinary thinking. The question of the beginning of the discipline, in this sense, oscillates with the question of being. This happened with Plato, Hegel, and Heidegger, as well as with Lautman and Badiou. We thereby find it legitimate to say that the itinerary in *Being and Event* falls in the same region of thinking that we are tracking, which recognizes the concept of discipline as the access of thought to being.

There is a sequence of thinking that began with Plato and was saturated with Lacan. Lacan posited a disjunction between thinking and being, with an empty lacuna in between. There is a certain limit to thought. At this point, Badiou proposes a change in perspective that the real of thinking—that one cannot think anything beyond this limit—should be taken as a positive category. Instead, we ask: What can we think of "nothing"? How maximally can we think of "nothing"? Badiou calls the being of such thinking, in the latter mode, the being of a truth. This gives us a sort of rhetorical way of establishing why the category of being would ineluctably come into the inquiry of the beginning: how else could we tie together discipline and the real, in the sense that once a discipline exceeds the point of impasse, the being of that discipline—that is, the consistency of that discipline—is conditioned by the real? The consistency of the real at this point is not merely the consistency of the obstacle of thinking, its point of impasse; it becomes the problem of the consistency of the discipline itself. It is the being of the discipline itself that is at stake. It is no longer a problem of logical consistency, but an ontological consistency.

What is the being of a discipline beyond its axioms? Badiou's generic procedure will show us that it is the being of its truths, that is, precisely the being of something within the discipline that cannot be directly derived from its axioms but is in conformity with them. In that sense, one way to rhetorically summarize the interplay of the question of the beginning of discipline with the question of being is this: the condition of the possibility of a discipline is its axiomatic origin. It is a set of axioms that carves a region of being to become a discipline. But axioms do not limit what sort of entities or truths *could* "consist" in that discipline. We referred to this aspect of axioms as being prescriptive and non-deterministic. If we call the axioms the formal condition of the being of the discipline, we can easily see that there is a material condition of the discipline as well. In this relation, we could then say that Badiou divided the real into cause and consistency in *Theory of the Subject*. In *Being and Event*, he proceeds to divide the real as consistency into two senses of consistency: logical or formal consistency, which is the axiomatic condition we just referred to, and ontological consistency, which refers to what "consists in." The logical/formal consistency is guaranteed by the axioms—the closure of an interiority—and the ontological consistency refers to the fact that, within this closure, a region of being (a discipline) is guaranteed by nothing but its own procedures.

The salient features of the formalization of axioms could perhaps be summarized then into two important aspects. The first aspect is the "object-less" aspect, that this formalization never presents a thing or things—neither objects of the world nor graphemes of the formal notations. The real of mathematics is within mathematics itself: mathematics is neither an extension of a higher-order logic (logicism) nor the representation of something else "outside" of mathematics (nature/realism, mind/intuitionism). What mathematics presents is nothing but the internal consistency of its own concepts. However, being removed from any possible object is precisely where the "knot" with idealism is cut, and where the access to being is made possible for thought. The mode of thinking condensed by mathematics is that which breaks away with the mode of thinking as consciousness:

> mathematics is precisely the thinking which has nothing to do with the experience of consciousness; it is the thinking which has no relation to *reality*, but which knots *letters* and the *real* together; a thinking faced with the void because it obeys the ideal of formalization.[46]

This is precisely the *ideal* of formalization that breaks away from *idealism* (as we have seen earlier in this section). This formalization empties out objects (and thereby consciousness) from thought, and by this evacuation of objects, it removes the distance between thought and the real: "Between the real and mathematical form there is nothing."[47] This aspect, which we termed the question of the beginning, is something that, although it has been with Badiou since from his structuralist era, found its full philosophical articulation in *Being and Event*.

The second aspect is the "subversive" aspect of formalism through axioms. The formalism allows thought's intervention to surpass the real of the discourse. Or, to say it in an opposite way, the real suppresses the impasse of formalism: the passing of the real by force.[48] This means breaking away from the structuralist dialectic that treats the real (of a given formalism) merely as an impasse, via what we saw in the historical dialectic as a forced passage of the real through a new formalization at the point where an impasse is reached. This aspect, which we termed the condition of novelty, is something that found its first theorization in *Theory of the Subject*, but this theorization finds a new ontological treatment realigned with the orientation that addresses the question of beginning.

Therefore, saying ontology = mathematics is another way to say that the science of being qua being is a formal theory that is both "object-less" and "subversive."

THE VOID

It is not an exaggeration to say that the concept of the void is one of the most crucial concepts in Badiou's entire program in *Being and Event*, if not the most important.[49] There is even less doubt that the concept of the void does most of the work in Badiou's edifice: from its unicity in standing for being qua being to its localization through the irruption of events, it is this concept that does the lion's share of the conceptual heavy lifting in *Being and Event*. Furthermore, for Badiou, the concept of the void is the focal point in the debate between philosophy and anti-philosophy.

As we saw, a situation is any multiplicity that is presented and therefore is a consistent multiplicity. To be presented and to be inconsistent are contradictory states (being presented is being under the law of count-as-one). The count-as-one structures a given content. Thinking of count-as-one as pure structuring will naturally create a conceptual divide: between consistency as the form of presentation and inconsistency as the matter (or the being) of presentation. Prior to the advent of the set theory, there was no way of thinking the pure matter without form. This was central to the great debate between Socrates and Parmenides. The claim here is that with the set theory, matter without form is possible; in the very structure of axioms, the thought of matter (the prior existence of elements: the axiom of Empty Set and the axiom of Infinity) has primacy over the thought of form (formation of the set—the axiom of Separation), and form in axioms are reduced to various operations and constraints on the matter (the axiom of Power Set, the axiom of Extensionality).

The law of the situation is that "the one is not" must be unthinkable. This is but "natural," Badiou emphasizes, possibly in the same breath as Husserl, who called natural those views that have a non-phenomenological interpretation of phenomena. Furthermore, and purely from the point of view of the interiority of the situation, inconsistency is "nothing," and this is different than saying inconsistency *is not.* "Nothing" and non-being are not the same. "Nothing" is the being which is not registered in a presentative modality. Conversely, there is the dichotomy of "oneness is" and "the one is." To understand the division between "inconsistency is nothing" and "inconsistency is not," we have to again pay attention to the division of being and existence in Badiou's theory: being is pure and inconsistent multiple, while existence pertains to the domain of presentation: everything that has already been presented. Everything that exists is the result of the operation of the count: what is not One does not exist (with a little refactoring of the Leibniz principle—see below). The point Badiou is making is that insofar as existence is inseparable from being, inconsistency is inseparable from consistency. Although being or inconsistency does not make it to the field of presentation or existence, what is, has its being in inconsistency. Conceding to the non-being of inconsistency is to concede to non-being altogether, which of course, is an absurd claim. But non-being is not what the presentation announces. It announces that inconsistency is a no-entity or no-thing since what is presented is always a consequence of being counted. The being of inconsistency is registered as "nothing" by the count. "Nothing" is not a term or a place within the situation, and yet every situation, by its mere existence, implies "nothing" —the unpresentable. That is why Leibniz's thesis is the ultimate law of presentation, and if it were not for mathematical ontology, the "being of nothing" would have remained undiscovered. But "nothing" is an unpresentable "something," solely derived from a logical/causal explanation: the one (a thing) is the result of the operation of the count, and for a thing to be counted, there must be a prior material for the operation. Inconsistency, in this sense, is the very being of consistency, and this inconsistency/being is registered as "nothing" in the field of presentation.

Hence, what is registered as "nothing" in the field of presentation is, at the same time, the marker of being of the presentation itself: the being of the one is registered in the situation as "nothing," and therefore "nothing" is what sutures the situation to its being, being qua being. Furthermore, since "nothing" stands for the inconsistent multiple, there is only one "nothing" despite infinitely many situations: all situations have one suture to being qua being. We will see how these latter features of "nothing" manifest themselves in the axioms of ZF. The oneness of the "nothing" implies that "nothing" is not localizable to any situation. No term in the situation refers to "nothing," yet "nothing" is, insofar as there is always a not-having-been-counted in every situation.[50] As such, *the void is the name of being*. This name is chosen as the name of the being of the situation. No discourse about being of the situation is possible using the terms that belong to the situation, for such terms have already been counted and are consistent. That is, the very being of the situation is unnameable, and hence the role of the void is to suture the situation to its being. Mathematically, this is presented by one of the existential axioms in ZF: the axiom of Empty Set, and according to the axiom of Extensionality, there is only one empty set. At the same time, there is an imposition of an absolute "unconsciousness" of the void: there is no conceivable encounter with the void. But ontology, which presents presentation inasmuch as it provides a theory of the presentative suture to being, requires a theory that presents the law of presentation based on the errancy of the void, which at the same time describes how the veridical non-encounter is grounded on this unpresentable errancy.

In a sense, ontology is nothing but the theory of the void. This statement is true on two levels. Ontology is the science of being qua being. However, at the same time, ontology is a presentation of the presentation, and it is subject to the effects of the count. As such, ontology cannot present what is unpresentable, that is, the inconsistency of being qua being. The only difference that ontology has with other situations is that, while in other situations, the law of the void is "unconscious," for ontology, this law is conscious. Having knowledge about the void as the being of all presentation, as the errancy that evades the law of presentation, is the only difference between ontology as a situation and every other situation. Therefore, although the void has no veridical status in any situation, in ontology, it finds the name empty set. In addition, in ontology, it is established that every situation can be built based on the void. In effect, axioms have only a single term to deal with, the empty set. The rest of the axioms are operational; many of them will allow the construction of more complex terms from this single term. This justifies why ontology should not have any entity other than the void to deal with. Having a term beside the void would mean that ontology distinguishes between the void and the other term, which would determine the void as an entity that is counted by the situation that is ontology. This would have given the void a determination other than the errancy. For this reason, ontology only has one term, the void, and shows how all other terms are different modalities of this singular term.[51]

We should remark here that the unicity of the void is *not* the same as grounding ontology on the One. The axiom of Empty Set prescribes the suture of ontology to an "empty multiple" and not to a substantial building block. This is clearly manifested in the way that the axiom of Empty Set is defined in ZF, using second-order logic:

$\exists x, \forall y: \neg (y \in x)$	there is some set x such that for all y, y does not belong to x

The use of the second order logic for the existential guarantee that ZF donates to the empty set is very important. The second-order logic means that the empty set is not a fixed set: a unique set that can be simply referred to "by name" in the first-order logic (such as the non-ZF axioms that make use of the Null set or Urelement, which is defined in first order). The semantic of using second order is precisely to avoid "fixing" the empty set and a "proper name for the one empty set."[52]

There cannot be two principles of being, only that of being qua being. So, if all situations are presentations of the same being, then one can conclude that a situation doesn't have any being other than the being of the void. So, ontology is distilled to one law: any situation is a composition of the void, either directly (e.g., the axiom of Power Set), or indirectly (the axiom of Extensionality).

The assertion that mathematics is ontology and the equation of the empty set with the void raises one difficult question. Axiomatic set theory is a regulatory and prescriptive theory for mathematics, and as such, its role is to support the rest of mathematical theory with an extensible architecture. How could the errancy of the void be inscribed in the axiom of Empty Set? In the axioms and certainly, through the rest of mathematical theory, the empty set is considered to be a perfectly consistent set. Assuming anything otherwise will compel us to conclude that the entire mathematical theory is fraught with errancy and inconsistency. Does this imply that mathematics is also fraught with errancy? This is apparently a bizarre consequence. How can we account for such a? We think the answer is in between philosophy and mathematics (which, in our case, is mapped to the theory of discipline and a discipline). If mathematics is ontology, then as ontology, it can certainly think of the void consistently. However, it is through philosophy as meta-ontology that we are able to think of the inconsistency of being qua being. The role of philosophy as meta-ontology, which inscribes the inconsistency of being qua being, an inconsistency whose encounter is absent from ontology as a situation, points to a division that is of enormous consequence in Badiou's program: through this division, we can think the event as a localized void in a historical situation. Ontology, as we know, prevents this, and thus ontology cannot think the new, but philosophy can, through a prior engagement with inconsistency inhered in presentations as the invisible but in-expugnable remainder of every count-as-one. To put this in another way, we can say that the presentation is not the same as the set, and the set is not the same as the situation; each of these categories belongs to a different discourse. Presentation is a philosophical category, the set is obviously a mathematical category, and the situation is a category that Badiou uses to refer to real concrete things.[53]

Badiou's theory contains mostly assertions that belong to philosophy, and the assertion that mathematics is ontology is one of, without mathematics being conscious about it. This interpretation conforms to the mathematical formalism, of which David Hilbert is the champion. Hilbert, in his 1899 book *The Foundations of Geometry*, sets a new vision of geometry as a pure system of manipulation of signs and symbols that heralded the formalist interpretation of the foundation of mathematics.[54] This new vision of geometry undertook to rebuild Euclidean geometry from scratch, no longer founded on appeals to self-evident propositions and basic intuitions but on manifestly contingent arrangements of axioms prescribing the operations that may be carried out on undefined symbols. In mathematics, there is no correspondence between reality and mathematical signs and relationships. What distinguishes the formalist movement is, first and foremost, a commitment to the autonomy and freedom of the axiomatic method (it is without any norm other than consistency), and the identification of mathematical propositions with their formal inscriptions (a sort of immanently mathematical materialism). The modus operandi for working mathematicians, to see as vain and obsolete any attempt to philosophize or concretize their work, is a stronger manifestation of the generic essence of what they do. To put it in an Althusserian light, the turn toward axioms espoused by Badiou in philosophy may be seen as a consequence of the prior scientific turn toward axioms.[55]

So, if the set is a mathematical category and the situation is a category that refers to concrete things, then what is the relation between the set and the situation? Badiou's answer is simple: none. There is no signification associated with the relation of these categories in his theory, except when one thinks in a meta-ontological register (i.e., philosophy). In that register, an existent, a situation, is thought of as a set. That would certainly mean for Badiou that there is no equivocation of a thing and a mathematical entity such as a set: a thing isn't a set, except when you are thinking of a thing in meta-ontology. This is a result right out of the axiomatic textbook.

This view of ontology is different from the classical characterization famously formulated by Descartes in 1644,[56] accordingly to whom ontology—or metaphysics—is the root of the entire tree of knowledge. In contrast, the theory of discipline teaches us that disciplines are self-contained interior regions of endeavors, in which human subjects participate, but don't "own" them. These regions are lateral and do not lean or otherwise depend on each other. For sure, disciplines influence each other, but, as we shall see later, it is only through the truths they produce that this influence takes place. Ontology as a discipline is neither the base, acting as the foundation of the rest of the knowledge, nor is it the structure of all structures, in contrast to other structures that are sutured to a particular region of the reality. There is no necessary or stable connection between ontology as a region and other regions. This fact, that there is no signification at stake between a set and a situation, witnesses the general property of disciplines, and at the same time recaptures the property that the only semantic relationship in mathematics is intra-mathematical. This directly relates to the creative dimension of axioms, as they allow disciplines to be creative without being foundational or grounded.

Let's then look at an example to illustrate this point. Badiou distinguishes two types of sets: natural and historical. Then, he makes an extra claim: the schema explored in mathematics under the name of the natural *set* speaks about a *situation* that we call nature. That is not to say the natural set is the same as a natural situation. Nor does it say that natural situations are completely determined by natural sets, which we know they are not (as natural sciences would definitely tell us). What this assertion means is that the ontology of natural situations is presented by the natural sets: the latter speaks about *what sort of beings are the multiples* in the former. The same holds for historical sets and historical situations. But, in comparing these two cases, there is one big differentiation that Badiou makes with respect to historical sets. The ontological schema of historical sets does not allow for historicity, which is another way of saying that there is no being of history, or history does not exist.[57] There is certainly an "effect" of history, similar to the effect of count-as-one, but unlike the count-as-one operation whose existential donation comes from the side of ontology, there is no accountability on the side of ontology for the effect of history. What accounts for the effect of history are "situational" categories, and by this, we mean those categories such as an event, intervention, fidelity, or subject that does exist in a historical situation, and they even have ontological schemas for their being (e.g., axiom of Choice for intervention and generic extension and forcing for operations of fidelity), but they exist as the effect of intra-situational operations.

This effect in our study bears a name: discipline. And hence the thesis that we would like to present in the section that follows: historical situations comport according to the logic we dubbed discipline, whose pure schema is marked by categories of interiority, novelty, and beginning. Also, the trans-ontological categories such as an event, intervention, fidelity, and subject are categories that should properly belong to discipline and not ontology, which is "a" discipline.

EVENT

The theory of events is likely the most complex part of *Being and Event*. There are two sources for this complexity that are intertwined. First, the being of an event is meta-mathematical, because, at its core, the event is a supernumerary concept. That means its definition and existence are disallowed by ontology. The event is a category outside of ontology, and this is ultimately why it is formulated by a matheme rather than a mathematical construction (an axiom or a theorem). The second source of complexity is that Badiou appeals to historical venues when articulating this concept: the French Revolution, Paris Commune, May 1968, and the trial of Socrates each fall under the historical, rather than formal, paradigm. Badiou's articulation of the event is, therefore, itself analogous to a rupture. *Being and Event* begins as a progression of formal arguments, and this methodical and didactic movement is suddenly interrupted by a poetic, historical, and non-formal account of the event.[58]

There is a stepping out from the realm of ontology into the trans-ontological or non-ontological, as well as a return to ontology from this latter region. This is clearly manifested in the flow of the book itself. Until the topic of infinity, we are within ontology. The event is introduced after infinity. And from the event, we come back to ontology and examine its impasses, a direct result of the manipulation of infinities (see figure 6.1).

From the perspective of our project, events take place within disciplines, and it is the main conceptual apparatus through which the point of the real of disciplines as cause and consistency may be theorized. In this regard, *Being and Event* tells the story of one particular discipline: the discipline of ontology and how it grapples with an event that bears the name of Cantor, ultimately formalized by the axiom of Infinity, which changed the entire map of mathematical sciences. Although what was developed by Easton, Gödel, and Cohen is now part of ontology proper, this particular story can only be told by a detour through Badiou's philosophy, which is uniquely suited for detecting those traces belonging to an event in mathematics. The role that the theory of event plays is twofold: at the surface, it provides a narrative of certain philosophical concepts that conceptualize rupture (cause) and shift (consistency) in all disciplines: that which provides the material (the site of the void), formal (the naming and intervention), efficient (subject and the procedures of fidelity), and final (truths) explanation for regional change. At the same time, it centers on one particular event in ontology: the advent of actual infinities. This authorizes us to say that the way the concept of the event appears in *Being and Event* makes it an ultra-one: it bears its own nomination; that is, the theory of event belongs to itself.

We have one categorical concern in this section, and that is the relation of events to discipline, which we will immediately turn to.

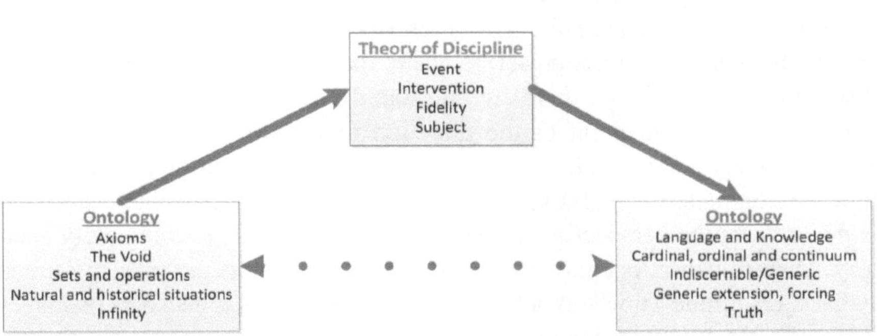

Figure 6.1 Event, Philosophy, and Ontology. *Source:* Author created.

Event and Discipline

We postulated that the theory of event marks a move from ontology to discipline, but it must also be complemented with the inverse move from discipline back to ontology. In our view, this postulate helps us to understand the structure of *Being and Event* better and, at the same time, allows us to extract important consequences with respect to the position of the categories, both ontological and meta-ontological. With respect to this postulate, we can think of two pertinent questions:

- How can we theorize this move from ontology to outside and back?
- What motivates this move with regard to the project that we have at hand?

We know ontology forecloses events. This is why events are defined neither axiomatically nor theoretically, but rather through the use of a matheme. Events, in fact, belong to situations which are only partially formalized as sets. In particular, we assert that they belong to only those specific historical situations we've referred to as disciplines.

The most important observation here is that the theory of events fits within the framework of axiomatic thinking. Our earlier introduction to Lautman showed that his "effective mathematics" addresses the theoretical practice within the activity of mathematicians: the structure of mathematical knowledge in relation to the concrete action of building the mathematical edifice. This progress takes place through the structural processes of liberation and saturation, resulting in mathematical creations that are mixes of opposite polarities. Lautman's vision involved a disciplinary grounded freedom of mathematicians to practice and develop their theories through the constant encounter and cross-pollination of their branches. This level of controlled dynamicity in mathematical practice is precisely the sense Badiou, within the framework of events, uses terms such as "encounters" and "decisions." Therefore, axiomatic thinking is no stranger to the points at which decisions must be made. While general theory prescribes the way in which generalities are developed, they do not prescribe where the points of dissociation in each discipline reside. That is, there are no grounds according to which we could determine either the point or the direction to which the discipline should move to at any given point. In fact, the methods of dissociation and participation are completely context sensitive—meaning non-formal: that a decision is made, from the perspective of the formality of axioms, is completely aleatory. Determination occurs only insofar as the encyclopedia of a discipline (what a discipline knows about itself) either forecloses a decision or makes it trivial—the ultimate paradigm of which are logicism and constructivism. This is precisely what the theory of events teaches us. At the point of an evental site, a decision is efficacious: making or not making a decision both lead to consequences. These consequences are not derivable from the already existing knowledge of the system, only from the space of something that must be decided on.[59]

Seen from this perspective, the theory of event is not much more than what Lautman has formalized as the methods of dissociation and participation, but expressed in a language that encompasses both ontology and what ontology disallows. In other words, Lautman's description of axiomatic systems is re-transcribed by Badiou through the evental schema. The placement of the theory of events as a part of the constitution of disciplines has an additional advantage: it derives more sense from the pragmatic and intuitional aspects attached to the discourse of events. These questions, more than being merely evental questions, according to Lautman, are presentative of axiomatic thinking. The concept of discipline donates regional sovereignty to disciplines only. Earlier, we discussed how the question of novelty is an intra-disciplinary question. Similarly, the question of whether this is a point of decision or not is only decidable within the discipline. Furthermore, the point of decision is something contingent: there are no founding theories to prescribe or anticipate the inflecting points within a discipline. At most, we can say that there are points at which a decision could or should be made. Where and what those points are, within each discipline, are indeterminable and non-axiomatic questions.

Disciplines are infinite multiples, but more precisely, they are historical situations. This means that there are sub-regions of the discipline that are not fully representable by what constitutes the encyclopedia of the given discipline—that is, what the discipline knows about itself. Here, it is the theory of event that finally untangles the myth of formal fidelity to the discipline—Badiou's observation from "Mark and Lack." We saw the first attempt to theorize this in *Theory of the Subject*, which deployed the grammar of SD and HD in order to show that there is a point of passage beyond what a discipline knows about itself. We are now at the stage to redeploy this point of the passage as an ontology based on the theory of consistent multiplicities.[60] The Platonism of this thinking resides in the ability to conjoin seemingly unrelated ideas which separately lacked the capacity to represent a part of the world. This conjunction leads to the emergence of something new, something that has not been in the world before—a cross-worldly and eternal truth.

Now let's come back to the topic that matters most to us in the discussion of events. In Part III, we tracked two encounters between a discipline and the real. These two encounters sanctioned both the multiplicity of disciplines and the real itself as multiple. We then journeyed through axiomatic thought, the question of beginning and ontology of consistent multiplicities. This itinerary established how the regionality of the discipline is precisely that of the set of axioms which circumscribes it. The localization of axiomatic consistency is, at the same time, the localization of the disciplinary real, and in that sense, the inquiry into the question of beginning not only reveals the unfounded and non-metaphysical nature of disciplines, it also establishes its multiplicity as particular regions of being: discipline is a situation, thereby its real is local, and this relativity of the real to discipline implies the multiplicity of disciplines. Therefore, it was quite legitimate for us to ask what the pure schema of this relative real is.

The second point of encounter concerning the real is its splitting between cause and consistency. We have claimed that, if a discipline exists, while it forces itself at the point of the real, it can also further "count on" the real. This counting on the real is precisely the counting on the ultra-one of the event, in the evental site as the singularity, functioning in the capacity of the real as the cause of the situation, insofar as it is at the edge of the void. At the same time, to count on the real means to extract some consistency out of the un-presented element of the site. For a discipline to be self-governing or an interiority, we have the requirement to split the concept of the real: a discipline draws its consistency from its real. That is, we want to think of the real relative to a situation as the possibility. The question of how exactly such a split would occur and function in a given discipline remains within that discipline, but knowing the pure form of this split pertains to the theory of events.

The theory of event explains why, at an evental site, decisions are efficacious. This theory, as elaborated in *Being and Event*, uses both mathematical and non-mathematical conceptions such as intervention and axiom of Choice for this explanation. But its main problem is transposable to what was covered in *Theory of the Subject*, where the spatial dimension was much more pronounced. In that work, Badiou thought about the evental site in the distinction between algebraic and topological "perspectives" of the same point, and not as "before and after." He did not say that the situation thinks of itself algebraically and then suddenly topologically. He instead examined the shift of perspective of the same point (of the real): this point, treated algebraically, is simply a reference to an outplace, but if we consider it topologically, what counts as a point in the system of placements is, in fact, a swarm of points. It possesses vicinities that are not necessarily included. So, what was a topological vicinity became the un-presented of an evental site. That is why the axiom of Foundation is the ontological schema of the real. If the foundation is the point of suture, it would be the real as the cause, and if the un-presented in the foundation is called upon as the generic, it is the real as consistency. In that sense, the role of ontology is to show us that there are disciplines—or historical situations—which, as founded situations, have a non-void singularity. That is, while the discipline is sutured to a specific presentation of being, its site is always sutured to the void. This is the point at which we leave ontology. The theory of "evental" disciplines tells us that the point of singularity needs to be divisible. This divisibility must present the two properties that we analyzed in *Theory of the Subject,* to be treatable both algebraically and topologically, and ultimately, ignoring the phenomenology of how this happens, this point of the real must be unfolded either as a suture or as a novelty. This is what discipline does. The event only comes as the name for this anamorphosis that is distilled from answering another question: the question of how we call the shift from a site initially considered merely as the cause (the void) to what can evoke consistency (the generic). It begins from the point of singularity and the perspective of non-presentation, but then the schema of the event allows for the addition of an independent assumption to the situation, according to which we can draw results that we were not able to previously.

This was precisely what we saw in Badiou's analysis of the prisoner's problem in "Logical Time". Badiou adds one simple assumption: "What if the other is stupid?" Adding this assumption does not disturb the logical consistency of the three procedures of R1, R2, and R3. It does offer, however, the possibility of one new deduction that avoids the last scansion and thus leads me to be the first prisoner the warden sees. This elicits the nature of an axiomatic economy: axioms are independent decisions. The case in point is the axiom of Infinity, the subject of the subsequent section. The rest of the axioms of ZF are neither strengthened nor weakened by the addition of this axiom. But the axiom of Infinity, as what bears the name of the event in ontology itself, suddenly opens up unconceivable vicinities to thought, making it the most important decision of contemporary mathematics and (as claimed by this work) altogether for contemporary thought.

Hegel has said, "Nature is weak and fails to exhibit the logical forms in their purity." This is due to the contradiction between the necessity of its forms (generated by the Notion) and the indifferent contingency and indeterminable irregularity. This contradiction is at its highest in the realm of "concrete individual forms" (aka Badiou's situations).

> The *immediately* concrete thing is a group of properties, external to one another and more or less indifferently related to each other; and for that reason, the simple subjectivity which exists for itself is also indifferent and abandons them to contingent and external determination. This is the *impotence* of Nature, that it preserves the determinations of the Notion only *abstractly*, and leaves their detailed specification to external determination.[61]

We think an analogy is to be made between the "impotence of Nature" and the "impotence" of ontology to treat historical situations. Akin to what Hegel is declaring about Nature, the "concrete individual," that is, the historical situation, realizes the ontological determinacy only abstractly. But, just as

Nature's impotence is the very birthplace of Spirit, the impotence of ontology is the very birthplace of truths, and the event is simply the schema for naming a localized region of a situation whereby the impotence of ontology declares itself. Therefore, the emergence of the concept of the event is the consequence of the impotence that makes ontology too weak to close but also too weak to propose.

Our proposal is thus: the thought of an event is already embedded in that of axiomatic thinking. Event formalizes the very process by which axioms are developed in condensation and saturation of the very practice within a discipline. Thereby, the concept of the event does not belong to ontology (which Badiou affirms) but belongs to discipline (what we propose). Ontology itself is a discipline, and as such, was subject to an evental rupture in and around the turn of the past century. In our view, this rupture is marked by the name of actual infinities and by the addition of the axiom of Infinity to set theory. As we have mentioned previously, this study is centered around the claim that the event of infinity is "the" event that defines our "epoch" as what comes after the epoch of the death of God.

It is time for us now to finally look at this evental rupture in thought: the axiom of actual infinities.

NOTES

1. Badiou, *Being and Event*, 482.
2. Fernando Zalamea, "Albert Lautman and the Creative Dialectic of Modern" in Albert Lautman's *Mathematics, Ideas, and the Physical Real*, xxiii–xxxviii. London, New York: Continuum International Publishing Group 2011.
3. Zalamea, "Albert Lautman," xxiv.
4. Zalamea, "Albert Lautman," xxvii.
5. Albert Lautman, *Mathematics, Ideals, and The Physical Real*, London, New York: Continuum International Publishing Group 2011, 41.
6. Zalamea, "Albert Lautman," xxx.
7. Zalamea, "Albert Lautman," xxxi.
8. Lautman, *Mathematics, Ideals, and the Physical Real*, 9.
9. Lautman, *Mathematics, Ideals, and the Physical Real*, 9.
10. In other words, meta-propositions take propositions as objects, and so accepting a set of meta-propositions also means determining all true (first-order) propositions.
11. Hilbert's masterful axiomatization of the Euclidean geometry in *The Foundation of Geometry* in 1899 is perceived as the hallmark of such a formalist effort. Hilbert's effort to axiomatize geometry, whose striking formalism makes use of no definitions—or definitions that are not reducible to mere abbreviations—is a formalism that neither borrows from a higher-level hierarchy of concepts, nor is trying to establish a hierarchy for other disciplines to follow it. It is merely a demonstration of

how formalism in a particular field of mathematics can revolutionize the practice of that field quite dramatically. Hilbert formalism and his "understanding of what constitutes a scientific concept is paradigmatic for Badiou" (Zachary Luke Frazer, "The Category of Formalization: From Epistemological Break to Truth Procedure" in Alain Badiou's *The Concept of Model*, Melbourne: re.press 2007, xxx).

12. Lautman, *Mathematics, Ideals, and the Physical Real*, 11.
13. Lautman, *Mathematics, Ideals, and the Physical Real*, 11.
14. Lautman, *Mathematics, Ideals, and the Physical Real*, 31.
15. Lautman, *Mathematics, Ideals, and the Physical Real*, 31.
16. Lautman, *Mathematics, Ideals, and the Physical Real*, 32–33.
17. Lautman, *Mathematics, Ideals, and the Physical Real*, 33.
18. Lautman, *Mathematics, Ideals, and the Physical Real*, 41.
19. Here of course by "type" Lautman has in mind the Aristotelian style of generalizations. The new type theory, for example in the case of Homotopy Type Theory (HTT) or univalent foundations, utilizes types that are not simply hierarchical, that is, a type can be determined axiomatically (in the case of HTT types are spaces). As such, the relation between axioms and types is more complicated in modern mathematics.
20. Lautman, *Mathematics, Ideals, and the Physical Real*, 41.
21. Lautman, *Mathematics, Ideals, and the Physical Real*, 190.
22. Leon Robin, Julius Stenzel, and Oscar Becker.
23. Zalamea, "Albert Lautman," xxxiii.
24. Lautman, *Mathematics, Ideals, and the Physical Real*, 190.
25. Lautman, *Mathematics, Ideals, and the Physical Real*, 191.
26. Alain Badiou, "The (Re)commencement of Dialectical Materialism" in Alain Badiou's *The Adventure of French Philosophy*, London, New York: Verso 2012, 163. Fraser in his amazing introduction to *The Concept of Model* after quoting the very same passage makes the following interesting observation: "Note that 'historical sets' translates the French '*théorie des ensembles historiques*,' an allusion to both the French term for 'set theory' ('*théorie des ensembles*') and to the subtitle of Sartre's *Critique of Dialectical Reason*: '*théorie des ensembles pratiques*' ('theory of practical ensembles')" (Fraser, "The Category of Formalization," xxxix).
27. Alain Badiou, "Infinitesimal Subversion," in *Concept and Form Volume One*, edited by Peter Hallward and Knox Peden, 187–208. London, New York: Verso 2012.
28. Badiou, *The Concept of Model*, 35.
29. We saw the appropriation of Frege's logicism by Miller and Badiou's critique of both in "Mark and Lack" in chapter 1. It is perhaps also useful to mention the debate between Hilbert and Frege over the necessity of definitions in axiomatic systems—this debate highlights the premodern stance defended by Frege versus the modern axiomatic thought that Hilbert represented. This debate, which was started with Frege and continued for some years, was over some fundamental issues such as: "how to understand the content of a mathematical theory, what a successful axiomatization consists in, what the 'truths' of a mathematical theory really are, and finally, what one is really asking when one asks about the consistency of a set of axioms or the independence of a given mathematical statement from others" (Patricia

Blanchette, "The Frege-Hilbert Controversy" in *The Stanford Encyclopedia of Philosophy*, Spring 201, Edward N. Zalta editor). One of these points pertained to the status and necessity of definitions in mathematics. As it is well known one of the striking features of Hilbert's axiomatization of Euclidean geometry is that Hilbert's presentation does not contain any definitions. This was a programmative decision on the side of Hilbert who did not think that a formal system requires any definitions. In his letters to Frege, Hilbert compellingly argues, definitions "are utterly impracticable in the field of mathematics." He writes, "only the whole structure of axioms yields a complete definition. Every axiom contributes something to the definition, and hence every new axiom changes the concept. A 'point' in Euclidean, non-Euclidean, Archimedian and non-Archimedian geometry is something different in each case" (David Hilbert—*Letter to Frege dated December 29th 1899, Frege Philosophical and Mathematical Correspondence*, University of Chicago Press 1980—cited from Fraser, "The Category of Formalization," xxvii–xxviii).

30. This also clarifies why the reading of Badiou's axiomatic beginning as a sort of onto-theological beginning is totally missing both points: of axioms and the formal system, as well as Badiou's specific engagement with them.

31. See Mark Kulstad, and Laurence Carlin, "Leibniz's Philosophy of Mind" in *The Stanford Encyclopedia of Philosophy*, Winter 2013 Edition, Edward N. Zalta (ed.).

32. Badiou, *Being and Event*, 23.

33. Badiou, *Being and Event*, 24.

34. In the language of *Logics of Worlds*, the presentation of being is an outcome of the inconsistency of the Whole. The appearance of pure multiples is that which guarantees the being of multiples: without them having to appear there is no guarantee for their being—and this guarantee is not a guarantee that is issued by providence. So, a being is a being insofar as it appears, despite us having the ability to think its pure being form its appearing. Badiou's point is that this (guarantee of being through appearing) is a consequence of the non-being of the Whole.

35. Badiou, *Being and Event*, 25.

36. Badiou, *Being and Event*, 25.

37. Badiou, *Being and Event*, 27.

38. Badiou, *Being and Event*, 26.

39. Badiou, *Being and Event*, 26.

40. Badiou, *Being and Event*, 28.

41. This comparison may be a little bit vague. The power of axioms is to articulate certain constraints independent of the objects which would satisfy them. But this can still be considered "deterministic": if two different objects satisfy those constraints, then the axioms "determine" the sense in which they are the same. But it cannot be determined from the axioms whether such an object exists. In that sense, axioms prescribe what it takes for objects to satisfy them, whereas generalizations only exist if those objects do exist and do possess certain characteristics pertinent to the generalizations in question.

42. Badiou, *Being and Event*, 29.

43. Badiou, *Being and Event*, 30.

44. Badiou, *Being and Event*, 30.

45. It is interesting that for Badiou it is not the category of the subject, but the category of the object that is the point of entanglement with idealism, "for the knot of idealism is not the category of the subject but that of the object . . . a materialist (and therefore objectless) theory of the subject is necessarily of the school of post-Cantorian mathematics" (From Badiou's 1987 seminar *Orientation de pensée transcendante* cited in Fraser, "The Category of Formalization," xxxv). We also saw the same statement by Lazarus regarding politics as a discipline: there is no object of politics (and for Lazarus, therefore, there is no subject of politics in an ordinary sense either; the 'subject' of politics for Lazarus is people: "people think"; see chapter 4). According to Badiou the same "non-objectal" feature of the discipline, already called out by Lazarus, is also true for mathematics post-Cantor.

46. Alain Badiou, *Infinite Thought: Truth and the Return to Philosophy*, London, New York: Continuum 2008, 67.

47. Badiou, *Infinite Thought*, 67.

48. Badiou, *Theory of the Subject*, 23.

49. Badiou, *Being and Event*, 114.

50. We could read here a certain Sartrean undertone of Badiou: "Nothing" is the situated name of the fact that the count is contingent, that it could have been otherwise—which, for Sartre, leads to the whole analytic of anguish and freedom. Unlike Sartre—and Lacan, for that matter—Badiou does not *a priori* link nothingness to the subject: the phantom of the count's contingency, the gap of nothingness, is not a subjective feature of a situation, but an "objective" one, not an access to the unstructured or the intromission of the real, but an effect of the *being* of structure. But without the decomposition of the real into its relative and absolute characters, only feasible in an ontology of situations, this claim could not have been made prior to Badiou. After Badiou, however, *further determinations* must be added in order to move from nothingness to subjective anxiety: in short, nothingness is subjectivized as anxiety *under certain determinate situations*—this result we already encountered in Badiou's critique of Lacan's "Logical Time."

51. It seems important to point out that these claims are within the world of ZF and should not be considered as overarching (say to other models of set theory). When we say the void is the only term, we of course mean the only term that is directly recognized by the axioms of ZF (other than ω). It is also important to note that there is also no axiom in ZF that says "there exists ONLY the empty set," or that all sets must be built with the empty set. This is mostly done from convention, but there can be other models where this is not true. "Void is the proper name of Being" could also be proven by the fact that every power set contains the empty set, rather than by every set built from the void. Those are two different claims. Although Badiou works with Von Neumann Universe so both are true, it might be worthwhile to distinguish them. Also, one can distinguish the void from another term by the axiom of Extensionality.

52. The very statement of the axiom in mathematical language already shows how the linguistic turn falters here: what is written in this axiom is almost automatically substantiated into a statement about the one just by being transcribed from the "there is some set x such that for all y, y does not belong to x" into "there is one empty set."

53. Another, and perhaps a better way, to account for this triad is the following: set, presentation, and situation are best thought of in terms of theory, model, and modeling relation: presentation is a meta-ontological concept, the set is a mathematical term (the consistent thought of presentation, even if mathematics does not know this) and the situation is the name of what counts presentations as set—it is the "fundamental metaphor" which abstracts from presentations just their "presented-ness" in order to correlate it to set. This seems like an extra-mathematical operation, but with the invention of category theory, which in some ways is a generalization of the model theory, this is no longer the case: no wonder that the term "situation" disappears from Badiou's work with *Logics of Worlds*, and is substituted by the term "world," which is both the modeling relation (situation) and the mathematical concept (set) taken together in the world/transcendental formalism.

54. David Hilbert, *The Foundations of Geometry*, La Salle, Illinois: Open Court, 1965.

55. Louis Althusser, *Lenin and Philosophy*, New York: Monthly Review Press, 1971.

56. René Descartes, "Principles of Philosophy," in *The Philosophical Writings, Volume 1*, Cambridge: Cambridge University Press 2006, 186.

57. "History is but empty symbolism" (Alain Badiou, *The Communist Hypothesis*, London, New York: Verso 2010, 236), and "we will allow that History exists only symbolically. In effect, it cannot appear. In order to appear, belonging to a world is necessary. However, History, as the alleged totality of human becoming, has no world that can locate it in an actual existence. It is a narrative constructed after the fact" (Badiou, *The Communist Hypothesis*, 238). There is a sense of history that belongs to truth—and that is precisely because the being of truth is post-evental.

58. In fact, the rest of the book is also structured in the same way, one can say that *Being and Event* in its architecture follows the theory advocates: ontology, event, intervention, fidelity and finally truth.

59. The triviality here is that the majority of these points pass by quite "silently"—witnessing the fact that what is rare is not the point of decision, but the decision itself.

60. This point elicits what we said before. The theory of event within the edifice of *Being and Event* is itself an ultra-one, not only providing the theory but at the same time disclosing one evental sequence within the discipline of ontology. In particular, the theory of event is coming to terms with the question of passage beyond the point of the real of a discipline.

61. George Wilhelm Friedrich Hegel, *Philosophy of Nature, Part Two of the Encyclopaedia of the Philosophical Sciences (1830)*, Oxford: Oxford University Press 2007, 22–23.

Chapter 7

Infinity

Axiomatic thinking, as a framework, allows us to think about the question of the real ontologically, without supposing the One or ground. In this way, Badiou answers Miller's question to Lacan regarding the latter's ontology. Regarding ontology either as the One or a foundation, Lacan was right to answer that psychoanalysis is "pre-ontological." Badiou shows us that this "pre" is a "platform" for axiomatic thought and, in particular, axiomatic set theory.

But while set theory establishes a framework to consistently and productively think the being of the real, there is only one axiom that directly addresses this real: the axiom of Infinity. Our question is this: What becomes of the real according to this existential axiom? This is the question that motivates the various topics of this chapter. The consequences of thinking the real according to the axiom of Infinity are of such a scope that we claim this axiom to be singular in distinguishing the contemporary age from what is declared by Badiou as the age of finitude or the death of God. But for us to establish this claim, a few things need to take place.

First, we need to understand what the thought of axiomatic infinity is and how it differs from the traditional thought of infinity from antiquity to today, which invariably falls into one of the following categories:

- Infinity as a modality of the One: the One (infinite) substance (Spinoza and all monotheist religions) or an absolute otherness: the source for the ethical injunction (phenomenology, Levinas).
- Infinity as a modality of finite: infinity as the inaccessible (Lacan, Hegel's bad infinite), infinity as the self-realization of finite (Hegel's good infinite).

In contrast to the above negative senses of infinity, we shall outline an affirmative definition of axiomatic infinity.

Second, we need to establish the relation between the thought of infinity and our own project: in relation to the path that has taken us from epistemology and the theory of discourse to the concept of discipline, the real, and the question of the beginning, a path which has yielded an axiomatic theory of multiplicity that touches the real—ontology. In that relation, we would like to establish the necessity of the axiom of infinity in thinking about the real consistently. Out of this thought of the real as infinity, as defined by the axioms, there comes the thought of the generic. This thought, as we shall see shortly, is the consequence of fully adopting the axiom of Infinity and is the upshot of a mathematical drama that witnessed two of its most brilliant heroes go to insanity, a drama centered on what was dubbed by Hilbert as the most important problem of mathematics—the Continuum Hypothesis (CH).

Previously, we saw how the thought of the event challenges the formal constraints of ontology, but we assert that this challenge does not extend to the axiom that Infinity exists. The underlying intuition of our thesis is that, ontologically, it is infinity and not the event which truly constitutes the power of Badiou's proposal. Badiou indeed inaugurates a new thinking of events. But in our view, this formal thought of events can be derived from infinity. The debate surrounding CH, which includes everything that is possible in Badiou's ontology in general, concerns the status of Infinity. Between Cantor and Gödel, what is at stake is the restriction of the power of this axiom. From Cohen onwards, the axiom of Infinity is capable of procuring the generic, which grounds a new thinking of the category of the subject. Our study will conclude with this topic.

We established in the previous chapter that axiomatic thinking decouples the question of beginning from the One and foundation. It allows us to think differently about what is at stake when something new begins. Yet events themselves neither found, totalize, nor inaugurate in the ontological sense. The properties of events are conditioned by the properties of the axioms: that is, it would be impossible to think of events without the thought of axioms. So, we ask the question: What is the evental element in mathematics which allowed Badiou to finally complete, in a positive sense, the issue that we have been tracking since "Mark and Lack" and *Theory of the Subject*? The axiom of Infinity will also alter the discourse on the problem of otherness and inaccessibility. Beginning with the axiom of Infinity, we have a new ground to rethink the problems of limits of finitude, discreteness, and discernibility.

If we are to take seriously the assertion that axioms capacitate the mode of thinking without objects, enabling a maximality of thinking, then we must be able to accept Cantor's theory of transfinite numbers from 1874. If the consistency of a discipline demands one to think consistently beyond its point of impasse, and this effort is only conceivable under an adequate theory of beginning un-sutured to the One, then the axiomatic thought of infinity in modern mathematics is paradigmatic for all disciplines. While axioms *enable* this thinking, in which the real as impossibility is always a requisite fallout, the theory of infinity *deploys* this mode of thinking in mathematics. Modern mathematics owes its richness to the axiom of Infinity. A mathematician who accepts this axiom cannot pretend that she is not a Platonist, nor can she fully chalk this decision up to pragmatism. That is, she has to accept that she is taking a stake in the maximality of thinking.

It is appropriate then, after Badiou, to say that the Eden opened by Cantor is not only for mathematicians, as Hilbert had gestured, but for all disciplines, and that "inquiries into its marvels" have only just begun.[1]

SENSES OF INFINITY

In this section, we provide a brief overview of the senses of infinity important to our discussion. As is known, infinity is one of the most ancient and important problems of philosophy. In particular, the question of the infinity of being has been a contentious matter throughout history. For the ancient Greeks, the thought of being always fell within a framework of finitude. For them, to be was to be finite. Infinity resulted in destruction and tragedy—thematically, what the Greeks called hubris. In fact, *apeiron*, from *a* (not) and *peras* (limit), literally limitless, was the only sense in which infinity was treated among them. Aristotle claimed *Infinitum Actu Non Datur*. That is, the Actual Infinity is an inconsistent notion. This opinion was shared by Leibniz, Berkley, Gauss, Cauchy, Poincare, Brouwer, Weyl, Luzin, and many other outstanding creators of classical logic and mathematics as a whole.

The thinking of infinity of being came into ontology through religion, specifically Christianity. But even then, the gap between being as infinite, which is the being of God, and the ordinary being of things that exist was an impassable chasm, only bridged by the miracle of Christ. The infinity of the natural world, on the other hand, was inaugurated by advancements in the mathematical sciences in the fifteenth and sixteenth centuries. For the first time, it dawned on the human mind that perhaps the natural world is itself infinite: this is the first time that ontology began to break with the Aristotelian dictum that Actual Infinity is a logical contradiction and cannot exist.

This development of infinity in the sciences is superbly laid out by Alexander Koyré.[2] For now, it is enough to say that ontological thinking has had a complex relationship with infinity, such that the statement "being is infinite" has the status of a radical decision rather than a logical necessity.

As fascinating as the ancient Greeks are, they teach us very little about infinity—not just because they see infinity as a source of hubris and malice, but more so because their version of finitude reverberates in our time with heightened force and sophistication. Therefore, we will review other notions of infinity that have been dominant throughout history. First, we have the notion of infinity as absolute otherness, the One—familiar, because it is the infinity of God in Spinoza's account. The second is the notion of infinity as a modality of finite, which is the subtler, more recent account that either sees infinity as inaccessibility—that is, the infinite chain of repetition of the finite—or the romantic notion of the infinity of the determinate being, which is again an attribute of the reflexive faculty of finite beings. Both of these senses are available in Hegel, famously depicted as *spurious* and *affirmative* modes of infinity. And the third sense is, of course, that of mathematical infinity, whose affirmative definition is provided by the axioms of set theory.

What all these notions of infinity have in common is their attempt to think of infinity consistently rather than as an imaginary construct. The infinity of the One finds its most consistent representation in Spinoza and infinity as a modality of finite in Hegel. But what is interesting about both is that they took the task of thinking about infinity to such an extent that there is nothing left outside—that is, the thoroughgoing monism of both of these great masters. In addition, despite their metaphysical commitments, their fidelity to this task, combined with their genius, enabled them to unearth much more about infinity than they originally set out to accomplish. For these reasons, and because they teach us a great deal about our own approach, we shall review their accounts of infinity first.

Infinity as a Modality of the One—Spinoza

Spinoza is definitely a philosopher of infinity—or perhaps we should say infinity is his philosophy. In *Being and Event*, Badiou dedicated one chapter to Spinoza,[3] and despite being critical, he praises the latter's ontology for possessing the profound insight required for any discourse worthy of the name. Similar to Badiou, Spinoza believes in truths that cannot be presented in any structured discourse. Also, Spinoza would be opposed to the dominant linguistic orientation, from pragmatism to other post-analytical philosophies, that assert that what cannot be spoken of is not true—or, in other words, truth has to measure up to language. In contrast, both philosophers adhered to a logico-mathematical form: in the case of Spinoza, geometry, and in the case of Badiou, set theory. In addition, Badiou is very intrigued by the passion according to which Spinoza participated in the sociopolitical debates of his time—especially in defense of freedom in the Dutch Republic, which was at the time consumed by religious and political debates, for which he famously suspended his *Ethics* project to write *Tractatus Theologico-Politicus*. So, to Badiou, who also uses a rigorous mathematical ontology to support a passionate political intervention, Spinoza is like a preamble.

For Spinoza, a *thing* is a mode of (an attribute of) God. The relation of God to a thing, insofar as its existence is concerned, is an immanent relationship. Being in God is the only relationship there is between things and God. So, for example, if we build new things out of other things, both this new thing and its constituents all belong to God. This structural relationship of belonging leaves no room for parts since everything there is, is in God. This amounts to a radical identification of presentation and representation. As Badiou remarks, to Spinoza, all situations have the same state—every composite thing is composed in a way such that the local state of a multiple isn't permissible in Spinoza, only the count-as-one of state, which is God.

Given that we now only have one state, does Spinoza also think that we only have one situation? The answer is no. The one and only substance presents itself according to infinite intelligible attributes. Each of these presentations is a situation that is also in God. The importance of this emphasis is that Spinoza moves the difference itself inside the One of God: if it was otherwise, then different situations could be counted separately, which is impossible. There are only modifications of a single substance corresponding to each of its infinite attributes, and an existent may be a combination of multiple such modes. For example, the human is an existent comprising two modifications of the substance, expressed as the attributes of extension and mind.

It should be mentioned that the above two "features" of Spinoza's ontology allow him to avoid the paradox of the set of all sets. The set of all sets, denoted as Ω, is impossible because for every set, there exists a power set that is in excess of that set. The set Ω implies a power set that is in excess of Ω, which is a contradiction (since Ω is the set of all sets). God in Spinoza can be considered equivalent to Ω. But since there are no inclusions permitted in Spinoza's ontology—God has no parts, only terms—everything that belongs to God is also represented in God. In other words, the "power set" of God is the same as God, and thus there is no set that is in excess of God. In this way, Spinoza appears to banish the void from ontology altogether successfully. The void neither belongs to a presentation because by not having any elements, it cannot contribute to the one-effect of the count as one of any situation—the one-effect of any situation to be determined by the causality as the count-as-one operator—nor is it included in any situation since to include for Spinoza, as we just saw, is the same as to belong—to be included requires to belong. This appears to be a deductive foreclosure of the void.

What is Badiou's strategy to confront this? Badiou hones into the (dis)similarities between substance/mode couple and infinite/finite couple. First, as Badiou remarks, attributes have nothing to do with ontological donation; that is, they are not existent; only the substance and modes of the substance exist. Second, if what exists is either the substance or the modes of the substance, and if the substance is absolutely infinite (notwithstanding the fact that the sense in which the substance is infinite is not clearly defined in this context), how does the infinite substance cause finite modes? How can we account for a causal chain between the infinity of the substance and the finitude of a mode according to the Principle of Sufficient Reason (PSR)? Here, Badiou argues that Spinoza will fail to establish a causal chain that can handle the absolute excess of the substance over finite things, and this failure is where the void is to be found.[4] The strategy is thus to disclose Spinoza's inability to account for the excess of the absolute infinite in a causal chain that grounds finite modes, thereby showing that Spinoza is far from foreclosing the void and banishing it from his ontology.

The question of how the infinite essence of God can beget finite modes is posed by Leibniz a few years after Spinoza's death, and since then, it has been the center of many discussions and interpretations. Spinoza himself only derives immediate infinite modes from the substance (*Ip21*) and mediates infinite modes from immediate infinite modes (*Ip22*). He, however, indicates clearly that finite modes do not follow either directly or indirectly from the absolute nature of the substance. The famous *Ip28*, which Badiou also cites, is the most crucial passage regarding the absolute rift between finite and infinite modes, as well as between finite mode and infinite nature of the substance.[5] The important point from this passage is that there seem to be two distinct causal chains: the causal chain that relates the self-sufficient substance to immediate infinite modes and the immediate infinite modes to mediate infinite modes ad infinitum, and the second chain that relates a finite thing to another finite thing that caused it, to a third thing that caused the second, ad nauseam. It is apparent from what Spinoza says that these two chains never reach each other and never connect. The causal chain that starts from God is a necessary causal chain, as the infinite modifications of God follow necessarily from the infinite nature of God, and since God necessarily exists, all the infinite modes (immediate and mediate) also necessarily exist. On the other hand, finite things are contingent. Spinoza does not believe that finite things necessarily follow God's nature or one of the infinite modes: this can be clearly shown by the disconnect between the necessary infinite causal chain and the causal chain pertaining to finite things.

So, the inevitable question is how Spinoza's claim (that whatever exists, exists in God) can be reconciled with the gap between these two causal chains. As was mentioned, Leibniz was the first who discussed (or discovered) this gap shortly after Spinoza's death.

There are several contemporary attempts to fill this gap, most famously by Curley[6] and Garrett.[7] We will not examine how successful, if at all, these attempts have been. On the contrary, what's important is, unlike Leibniz and the contemporary commentators, for Badiou, this gap is an unavoidable part of any ontology. Spinoza was likely much subtler than his critics and apologists since he makes no attempt to eliminate this gap—and according to Badiou, he was very much aware of it at all times. In fact, similar to the infinite causal chain of *Ip28*, Spinoza has in mind two other infinite chains: the chain of bodies, which in its infinite totality gives us the conception of Nature, and the chain of eternal thoughts, that which gives us the conception of an infinite understanding of God. But it is clear that the totality of the chain, which perhaps can be best conceived as the Kantian regulative categories of Nature and God, is not the same as the totality of the One, which is an existential category: the presentation of an infinite chain of determinate finite things (bodies or thoughts) is not the same as an existential deduction from the necessary existence of the one substance. This is so because, for Spinoza, everything that exists is either the substance itself or the modes of the substance: the substance necessarily exists, and in *Ip21*, *Ip22*, and *Ip23*, Spinoza indicates that the infinite modes also necessarily exist. The existence of things is thus donated on two bases: they are either derived from the nature of God directly (*Ip21*), or they are modifications of something that is derived directly from the nature of God (alluded to in *Ip22* and *Ip23*). The modes that are modifications of what is derived directly from the nature of God are infinite—since modification of something infinite is still infinite (and necessary since everything that is grounded by the nature of God necessarily exists). So, again, the gap pertains to how infinite modifications of modes that are originally dependent on the nature of God can ever lead to finite, determinate beings. If all presentation is of finite things, then there is an absolute gap between the multiplicities of infinite modes that are not presentable (because there is no deductive link between the infinite modes and finite modes) and the domain of presentation. Given that everything that exists must be either based on the nature of God or the modifications of infinite modes, the being of what is presented is ultimately based on what is not presentable. This gap is exactly what Badiou calls the void.

But, the notoriously hard problem of deducting finite from infinite does not seem to have bothered Spinoza at all. He left open two deductive sequences, two discourses, whose real is the gap that resides between them. This gap is unthinkable by the great constructivist mind of Leibniz—and following Leibniz, everyone who, in the specter of the real, attempted to take shelter under the safeguard of a foundation. Interestingly and thankfully, Spinoza was not one of them—something that perhaps attests to his greatness. What does this gap tell us? That we need to push the task of thinking beyond this point of impasse, not by eliminating it, but just conversely, by deploying it productively and forcefully, which in the axiomatic framework is distilled to a decision. This was obviously impossible until Cantor and modern axiomatic thinking, but Spinoza's encounter unequivocally belongs to the same evental sequence.

Infinity as a Modality of Finite—Hegel

The Affirmative Infinity is the first appearance of the concept of systematic totality in *Logic*. It belongs to the speculative nucleus of the Dialectic of One, the first complete revolution around the circle, anticipating the categories of Ground, Disjunctive Syllogism, Absolute Idea, and Absolute Spirit that are yet to come in *Logic* and if we extend this to Hegel's *Philosophy of Nature*, this systematic totality will manifest in Animal Organism, and in *Philosophy of Mind*, domains such as Free Mind, State and finally Philosophy itself. The point here is this: like all the categories of *Logic*, Affirmative Infinity demonstrates a way in which being can be thought. Here, the truth of being manifests as a type of infinity that is internal to being—being, qualitatively, is infinite, and as such, infinity will manifest itself in various states of being, for example, as Animal Organism, State, or Philosophy. That means, during this process, several states of being in nature and in history, albeit of a finite constitution, possess in their truth the character of Affirmative Infinity (the most famous example of a non-logical category is, of course, the Hegelian State).

There is an important decision at stake here. One of the biggest critiques that Hegel makes of Spinoza is that the latter never derives infinite attributes and infinite modes (let alone finite modes, which we saw previously) from the infinite nature of the substance. It is true that Spinoza just posits this fact without any proof, and while Hegel admires Spinoza's intuition in thinking determinacy (attributes) immediately subsequent to pure being (substance), he is critical of this absence of proof: in Hegel's mind, Spinoza foregoes PSR (yet again), by not providing any ground for why determinacy must emerge from pure being—something that is brilliantly manifested at the very beginning of the *Logic*. The point, however, is this: speculative ontology deduces finitude as the necessary outcome of the determinacy of being and as logically prior to the category of (affirmative or spurious) infinity: in a sense, the finite is the condition of possibility of infinity, and infinity is the truth of finite. This is one of the most amazing and novel parts of Hegel's logic: not only is there a logical (and ontological) continuity between finitude and infinity, infinity is literally derived from the finite. Here, Hegel's logic is in absolute opposition to the metaphysical and religious traditions surrounding the discourse of infinity, where there is a clear rupture between finite and infinite. We saw this most clearly in the gap between the two sequences of deduction in Spinoza, between those finite and infinite modes. Moreover, it is the infinite that is usually the condition of possibility for the finite. In contrast, for Hegel, the thought of a pure being does not immediately yield the thought of the infinite. It is from pure being's determinacy that the Finite "foams forth" prior to thoughts of bad or good infinities.

Hegel's orientation toward infinity here is even more amazing than Spinoza's. While Spinoza leaves the gap between infinite and finite wide open, Hegel attempts to close this gap: and not from the usual side—which is infinite—but from the side of the finite. It is, therefore, important for us to understand how successful Hegel was in closing this gap. This takes us a little bit back to the genesis of the idea of determination in Hegel.

As Badiou points out, for Hegel, the conditions of "determinateness" are two:

1. A determinate thing is determined insofar as it is at the same time what it is not: not just an "other," that is, another determinate thing, but Other, the whole space of otherness. This is the result of the operator of negation, that is, a moment of determinateness. For every determinate being, nothing, which is the otherness, or Other of being, is to be part of being, and this is the main difference between being and determinate being in Hegel's logic: determinate being is the being that has the nothing, that is, the Other of being, as a part of itself. There is no being without having nothing as a part of it. This inscription of Other inside every other is where infinity originates in Hegel.
2. The oneness of the same is thus a mark that is possessed by the same. It is already also inscribed in the same. This is a consequence of the previous point: what makes a determinate thing at the same time makes it One. This is the same as saying that count-as-one is inscribed in the presentation, as opposed to the assertion made by ontology that presentation is the outcome of the count-as-one and the latter is squarely outside of the former.

Thus, we have two notions at play here: we have the notion of a limit (*Grenze*), which is the proper boundary, its count-as-one, according to which a thing separates itself from everything else, the Other, which is, as above lays out, is also inscribed inside every determinate thing. We also have the notion of a frontier (*Schranke*), the space that is laid beyond the thing.[8]

To Badiou, ontology puts the count-as-one outside of the situation. To guarantee the oneness of the count, we have to have a meta-count-as-one of the situation, that is, to recount the situation in the state of the situation. In this sense, count-as-one is the Other of the situation, and the state is the internal mechanism through which the situation can hold on to the oneness in the absence of the Other. Conversely, Hegel puts the count inside the one:

a determinate thing, that is, a situation, is determinate insofar as it inscribes within it the Other.

This points to a clear difference in approach. Axioms provide a framework for a subtractive ontology: the ontology that does not think about the genesis of being but only its presentation. For axiomatic set theory, the being-there of multiplicities is an axiom—in particular, the axiom of Separation—it is a primitive of thought. For Hegel, the being-there is a generative fact; it is not given by any means. The only thing that is given to Hegel is being as such. He has to derive being-there as well as its modes of finitude and infinity from pure being. ZF has the same rigor and power, whether or not mathematics assumes that being qua being is infinite or finite.[9] That is why the axioms only speak about what is given and not their "sizes."[10]

But Hegel's program is different and has a different priority. Hegel is committed to a generative ontology. For him, the one of the count cannot be outside of beings. For this reason, he derives the determinateness of being and the thought of something, from the inclusion of nothing in the being of Something. This results in three things:

1) What makes Something one, the limit, is not outside of Something.
2) What the "not" of Something is, the frontier, is inside of Something.
3) Positing Something will result in positing of the Other.[11]

This automation of positing, passing-beyond, and re-positing, generates the operator of infinity: all this from saying "being-there is one of being." Hegel, with a special genius, set out to co-engender the finite and the infinite on the basis of the point of being alone. Infinity becomes an internal reason of the finite itself, a simple attribute of experience in general because it is a consequence of the regime of the one, of the between in which the thing resides, in the suture of its being-one and its being. Being *has* to be infinite: "The finite is, therefore, itself that passing over of itself, it is itself the fact of being infinite."[12]

This immanence of passing over to the constitution of a finite thing is what also constitutes the move from the finite to the (bad) infinite: the bad infinite is the repetitive steps of passing over finite beings, ad infinitum. Badiou remarks that this resulting (bad) infinite has no "being," and thus, he calls it the void.[13] As such, the (bad) infinite is not the being of the "finites" that pass over to each other; the infinite is thus "the passing" over of the "finites" to each other. The being of the (bad) infinite is where one finite being passes over to another finite being. That point of passing over does not have a being and was not given a being by Hegel, and that is why Badiou calls it void.

The question is to see how Hegel moves from this immanent definition of the infinite, which is bad, to a definition that, while it does not merely represent this repetition, at the same time remains immanent. Badiou mentions this contention nicely in the following passage:

> After all the bad infinity is bad due to the very same thing which makes it good in Hegelian terms: it does not break the ontological immanence of the one.[14]

It will be interesting to see how Hegel can maintain the immanency of the structure within the presentation at the same time that he has to step outside the chain and reach a global level that will allow him to define the good infinite. Before we consider Badiou's treatment of the move from finite to infinity, let's have a quick recap of what *Logic* is telling us about this move. Being proves to be Finite before it proves to be Affirmative Infinity.

> In keeping with this determinateness, the finite stands opposed to the infinite as a *real determinate being*; they stand thus in qualitative relation, each *remaining* external to the other; the *immediate being* of the infinite resuscitates the *being* of its negation, of the finite again which at first seemed to have vanished in the infinite.[15]

In the conception of Finite, to point beyond itself, there seem to be two distinct moments here in Hegel, both of which are packed in the above passage from *Logic*: when Finite ceases to be, and its Other comes to be, the being of this Other is nominally infinite insofar as it transcends the finite. But whatever transcends the finite is not truly infinite, because it depends on what is being transcended. Therefore, it is not truly infinite by quality, but only nomination—and that is what is perceived to be "bad" about this infinite. This seems to be definitely one sense that we can deduct from Hegel's derivation.

> The infinite as thus posited over against the finite, in a relation wherein they are as qualitatively distinct others, is to be called the *spurious infinite*.[16]

But there is yet another sense. The other sense is the sense of the algebraic alteration of Other and Something to each other. Other, what we just called "bad" infinite, is determined itself by Finite, so as Finite posits Limitation, Limitation re-posits Finite, ad infinitum. This sequence of alternations is logically unceasing and therefore gives rise to the thought of an infinity of a sequence that does not halt. This thought of infinity, as an unending sequence, itself is a thought of "bad" infinite again, but it is certainly not the same thought of "bad" infinite as the Negation that transcends Quality, or as Limitation that transcends Finite or Other that transcends Something.

Now, why is this important? Because one of these infinities point to Affirmative Infinity as a "deduction" from finite, and the other as the space of "finiting."

As we saw a bit earlier, the mutual determination of finite and bad infinity gives rise to a thought of infinity that is self-related—and thus, its determination is within it. In that sense, it is a return to Determinate Being, but instead of this determination to be unrealized and in itself, it is now fully realized and for itself. This makes this thought of Determinate Being that is for itself the thought of Affirmative Infinity. In that sense, Affirmative Infinity is the Finite that finds its Ought fully realized and returned to it. This is the actual deduction in *Logic*.

But, as we've said, there is something else, something that in Hegel's mind actually leads directly to the realization of Notion. Right under the heading of *Affirmative Infinity* in *Logic*, we find this crucial passage:

> In this alternating determination of the finite and the infinite from one to the other and back again, their truth is already implicitly *present*, and all that is required is to take up what is before us. This transition from one to the other and back again constitutes the external realization of the Notion. In this realization is *posited* the content of the Notion, but it is posited as *external*, as falling asunder.[17]

In the quote above, it looks like Hegel is pointing yet to another way of deriving Affirmative Infinity, which is not the result of the unity of finite and infinite: the "finite-infinite" and the "infinite-finite," which is the way the rest of the section reads. As if there are two matrixes in Hegel regarding how to deduct Affirmative Infinity: one points back to Finite and allows it to find its true "infinity" in the mutual determination of its Other, and the other is to totally "step out" of this reciprocal determination of the finite and (bad) infinite that alternates back and forth, a stepping out that implies a truth *present right before our eyes*—we only have to "take up what is before us"—an "external realization" of the Notion, the truth of which "falling out" of the two.[18]

Right after the above quote—a momentarily "stepping out" of the cyclic alterations—Hegel dives back into it and invokes the unity of finite and infinite—he wants his unity back. But that momentary glance outside is what Badiou captures and holds on to.

> Hegel's stroke of genius, if it is not rather a matter of supreme dexterity, is to abruptly return to pure presentation, towards inconsistency as such, and to declare that what constitutes the good infinity is the *presence* of the bad. That the bad infinity is effective is precisely what its badness cannot account for. Beyond repeating itself, the something detains, in excess of that repetition, the essential and presentable capacity to repeat itself.[19]

What do we have when we step out of the chain of alterations? We get the pure presentation: whence the truth of this alteration that is "*present* in itself" (Hegel's emphasis). The space of presentation is an objectless space where the tireless objective oscillation plays out. Badiou thus turns the tables completely: what Hegel is telling us is that the tireless repetition of the existence of Something and Other is the "implicit" indication of presentation itself. In fact, what Hegel is telling us is that Affirmative Infinity stands for the "field" of presentation, and it is because of the infinity of presentation that the existence can repeat itself, ad infinitum: "the good presence of the bad operation."[20] In addition, and even more importantly, the definition of infinity is coming from a register that is "outside" of the register of repetition. The repetition is the property of finitude, and the conception of finitude neither logically nor ontologically can be used for deriving the infinite. At most, it derives the conception of the indefinite. In that sense, Hegel's Spurious Infinity is not even an infinity; it is indefiniteness.

This is an amazing outcome. The "moment" captured by Badiou in the deduction of Affirmative Infinity seems to be adding one more thought to what Spinoza had given us before. We saw how Spinoza—by breaking away from all the rules of discourse, what he so sternly held in the highest regard—implicitly affirmed that the causal chain of finite and infinite are two disjoint sequences. They are separate and independent, and there is no continuity between the two. In the middle of a strict discursive fidelity, Spinoza gives us a glimpse of a caesura with the law, as perhaps the highest moment of his genius that anticipated what is to come. Within this sudden break introduced by Spinoza, this lacuna that is now open, Hegel is adding a thought: the consistency of the repetitive operation of presentation is due to the infinity of the presentation itself—the infinity of the "field" of presentation is the condition of possibility of this repetition. It is as if the Hegelian answer to the Spinozist *aporia* is that the two sequences identified by Spinoza are never meant to reach each other because one is within the other.

So, while the "conscious" Hegel is a step backward from Spinoza, suturing the finite and infinite—to the demise of the infinite and the affirmation of the finite building up toward a narrative that becomes the core of what we named in chapter 3 the discursive idealism—the "unconscious" Hegel lifts Spinoza up, and that is what sets the stage for us to now look at the full disclosure of the thought of infinite within axiomatic thinking.

The Axiomatic Infinity

Modern mathematics begins around the end of the nineteenth century. While this modernization has many facets—from foundations, and problem-solving techniques, to theory building, and so on[21]—the most important, from the perspective of the current study, is the invention of set theory by Georg Cantor in 1870, the consolidation of its axioms by Ernst Zermelo in 1908, and later its improvement by Abraham Fraenkel in 1922, in what is now referred to as Zermelo-Fraenkel axioms (ZF) with the axiom of Choice (ZFC). The thought of infinity we are tracking finds its axiomatic expression in ZFC.

The first lesson that modern mathematics teaches us about infinity, in contrast to what common sense may tell us, is that infinity is a simpler thought than the finite.[22] Problems about infinity are often easier than problems about the finite. In several cases, there exist solutions for a sequence with infinite elements, whereas comparable sequences with a finite number of elements do not have a given mathematical solution.[23] Dedekind was the first who thought that infinite numbers precede finite numbers in construction.[24]

The second important lesson about infinity that modern mathematics teaches us is that infinite is not the same as indefinite. We saw how the earliest conception of the infinite by the ancient Greeks literally meant "what has no limit," or simply limitless. This negative conception was dominant in mathematics for quite some, and despite serious philosophical, metaphysical, and theological ruminations about the conception of infinity—such as the ones we saw from Spinoza and Hegel—the use of infinity in mathematics was frowned upon by serious mathematicians.[25] It is easy to see why: the conception and construction of the infinite was conceived through the finite. It is obviously futile to use the concept of the finite number as a paradigm for the construction of the infinite number(s). This is something that was known since Zeno's paradox. Seen from the side of the finite, the infinite is that which is not reachable. That is, defining the infinite based on the finite only results in a conception of the infinite as indefinite or inaccessible. Yet, the same conception of infinite as indefinite also works in reverse. Consider Euclid's proof that there is no biggest prime number. The essence of his proof is to show that, no matter how large a prime number is, we can always construct one bigger. That is, the indefiniteness of the set of primes proves (by contradiction) that there is no largest prime.

If we look closely into the problem of indefiniteness, we see a similarity with Hegel's bad infinity—the problem of repetition. Yet the concept of indefinite as the paradigm to think infinite prevailed for centuries until logical inconsistencies began to appear. Galileo's famous paradox asked whether natural numbers (0, 1, 2, 3, . . .) are "more numerous" than their squares (0, 1, 4, 9. . . .). At first glance, we might conclude that natural numbers are "more numerous" than perfect squares (or than even numbers, or integer multiples of some natural number, or prime numbers, etc.) since there are infinitely many natural numbers which are not square numbers. However, Galileo points out that each natural number corresponds to only one square number and vice versa, according to a bi-univocal (i.e., one-to-one) correspondence:

Table 7.1 Galileo's Paradox—Natural Numbers and Their Squares

0	1	2	3	4	5	6	7	8	9	. . .
0	1	4	9	16	25	36	49	64	81	. . .

Source: Author created.

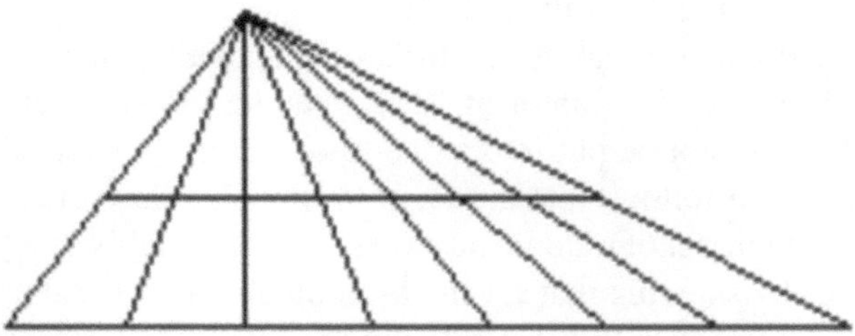

Figure 7.1 Comparing Two Unequal Line Segments. *Source:* Author created.

Now, two *finite* sets in one-to-one correspondence have the same number of elements. By extending this criterion to infinite sets, we should conclude that natural numbers aren't more numerous than the perfect squares or the even numbers, and so on, contrary to the Aristotelian principle, according to which the whole is greater than its parts.

There are such paradoxes even in Geometry. For example, consider two unequal segments. It seems the longer one contains more points than the shorter one. But trace from an external point some straight lines intersecting both of them as depicted in figure 7.1.

As we can see, each point of the shorter segment corresponds to one, and only one, point of the longer. Therefore, it seems that both segments contain the same number of points. Infinitesimal calculus and the notion of infinite sums also created an issue with the conception of indeterminateness. Infinitesimal calculus brought almost the whole physical world within the scope of mathematics, whereby it solved problems of change and motion. Powerful methods were developed to calculate or prove the divergence of infinite series of operations with numbers.

These paradoxes, and especially the methods thus deployed, showed the inadequacy of the treatment of the infinite as indefinite. Simply put, there was a need for a conception of infinity that allowed its treatment as a unity, as opposed to a property of a limitless series. The dilemma was precisely the prevailing thought of the finite, and the solution was to disentangle the infinite from the finite.

Two developments made a decisive change: Dedekind's definition of the infinite and Cantor's uncountability theorem. The remarkable orientation introduced by Dedekind is to effectively make Galileo's paradox into a definition: A system S is said to be infinite when it is similar to a proper part of itself. On the contrary case, *S* is said to be a finite system. This was a remarkable turn of events. Dedekind's genius was precisely an example of thinking beyond the point of the impasse of the number system. This positive definition of infinity by Dedekind points toward the era of axiomatic Infinity. It is a conception of the infinite that is not related to the finite, and, thereby not sutured to the thought of the indefinite.

The second decisive break is Cantor's uncountability theorem. Uncountability was discovered by Cantor in 1874 when he showed that the set **R** of real numbers could not be put into a one-to-one correspondence with **N**, the natural numbers. (It follows that the set of irrationals is also uncountable, and hence, the countable set of rational numbers has uncountably many gaps.) The proof proceeds by showing that given any countable set of real numbers, one can find a real number not in that set, and therefore no countable set covers all of *R*. What does this result tell us? It tells us that the set of real numbers, which is an infinite set, is a larger infinity than the set of natural numbers, which is also another infinite set. That is, we have multiple sizes of infinity. This was a definite blow to the conception of infinite qua indefinite—we now have a way to compare different infinities together. The profound consequence of this discovery is its effect on the very notion of differentiation. As we will elaborate in the next section, since Aristotle, the concept of differentiation was coupled with determination, and the latter with the concept of finitude: that is, a determination is only possible for finite things—there is no infinite determination—and without determination, there is no difference. What Cantor proved was the possibility that infinity, considered an "indeterminate" notion at the time, can be different from another infinity. With the uncountability result, we can now distinguish between two infinities. So, if a thing has an other, this does not mean that this thing must be finite.

There are profound consequences of this change. The first is that you can now derive a determinate difference from the indeterminate. This brings into question the very grounding of thingness (unity) in finitude, which has been the unquestioned assumption in all metaphysics. The second profound consequence is the problem of multiple infinities. Cantor proved that countably infinite sets have a cardinality different from that of irrational and real sets.

The key point is to understand that infinity is structurally different than the finite. This illustrates a crucial "ontological law" of infinity: infinite sets cannot be constructed by using the existential operations of set theory. We cannot build infinite sets from finite sets, nor can we build infinite numbers from finite numbers. The two never meet. This much Spinoza taught us by his hesitation. The logical properties of infinity aside, the existence of infinite sets cannot be derived from the existence of finite sets—a lesson mathematics learned not in an easy way, as was witnessed by the attempts of mathematical elites such as Dedekind and Frege.

Dedekind's attempt ultimately fails at the same point as did Frege's: in the transition from concept to assertion of existence. And at the root of the affair is the same thing: Frege and Dedekind both seek to deduce from "pure logic," or thought as such, not just the operational rules of number, but the fact of its existence for thought. Now, just like the empty set, or zero, *the infinite will not be deduced*: we have to *decide* its existence axiomatically, which comes down to admitting that one takes this existence, not for a construction of thought, but for a fact of Being.[26]

This is precisely where axioms will be able to help us. The issue is not only the chasm between infinite and finite, the chasm that Spinoza left open and Dedekind tried to close (despite his recourse to Spinoza for his proof). The true issue lies in the attempt to deduce a positive existence of the infinite. Dedekind was not able to accomplish what the "unconscious" Hegel does in deducing Affirmative Infinity: to recast the issue in a different register, which is the sign of axiomatic thinking in our view. Dedekind himself was no stranger to the axiomatic register of thinking, as his definition of the infinite shows. But when it came to the existential seal, he, for some reason, did not follow the same route and instead attempted a proof.

The problem is not only that infinity is logically different from finitude. It is ontologically different too. While finite sets could be constructed using the existential operations of set theory, the axiom of Extensionality, the axiom of Power Set, and so on, infinite sets cannot be constructed; they have to be assumed. As Badiou remarks, it was an ontological decision to posit the first infinite ordinal set $\boldsymbol{I}$ in the axioms of set theory: It guarantees the existence of at least one infinite set, namely a set containing the natural numbers. It was first published by Zermelo as part of his set theory treatise in 1908. The following is the formal expression of this axiom:

ZFC axiom of Infinity: $\exists \boldsymbol{I}\ (\varnothing \in \boldsymbol{I} \wedge \forall x \in \boldsymbol{I}\ ((x \cup \{x\}) \in \boldsymbol{I}$

The axiom states, there is a set $\boldsymbol{I}$, such that the empty set is in $\boldsymbol{I}$ and such that whenever any x is a member of $\boldsymbol{I}$, the set formed by taking the union of x with its singleton $\{x\}$ is also a member of $\boldsymbol{I}$.

The Impact of the Axiom of Infinity

Could mathematical thinking fail to "turn the table" and pursue its efforts in the same direction that Dedekind, Peano, and Frege envisioned: To strive for proof, or else to do away with infinity? Absolutely! There was nothing that could seal the fate of actual infinities in mathematics. It was a pure decision, and that is why it has the evental character: the character to "begin" by seeing things in a "different register"—that is precisely what we mean by *de-suturing the beginning from the foundation.* All axioms of set theory are independent of each other. But, the mathematics whose models use structures of infinite sizes could not have truly "begun" without this particular axiom. In fact, mathematics without infinity is only a mere collection of algorithms—in some sense. It would be reduced to Peano Arithmetic (PA). It is not only the case that, without axioms, we could not think of infinity—as something different than indefinite, infinitesimal, or limit—but in fact, *without the axiom of Infinity, mathematics would not be a discipline that could think of its own novelties through axioms.* In what follows, we shall try to justify this statement by providing a brief summary of the impact that the axiom has had in the development of modern set theory and axiomatic thinking.[27] In essence, it has influenced the construction of new sets that allow us to prove or disprove statements that are derivable in formal theories based on ZF and in ZF's extensions.

When we speak of structures of sets, we speak particularly of the von Neumann universe, or a hereditary well-founded set. A hereditary set (or a pure set) is a set whose elements are all themselves hereditary sets. That is, all elements of the set are themselves sets, as are all elements of the elements, and so on. For example, the empty set is a hereditary set, and thus the set $\{\Phi\}$ containing only the empty set is a hereditary set. Similarly, the set $\{\Phi,\{\Phi\}\}$ is a hereditary set. The inductive definition of hereditary sets presupposes that set membership is well-founded. A von Neumann universe, or a von Neumann hierarchy of sets, denoted V, is the class of hereditary well-founded sets, and is defined inductively as follows:

$$V_0 = \Phi, \text{ For any ordinal } \beta, V_\beta = P(V_{\beta-1})$$

Every ordinal is either zero, a successor, or a limit ordinal. If α is an ordinal, the successor ordinal of α is the first ordinal that is greater than α: $S(\alpha) = \alpha \cup \{\alpha\}$ (von Neumann definition of ordinals). A limit ordinal is neither zero nor a successor ordinal. An ordinal number λ is a limit ordinal if and only if the following is true:

$$\forall\beta\ (\beta \text{ an ordinal}) < \lambda, \exists \text{ an ordinal } \gamma \text{ such that } \beta < \gamma < \lambda$$

For example, ω is a limit ordinal since for every natural number $m < \omega$ there exists a larger natural number n such that $m < n < \omega$. If λ is a limit ordinal, for all $\beta < \lambda$, we will have:

$$V\lambda = \cup_{\beta<\lambda} V\beta$$

The sets in V are divided into the transfinite hierarchy $V\alpha$, called the cumulative hierarchy, based on their ranks. Therefore: $V = \cup\alpha\ V\alpha$.[28]

Table 7.2 Formulae for von Neumann Universes

0	Successor ordinal β	Limit ordinal λ
$V_0 = \Phi$	$V\beta = P(V\beta_{-1})$	$V\lambda = \cup_{\beta<\lambda} V\beta$

Source: Author created.

$V\omega$ is the model for Peano Arithmetic (PA) in ZF without the axiom of Infinity (<ZF-Infinity>). This means two things. First, that $V\omega$ is closed to PA, which means using the axiom of Power Set and the axiom of Extensionality, we cannot produce sets that do not belong to $V\omega$. Furthermore, we can prove the consistency of PA in $V\omega$. But what about ZF (which includes the axiom of Infinity)? What universe can be the model for all ZF? If such a universe exists, it must be of the form $V\kappa$ for some cardinal κ. The cardinal κ is then called inaccessible, because $V\kappa$ cannot be reached from below. The cardinal ω is inaccessible in the same sense, for which we have the axiom of Infinity, but in ZF we cannot prove the existence of an inaccessible $\kappa > \omega$, which means that the assertion $\kappa > \omega$ exists amounts to a new axiom in ZF, stronger than the ordinary axiom of infinity.

1. Axiom of Inaccessible Cardinal (AIC): For every cardinal μ, there is an inaccessible cardinal κ which is strictly larger, $\mu < \kappa$.

AIC is important because the consistency of ZF is only provable in ZF + AIC, not ZF alone—or in other words, for proving the consistency of ZF, the ground model of $V\kappa$ is necessary. Furthermore, there exist a slew of other statements whose consistencies are only provable if ZF itself is consistent. This is called relative consistency, and can be stated in the following form:

If ZF + Axiom *A* is consistent, then ZF + Axiom *B* is consistent

Gödel's and Cohen's proofs are examples of such statements that involve the axiom of Choice (AC) and the Continuum Hypothesis (CH):

2. If ZF is consistent, then ZF + AC + CH is consistent (Gödel 1938).
3. If ZF is consistent, then ZF + ¬AC is consistent (Cohen 1963).
4. If ZF is consistent, then ZF + AC + ¬CH is consistent (Cohen 1963).

After Cohen another remarkable relative consistency statement was proven by Robert Solovay, as follows:

5. If ZF + AC + AIC, then ZF + "every set of reals is Lebesgue measurable" is consistent (Solovay 1964).

The remarkable aspect of this statement is that it effectively adds another axiom of infinity, which has to do with the existence of "measurable cardinals": a new axiom for which we need the axiom of the existence of inaccessible cardinals, neither of which is provable in ZF alone.[29]

Zermelo's AC is also an axiom that is dependent on relative consistency. The axiom's roots lie in the infinite process so prominent in nineteenth-century analysis. The assumption underlying the process was not, however, made explicit until Cantorian set theory. The decisive question was this: Can every set be well-ordered? Well-ordering was originally treated as an assumption by Cantor in 1883 (he famously proposed well-ordering to be a valid "law of thought"), but a number of Cantor's contemporaries did not accept that as a suitable explanation, and a decade later, Cantor was still looking for a proof for the well-ordering statement. In pursuit of proof for well-ordering, Zermelo formulated AC in 1904. The central idea in AC is the existence of a choice function. But such a choice function is trivial if the number of sets to choose from is finite. Even if infinitely many sets were collected from the natural numbers, it will always be possible to choose the smallest element from each set to produce a set. That is, the choice function provides the set of chosen elements. However, no choice function is known for the collection of all non-empty subsets of the real numbers (if there are non-constructible reals). In that case, the axiom of Choice must be invoked.[30] In this case, the existence of the function of choice is only guaranteed axiomatically. As we mentioned above, in 1938, Gödel proved the relative consistency of AC and CH with ZF, and Cohen proved the independence of ZF from AC or the consistency of ZF + ¬AC.[31]

But pursuant to the problem of measurability in 1964, an old problem was shown to be related to the axiom of measurable cardinals: the problem of determinacy. In 1964, two Polish mathematicians, Jan Mycielski and Stanislaw Swierczkowski, proved that all determined sets are Lebesgue measurable. This basically states that, for example, all sets of real numbers are determined. This result is so powerful that it called for its own axiom: the axiom of Determinacy (AD). AD is a much stronger axiom than AC, and as an alternative to AC, makes the sets of reals to be "better behaved (at least as far as measurability is concerned) than under the axiom of choice."[32]

To prove the consistency of ZF + AD, there is a need for an axiom of infinity much stronger than just the existence of measurable cardinals. The corresponding large cardinals were introduced by Woodin in 1984 and are called Woodin cardinals. If they exist, they are much larger than the smallest measurable cardinal. And to prove the consistency of AD, we need to assume that there are infinitely many Woodin cardinals. With this massive assumption, together with AC, Woodin in 1985 proved that:

6. ZF + AD is consistent if and only if ZF + AC + "there are infinitely many Woodin cardinals" is consistent (Woodin, 1984).[33]

There is one more advancement that we would like to add to the above list: the existence of a self-similar $V\alpha$. In the pursuit of the minimal large cardinal assumption that implies AD, other large cardinal assumptions were studied that proved to be even stronger than Woodin cardinals. The study of the self-similar $V\alpha$ showed that the first ordinal not fixed by the self-similar function f: $V\alpha \rightarrow V\alpha$ is greater than infinitely many Woodin cardinals. Therefore, the existence of a self-similar $V\alpha$ is a large cardinal axiom, even stronger than the existence of infinitely many Woodin cardinals:[34]

7. Axiom of Self-Similar $\boldsymbol{V\alpha}$: For a rank α of a von Neumann universe $\boldsymbol{V}$ there exists a function f: $\boldsymbol{V\alpha} \rightarrow \boldsymbol{V\alpha}$ that preserves all properties that can be defined in the language of ZF (Dehornoy, 1995).

The latter axiom has an intriguing property, aside from providing yet another axiom of infinity. The above axiom led to the discovery of braid ordering. Braid ordering is a branch of mathematics dealing with finite sets that can be connected (braided) together in a certain way. Yet, by studying the properties of the linear ordering of braids, Patrick Dehornoy started defining the properties of huge cardinals by using the axiom of self-similar $V\alpha$. Braids are essentially finite objects, and therefore the theorem that is named after Dehornoy is about finite sets. In particular, Dehornoy's theorem can be transcribed as a theorem in PA (which, as formerly mentioned, is in ZF–Infinity). Yet, he discovered his theorem through a study of large cardinals. This was the first time, perhaps, that a mathematician used the properties derived from an axiom of Infinity to discover theorems about finite sets. More recent efforts and research projects have since followed Dehornoy's path.[35] This is important to our (and Badiou's) stance on the finite as a regionalization of the infinite.

Another interesting aspect revealed by the search for ever-larger infinities is this: the larger the infinity we find, the more we are able to prove things. This means that provability in ZF is a direct correlate of the size of infinities. This may be rather intuitive if we think about the sequence of statements we reviewed above (sequences 1 to 7).

> If κ is a cardinal so large that $V\kappa$ is a model of ZF and $\lambda > \kappa$ is a cardinal such that $V\lambda$ is also a model of ZF, then the existence of λ implies everything and more than is implied by the existence of κ. This is because $V\kappa$ is a model of ZF + "κ does not exist" (since κ is not in $V\kappa$), whereas $V\lambda$ is a model of ZF + "κ exists" (since $\kappa < \lambda$ and hence κ is in $V\lambda$). Thus, one might begin to hope that any sentence not settled by the ZF axioms might be settled by a suitable large cardinal axiom.[36]

This may also be justified by the fact that the consistency of PA is not provable in a model based on <ZF–Infinity>, but by $V\omega$ (the model for ZF, which includes the axiom of Infinity). That means the ordinary axiom of Infinity is strong enough to prove the consistency of PA. In addition, the consistency of ZF itself is not provable using the axiom of Infinity alone. This axiom is not strong enough to prove this consistency; we need AIC, the axiom of inaccessible cardinals, and the model $V\kappa$ to prove the consistency of ZF. When assuming the model $V\kappa$, we are able to prove statements about AC and CH as we have formerly seen. If the cardinals are large enough (as in the self-similar ordinal we saw above), we begin to derive properties about ordinary integers that the model of <ZF–Infinity> was unable to support, nor was the models "smaller," such as the models supporting AD or measurable cardinals. This hints at the fact that for large enough cardinals, we begin to discover properties for finite sets that were not possible, further supporting the claim that finite sets can be thought of in a much richer way should we approach them from the standpoint of infinite sets.

Finite as the Regionalization of Infinite

The reader recalls our attempts in part II to illustrate Badiou's formulation of the alternating movements between materialism and idealism in modern European thought (see figure 3.6). In that figure, we captured the idealism of our time as the "discursive idealism," whose main category is repetition. Back in part II, however, we still did not have the proper constructs to fully capture what is ontologically at stake between the idealism and materialism of our time. We believe that now we can restate this encounter in more precise terms that belong to ontology: the primacy of finitude and the derivation of infinity from finite (discursive idealism) in contrast to the primacy of infinitude and the derivation of finitude as a regionalization of infinity (ontology of pure multiples). Furthermore, we mentioned that in our time, the operator of the shift from idealism to materialism is regionalization. But, again, we did not have a clear idea of what this operator is and how it operates. We are now able to say that the main category of the ontology of pure multiples is the existence of infinities, and the operator of regionalization is the ability of thought to derive an existent finitude from ω_0 the infinity whose existence is posited axiomatically. Ontology has been able to show that not only is this derivation necessarily the case (as in the concept of an infinite chain of succession of finite numbers, which we shall see shortly), but also, with the existence of infinities that are much larger than the first infinite ordinal, we are able to think the being of finite things through properties that were not accessible to thought without the axiom of Infinity.

Another important development in mathematics using the axiom of Infinity that illustrates this regionalization very vividly is the theory of cardinals, which is at the heart of the set theory. The theory of cardinals concerns the question of the "size" of multiples. As we know, multiples are sized by putting them into a one-to-one correspondence to the set of ordinal numbers. The concept of correspondence is expressed as a function which, in set theory (using the axioms of Separation and Extensionality), simply results in another multiple by choosing one element from one set and another element from another set based on the following three rules:

- for every element of α, there corresponds, via f, an element of β;
- to two different elements of α correspond two different elements of β; and
- every element of β is the correspondent, by f, of an element of α.

The function of correspondence gives us an existential procedure of comparison between two multiples, whether finite or infinite. Using this method, we have a way to know what it means for two multiples to be of the same quantity. Furthermore, since any set can correspond in this way to any other, so are natural multiples or ordinals. They can be compared to any other (ordinal or non-ordinal) sets since there is well-ordering among ordinals; that is, we know which ordinal contains what other ordinals, and thereby which ordinal is the bigger among an arbitrary number of ordinals. We can then assign ordinals to multiples. Let's call this a "size class." A size class comprises ordinals that can be corresponded to a given multiple. This means that any multiple can correspond to a size class. The opposite is also true. Every size class of multiples finds corresponding ordinals. What we are asserting here is that, first, multiples, finite or infinite, are all subject to the operation of size. Second, the operation of size amounts to a one-to-one map of a given multiple to, at least, one ordinal. The concepts of bigger and smaller are the correlates of the operation of size. Third, we should remember that ω_0 is the first infinite ordinal; there exist ordinals that are far bigger than ω_0. For example, following Cantor, the size of R is 2^0, which is a size bigger than ω_0. Thus, the third assertion is that ordinals can measure the size of multiples that are bigger than ω_0. For example, there exists at least one ordinal ω_n such that $\omega_0 \in \omega_n$ and ω_n measures the size of R (in this case $\omega_n = 2^0$). Fourth, we know that we cannot count R, but we can size R, as all multiples can be sized. In other words, counting and sizing are two different operations on the multiples. Counting is an operation that is limited by an infinite ordinal ω_0—anything that is countable must be in a one-to-one map to an ordinal that is a member of ω_0. But there are sets, ordinal or otherwise, that are bigger than ω_0 and thus are not countable: the continuum is one of those sets; that is, the continuum is not countable (and it is not an ordinal). However, those multiples, although not subject to the operation of counting, are still subject to the operation of sizing. The size of those sets also—similar to the countable sets—is measured by a one-to-one map to a size-class of ordinals. Another way of saying this is the following: if S is a countable multiple and λ is the smallest ordinal that measures the size of S, then $\lambda \in \omega_0$.

Therefore, any multiple S is subject to the operation of size, call it f, an operation that puts S in a one-to-one correspondence to at least one ordinal. Regarding the finite sets two immediate observations can be made. The existence of f is obvious, and counting—which as we have seen for finite sets is the same as sizing—is intuitively performed by applying f that corresponds a finite multiple to an integer (which is an ordinal). Furthermore, there is only one ordinal that measures a finite multiple, and that is again the ordinal that represents an integer. These two observations are rather more complicated when it comes to the size of infinite multiples. It is far from obvious that a correspondence function should exist for infinite multiples, and it is only by applying the axiom of Choice that we can guarantee the existence of such a function among infinite multiplicities. Second, an infinite multiplicity can be put in correspondence to an infinite number of ordinals (as we just saw in the concept of a size-class). We can say now what measures the actual size of an infinite set is then the "smallest" ordinal in a size-class—and here, by smallest we don't mean in size, as all those ordinals are of the same size (because they are in the same size-class), but by the membership relationship. Set theory names that smallest ordinal in a size-class a cardinal number (or a cardinal). This means:

> For a multiple S and an ordinal λ if $f(S, \lambda)$ and $\forall n$ if $(f(S,\omega_n) \wedge \lambda \in \omega_n)$, then λ (the smallest ordinal) is called a cardinal, and it is what sizes S.

This provides us with a definition of cardinal using the concept of ordinal. In this sense each cardinal provides a "name" for the measuring scale of the size. In this way cardinals are essentially ordinals that may be a lot larger than ω_0. We saw, when the question of "size" of infinite multiples comes to the picture, size designates a class that many (potentially infinite) multiples may belong to. This means that all those multiples have correspondence with infinitely many sets—which in turn also means that those ordinals have correspondence with each other. Thereby in a size class there may be infinitely many ordinals. But since ordinals are well-ordered we can always know which ordinal is the smallest ordinal (by relying on the property of belonging). The aforementioned definition states that the smallest ordinal in the class is a cardinal.

From this very brief summary the following results will ensue:

1. All members of ω_0 are cardinals—that is all finite ordinals are cardinals.

2. Accordingly, we now have a criterion by which we would know what multiple is infinite or finite: an infinite multiple is a multiple that its size is measured by a cardinal greater or equal to ω_0, otherwise it is a finite multiple.
3. ω_0 itself is a cardinal as it is the smallest infinite ordinal.
4. So, being is quantifiable.
5. The quantification of being is uniform from finite to infinity.
6. Infinity is no longer only thinkable in ontology in qualitative terms. It is possible in ontology to think infinity according to quantitative terms.

The fact that we can "size" infinity is a very profound ontological property of being. Being is quantifiable. Infinity is no longer a mere quality—a jump from one quality of being as finite to another, through a chasm that only metaphysics and mysticism could provide a narrative for. Being is uniformly quantifiable, and the same scale that quantifies finite, also quantifies infinite. But this scale is much broader for infinite beings, of which the quantification of finite being is a special case. Finite in this sense is precisely a regionalization of infinity.

Axiom of Infinity and the Axiomatic Thinking in General

We started our inquiry into axiomatic thinking as a way out of the impasses of thinking the beginning. We showed how the thought of beginning without the One and foundation is possible only through the axiomatic mode of thinking, and just now we have presented the existence of the smallest ordinal number of size infinity as an axiom within the larger landscape of axioms. But now we are able to recognize that, at the same time, the very thought of beginning without any presuppositions is itself a thought that has a deeper commitment to infinity. In other words, if there is no commitment to infinity, the use of axioms, while certainly powerful—such as in constructivist programs like intuitionism—is merely a matter of convenience and not necessity. However, with a commitment to infinity, there is no way but to start with axioms; axioms become the necessary mode of thinking and the necessary method of beginning. In this sense, the axiom of Infinity is not only one axiom within the axiomatic system, it exemplifies how every true beginning necessitates axioms, and thereby it is the finest achievement of axiomatic thought. This leads us to believe that there is a reciprocal relation between the thought of (actual) infinity and the thought of axioms as the beginning. In this sense, we say that *the axiom of Infinity is what the axiomatic thinking has to show for its commitment to thinking maximally*.

A clear example of the kind of invocation of axioms that the thought of infinity has evoked surrounds the heated debate on the Continuum

Hypothesis. We saw earlier that the schema for ordinal numbers under ZF allows two types of ordinals (aside from the empty set): as the successor (as per the definition of ordinal we saw earlier) and as limit (as per the axiom of Infinity). The limit ordinal is naturally an infinite multiple: the very definition of the limit ordinal is borrowed from the structure of the infinite multiple—that is, the limit ordinal and the axiomatic definition of the infinite multiple are structurally the same.

The first and the only axiomatic limit ordinal is ω_0. But we can build ordinals that come after ω_0. The successor ordinals to ω_0 are built in the same way as any other successor ordinal:

$$s(\omega_0) = \omega_0 \cup \{\omega_0\}$$

This process of succession yields infinite ordinals each in a one-to-one correspondence to the first limit ordinal. The question, however, is whether there exists a next limit ordinal, which is also the successor cardinal to ω_0 say ω_1. What does ZFC say about this next cardinal after ω_0?

Thanks to Cantor we know $P(\omega_0)$ is a multiple that does not have a one-to-one correspondence to ω_0 nor with any of the successor ordinals of ω_0. Thereby, there must be a bigger cardinal that could size $P(\omega_0)$. This bigger cardinal is, of course, a limit ordinal also. Let's call this cardinal $\omega\alpha$. Therefore, ZFC guarantees the existence of at least one cardinal which is bigger than ω_0. But the latter process can be repeated this time for $P(\omega\alpha)$. Let's say that the cardinal resulting from this power set operation is $\omega\beta$, where of course $\omega\beta > \omega\alpha$. By repeating this exercise, we can land on ever-greater cardinals:

$$P(\omega_0) = \omega\alpha,\ P(\omega\alpha) = \omega\beta,\ P(\omega\beta) = \omega\alpha, \ldots \text{ Where } \omega_0 < \omega\alpha < \omega\beta < \omega\alpha < \ldots$$

The above sequence will, of course, generate the well-ordered sequence of infinite number of cardinals. The issue that this sequence does not address is whether this sequence exhausts all cardinal numbers. In other words, are there cardinals between ω_0 and $\omega\alpha$, and between $\omega\alpha$ and $\omega\beta$, and so on. Let's state this in a different way. We know that there is a limit ordinal that is greater than ω_0. Let's call ω_1 the first limit ordinal that is greater than ω_0, and ω_2 the first limit ordinal greater than ω_1, and so on. The latter question can be reformulated in this way: are these multiples (begotten by the successive applications of the power set operation over the limit ordinal ω_0 and its set of subsets) actually the same as the cardinal numbers that are immediately bigger than ω_0, that is, ω_1, ω_2, ω_3, and so on? Yet, more specifically, is the following the case?

$$\omega_1 = \omega\alpha,\ \omega_2 = \omega\beta,\ \omega_3 = \omega\gamma, \ldots$$

The above question, dubbed the General Continuum Hypothesis (GCH), is the general form of Cantor's CH (Cantor's CH only asks the first equation in the list). Hilbert was quite right when in 1900 he put GCH (or in particular CH) as the first, that is, the most important, of twenty-three problems to which mathematicians should pay heed.

But why is this problem important from our point of view? It is important because this problem takes to task the commitment of thought to the axiomatic mode of thinking.

A century after the advent of the set theory by Cantor, and as a generalization of the work of Paul Cohen that we will see later in this chapter, William Easton in 1970 discovered a result that is now referred to as the Easton theorem (or the Cohen-Easton theorem). According to this theorem, the cardinal $\omega\alpha = 2^0$ could be equal to any one of $\omega_1, \omega_2, \omega_3, \ldots$ and nothing in ZFC could prevent this. In general, the Easton theorem proves that the notion of succession is undecidable when it comes to infinite cardinal numbers: it is impossible to say whether a cardinal is a successor of another cardinal; or given a cardinal, what its successor cardinal is. Although there are several ways to build higher size classes of ordinals, it is not possible to show whether a given construction yields a successor cardinal. In particular, the cardinal 2^0 could take any infinite cardinal above ω_0. This means that an undecidable gap separates an infinite set and its power set, a gap that cannot be sutured by ontology. This also means that the values of various infinite cardinals can, depending on our choice and compatibility with the rules of ontology, take a variety of values within the sequence of ordinals. It is impossible for an infinite situation to determine how much it is exceeded by its state. In other words, it is not possible for ontology to determine the complete identity of the representations collected by a state.

Badiou refers to this as the errancy of ontology. As the pronunciation of Being, ontology is unfaithful to Being, as it cannot pronounce a measure that measures the excess of the representation over the presentation: there is nothing in the presentation that could determine this excess.

But this unfaithfulness of infinity is precisely what evokes the axiomatic modality of thought to yet again shape the destiny of thinking. This evocation brings to fore in our time, more than any other, the question which has awaited us in the guise of the Continuum Hypothesis: Are we espousing the axiom of Constructibility toward a path of measured representation, or the Generic toward the path of unmeasured excess, toward truth and subject? Ontology cannot make this decision, as each of these orientations makes Being intelligible in a different sense. While we will examine these axioms in subsequent sections of this chapter, what is important for us here is the very act of convocation: the convocation of a decision by the thought of actual infinities.

We saw in the previous chapter that our interpretation of axioms already includes the thought of events, and in particular, the axiom of Infinity is what bears the name of the event in ontology itself, as it opened up unconceivable vicinities to the thought. We now see that CH manifests the same evental character when we operate in fidelity to the event of infinity: the thought of actual infinities convokes a decision. What this convocation really means is that under ZFC we need to involve the axiomatic mode of thinking anew; that we need to decide on the orientation according to which we can render the *aporia* of CH thinkable. We saw <ZF-Infinity> was an orientation of thought consistent with axioms and provided the model in support of Peano arithmetic. But the concept of succession in that arithmetic caused by the thought of infinity failed to draw any consistency from this thought. On the other hand, ZFC, as we just saw, renders a model of thinking that draws new consistencies from the thought of actual infinities. We can now understand the event as cause to be the foreclosure of the maximality of thinking, and thereby is the foreclosure of axiomatic thinking itself. Whereas, event as consistency is fidelity to thinking maximally. The problem of CH calls for a similar reorientation, and the reader no doubt can anticipate that one side of the problem of continuum amounts to the foreclosure and the other side to the fidelity to the axiomatic thinking. We shall shortly look at this encounter in the present chapter.

INFINITY AND THE REAL

Now it is time for us to look at how the thought of the infinite influences the category of the real. We have spent a great deal of time thinking through the various ways that this category was theorized by Lacan and Badiou, and in particular in relation to the question of cause (Lacan/Miller) and consistency (Lacan/Badiou). The immediate sense in which this relation should be investigated is the sense in which the real is thought as the inaccessible. This sense is conformant with the thought of the real as cause. So, the first question is this: How does the idea of infinite, or more explicitly the axiom of Infinity influence the thought of the inaccessible?

The Seal of Infinity

As we saw, the thought of Actual Infinity was something that was banished from both philosophy and mathematics as an incoherent thought, since at least (if not earlier than) Aristotle. Aristotle deemed Actual Infinity (AI) as a contradiction since for him AI implies three things. First, its actuality implies that that something is already there. Second, its infinity implies an operation of passage from what is covered by it to what is yet not covered by it. Third, in affirming the previous operation, we can confirm that there is still-more.

For Aristotle, what is posited must always be posited with a limit. Hence, from the get-go, AI is a losing proposition. The contradiction lies in the fact that being-there for Aristotle means having a limit, which contradicts the still-more of the indeterminate being. Being-there already implies exhaustion of the thing in question. That is, if there is still-more, the thing is still not there—it is not yet (fully) presented. This harkens back to the problem of unity of the infinity that we spoke about. The passage of still-more fails to recognize the unity of the infinity.[37] The issue seems to be then with the "assumption" of the operation of passage about what is already there. For the operation of passage to work, there must be something "already," what is there at the beginning. What form of being is this that donates the assumption of already? And how could the operation of passage ever contribute to this donation? The answer is that it cannot (and hence the "impotence" of which Badiou speaks). There is no way that the operation of passage can "know" that what it is passing through is something of what is already there, unless this operation is being performed from "already inside" what is. This reminds us Badiou's reference to Lacan's *Seminar XXII* regarding the consistency of the sequence of numbers that Peano had postulated:[38]

> It must be admitted that there is in this sequence of numbers a consistency such that one has the greatest of difficulties not holding it as constituent of the real. All approach to the real is woven for us by the number. But to what is owed this consistency that there is in the number? It is not natural at all, and it is indeed this that makes me approach the category of the real as knotted to what I am also lead to give a consistency, the imaginary and the symbolic.[39]

In this seminar, Lacan introduces a way of progressively creating shapes that are linked together through Borromean knots. He discovered in his own experiences that the progressive addition of shapes to the figures above follows a rule:

> In making use of rounds of thread in composing these Borromean chains, I illustrate something that is not without relation to the sequence of numbers.
>
> You know how, by means of some axioms, Peano articulates it. It is the function of the successor, of the n + 1, that he stresses as structuring the whole number—which supposes nonetheless to start with one that is not the successor of any, which he designates by zero. All that these axioms produce will be from then on, conforming to the arithmetic requirement, homologous to the series of whole numbers.
>
> The knot is something else. Here in fact the function of the plus-one is specified as such. Suppress the plus-one, and there is no more series—by the sole fact of the sectioning of this one-among-others, the others are liberated, each as one.

> This could be a wholly material way of making you feel that One is not a number, although the sequence of numbers is made of ones.[40]

What is at stake in all these situations (Aristotle, Peano, and Lacan) is the same argument, and admittedly Badiou is untangling this argument with such veracity that it has made the corresponding passages among the hardest in *Being and Event*.[41] But the underlying idea is very simple, as most profound philosophical insights are. The issue here is that Badiou has to untie the presuppositions that have been packed with these arguments, and that is perhaps what has made his presentation somewhat difficult to follow in this section.

We choose to provide an example to illustrate his point. In computer languages, there are several techniques to design algorithms that perform repetitive tasks. One of the techniques is referred to as the while-loop. The idea is that while a certain condition is still true the algorithm inside the loop performs certain operations. The operations end when they reach to the bottom of the loop and that is where the algorithm evaluates the condition, and if it is true, it then goes to the top of the loop and runs through the same operations. The point is that none of the operations inside the loop "knows" anything about the condition of the loop itself; they simply run "inside" that loop (see figure 7.2).

The while-loop is precisely what is missing from the assumptions: that the operator of passage is executing inside something that is already there; there is no way for the operator of passage to know that what it collects (or performs) belongs to the same being, the same as the operations inside the while-loop that have no idea about being part of an algorithmic construction. The language Badiou chose to differentiate the reason for no exhaustion, whether the operation is "empirically incapable of reaching the limit," or "the exhaustion, rather than being empirical, is one of principle" lends itself so well to the analogy of a computer program. The infinite loop is a loop that runs because its condition always computes to true (Badiou's matter of principle, such as the one in the example above)—for example, loops that read from an I/O buffer—versus when a program is hanged, due to reasons not related to its algorithm (e.g. stack overflow, thread starvation, and so on which in Badiou's terminology are categorized as empirical reasons).

```
do{
   //get some input.
   //if the input meets my conditions, break;
   //Otherwise ask again.
} while(true)
```

Figure 7.2 An Example of Do-While Loop in Java. *Source:* Author created.

The ideological presupposition in Aristotle's assertion and somewhat in Lacan's expression of wonder about the consistency of the sequence of numbers is that both have assumed that the operator of passage (Aristotle) or successor operation (Lacan) is executing outside of the limit of what is already there, without realizing that the existential guarantee for the operator to function comes from it executing inside something else (inside a while-loop).

This is what Badiou calls the second existential seal: "a second existent (besides the 'already') which acts as cause of the failure of the procedure of exhaustion: that is, a multiple which is supposed such that the 'still-more' is reiterated inside it."[42] This existent, like in the case of the while-loop, is the Other of the operator of passage. While each term of the infinite multiple as finites are counted and exhausted—similar to Hegel's tireless positing of Finite—they make an other of each other, the whole loop guaranteed by the existence of an Other.

> The second existential seal forbids one from imagining that the infinite can be deduced from the finite. If one terms "finite" whatever can be entirely traversed by a rule-thus whatever, in a point, subsumes its Other as an other-then it is clear that infinity cannot be inferred from it, because infinity requires that the Other originate from elsewhere than any rule concerning the others.[43]

Whence is this Other originated? We have already seen the answer to this: from the axiomatic thought, whose mathematical variance we reviewed in the previous section. Once we posit, through the axioms, the existence of one infinity, ***I***, then it follows that being is also infinite: the existence of one infinite being suffices for us to say that being qua being is infinite. Therefore, axioms not only posit the *being of infinity* but also the *infinity of being*.

Infinity and the Names of the Real

Now, coming back to the discussion of infinity and the real, we chose to present the impact of actual infinity on the thought of the real through a classification according to the three concerns or themes introduced earlier in this work. Accordingly, by thinking the real in the presence of AI we can classify the real under three names—three ways in which the encounter with AI changes what we think of the real. At the beginning of this study, we showed how there have been three concerns that permeate Badiou's entire oeuvre from the mid-1960s to his mature work in *Being and Event*. In the introduction of this work we named them interiority, novelty, and beginning—each being a central question to the works we have picked as our guide. Interiority was the central concern in "Mark and Lack," novelty in *Theory of the Subject* and beginning in *Being and Event*.[44] These three questions are interwoven in a single logic we have called discipline. The last chapter of this work will provide a formal definition of what a discipline is, but for our purposes here, we shall remark that a discipline is a region of being that owns its proper content, which is determined internally (the interiority of a discipline), is historical, which means it produces new content from within itself (the novelty in a discipline), and is regionalized based on a set of axioms (the beginning of a discipline). Therefore, a discipline is a region of being whose genesis is axiomatic (form), whose material is immanent (matter), and whose mode of being is historical (produces newness). Infinity is a central aspect of all these three dimensions: discipline is not bounded, it's being is infinite; its infinity is not due to circularity but periodized historicity; its beginning is not grounded (prior principles) or caused (the substance). Hence, discipline is only thinkable when thought is committed to the existence of actual infinities. Using this conception of discipline and its three "dimensions" we could theorize the encounter that the category of the real has with infinity in accordance to these three dimensions. But before doing that, let's review what we've learned through our previous encounter with the category of the real near the end of part II.[45]

There is a way in which the three registers of the real/symbolic/imaginary work together. Lacan famously models the mechanism responsible for this "working together" of the three registers using the topological notion of the knot, as we have seen in part II. Based on this account, knot is itself in the real, as we saw Lacan said, but its effects register itself in the imaginary as consistency. So, in the end, consistency is an imaginary construct. We again saw in part II how Badiou "turned the table" (this means to manifest an evental character). What if we think of the real as supporting consistency within a topological order, while the knot is the algebraic representation of such topological consistency? We suggested that this "turning the table" revamps the entire semantic of the three registers. This revamping follows the formulation of late Lacan the closest: the formulation that does recognize the "gaps" in the symbolic. This formulation maps consistency to the real, the knot to the symbolic and the consistent Other to the imaginary. This is the materialist reversal of the three registers according to which it is neither the symbolic (language) nor the imaginary (ideology) that is responsible neither for the consistency nor for the novelty; but the real (being) is. This materialist reversal teaches us that a (if not the) main function of the symbolic and imaginary is preservation, the preservation of the consistency of the status quo: the symbolic performs this through the law and the imaginary through the ideals such as the consistent Other: God, Man, and Science. But what is indeed responsible for both consistency and novelty (that is a new consistency) is neither the symbolic nor the imaginary to begin with: it is the real. This materialist reversal was what we saw at the end of the second part of this work: the theory through which *Theory of the Subject* attempted to establish a new relation between the subject and the real, in support of consistency. The model of this theory however is centered around "the real of the subject," of the political subject, as the guarantor of consistency. At the end, the real received two determinations in excess of what its Lacanian genesis had formulated: the real could be thought algebraically as well as topologically, and accordingly could be split to cause and consistency. In addition, the real is a category of the subject.

The claim here is this: in thinking the real in relation to discipline and its three dimensions of interiority, novelty, and beginning, we could now pick up more determinations, and we will be able to show that the prior subjective, causal, and consistent determinations are mere regionalization of the real when we think in the disciplinary register. We give each of these determinations a name. But before proceeding to see what these names are, we should remark something crucial. These names should not be perceived as an attempt to diffract the real by yet another three new labels. There is, of course, an internal logic that exists between these three names that together constitute the anatomy of disciplines. A discipline is built of these three moments. These moments together constitute the so-called "life" of the discipline. Furthermore, as the foregoing has shown, discipline broadens the logic of structure.[46] The Lacanian genesis of the category of the real, which pertains to structuralism in general and structural linguistics in particular, gives two names to the real: the cause and the inaccessible. The movement from structure to discipline, while still preserving the category of the real, will allow us—using the integrated logic of the discipline and its three moments—to find these new names for the category of the real. But, as the proceedings will show, similar to the thought of structure that is attainable by having certain constraints over the discipline (considering the discipline only by its structural logic), the names of the real in relation to structure could also be derived by having constraints over the names that the real has in relation to the thought of discipline.

Now, what are these names? In encounter with the dimension of interiority, the real finds the name consistency, in encounter with the dimension of novelty, the real finds the name truth, and in encounter with the dimension of beginning, the real finds the name condition. In the remainder of this section we will briefly look at these names and how they capture the encounters of the real with each of these dimensions.

The name of the real as consistency speaks to the register of thinking that belongs to the interiority of the discipline. In this register, the real is thought directly in relation to the axiom of Infinity. With the axiom of Infinity, in effect the real can be thought in the modality of existent infinity (or infinities). This means as existent infinity, the real is thinkable in an affirmative register. That is, unlike the modality of inaccessible in which the real could be thought only negatively as that which is unreachable and forever defaulting the reign of the signifier; it can now be thought about positively. In this modality, the real finds the character of consistency in contrast to which the modality of inaccessible acts as the cause—as the traumatic and absent cause that grounds the compulsive repetition. In this character, the real is not thought as the absent core of the symbolic, but in fact as that which renders the very possibility of being able to think symbolically: the consistency according to which the metonymy of progression can be exercised (i.e., the loop within which the operations keep running). This is what was missing from Lacan's (and Peano's) theorization: the assumption of the actual infinity—the consistency of the successor operator—the fact that it always "falls" on another number—is because it is "within" an infinite space of an operation; an actual infinity called the first limit ordinal ω_0. As the profound result of having access to this first infinite multiple, we are allowed now to think in a reverse order: we can think infinite succession as a property of the actual infinity where the algebraic operation of succession is "meant" to work due to the existential seal that Dedekind has instrumented within the definition of the infinite. So, what was previously a stretching out from algebraic to topological and from structural to historical dialectic with having access to ω_0 is now obverted. We now have access to a greater universe and we are all the poorer for thinking finitely. Since we have a consistent reason to begin with infinity and from this perspective, we can always already treat the real as cause and consistency. We don't need to choose between the two, because we have a richer concept. We can ask now: What are the constraints of a situation within the ZFC universe for which succession yields an inaccessible limit? The answer is, for an arbitrary finite ordinal set W and a succession operator s, as follows: $s(W) = W \cup \{W\}$.[47]

To such an ordinal set, ω_0 appears as inaccessible, and this is the sense in which infinite could be perceived as cause. But for this, we had to add extra determinants to arrive at such a finite ordinal. From the perspective of our axiomatic substructure, we are now at a prior level of determination than the one we need to choose between cause and consistency of the real. This is where ontology is happening. This clearly shows us why the resolution of the problem of the real required the resolution of the question of beginning through axioms. Furthermore, if we think the real of every situation as the guarantor of its consistency, then the knowledge that represents a given situation can be perceived as a constraint to the consistency of its real. The real in this case is no longer the impasse of the consistency of the situation, but what underwrites it. So, unlike the Lacanian model that we are always within the symbolic (the knowledge) and the real is what disturbs the symbolic consistency (it makes hole in the knowledge), it is indeed the symbolic (and the knowledge) that gets in the way of consistency that the being of the situation always already harbors.

The name of the real as truth speaks to the register of thinking that belongs to the novelty in discipline. The real as consistency is still thinkable within the so-called constructible universe of sets. What we call <ZF+Constructibility>, five axioms of ZF with the axiom of Constructibility. As we shall see shortly in this chapter, Gödel showed that GCH is a theorem of <ZF+Constructibility>, using the constructible universe ***L***, an inner model of <ZF+Constructibility>. Paul Cohen in 1963 proved <ZFC+¬CH> is consistent in ***L***. This result in particular showed that situations have regions that are completely indiscernible/unnameable—regions that the knowledge of the situation is incapable of accounting for—what Badiou (following Cohen) calls the generic. The concept of the generic gives the real a totally new conception as the indiscernible. What does this mean for the concept of the real? Had we stopped short of indiscernibles, we would have advanced from Lacan, but still left some elements that Badiou developed within *Theory of the Subject* out of the picture. We would have known that we are authorized ontologically to widen our reach beyond what Badiou calls structural dialectics, but we would not have been allowed to think what accounts for the internal historicity of a discipline: because we are still constrained by discernment. With indiscernibles, the real could be thought in terms of indiscernible consistency that comes to existence via the work of the subject—the truth. The theory of the generic is what Badiou had been in search for since his days with *Cahiers*, a theory that, while it sees consistency as what disciplinary progression engenders, it does not see this consistency either as being caused or constructed by what there is, but as a novelty that comes to existence ex nihilo. This is a consistency that is truly indiscernible insofar as the governing laws of the situation are concerned. A subjective act will bring this novel consistency to existence, but a consistency whose being is independent of the subjective act and whose existence is guaranteed only when we assume the axiom of inaccessible cardinals (axiom 1 in the previous section), which is a requirement for the consistency of ZFC.

The name of the real as condition speaks to a register of thinking that belongs to the beginning of the discipline. In this register, the real is thought as the guarantor of existence of multiple disciplines: Badiou has named four, which he calls the conditions of truth, each of which a discipline in its own right: science, politics, art, and love. In this register, and in the encounter with the question of beginning, the real is a particular way in which the very regionalization of being takes place by thinking axiomatically. This axiomatic regionalization of being creates traces in the real within the presentation of the being qua being that Badiou calls a situation, such that it makes the real the correlate of that particular region. In this sense, the regionalization of being divides the real, and this division is the guarantor of the existence of multiple disciplines.

According to these encounters, the real finds three names/determinations. The real as consistency is a direct outcome of the existence of infinities. As such, the foregoing remarks should have made this name abundantly clear. The rest of this chapter will focus on the name of the real as truth. This is the name under which the real finds its way yet again to the category of the subject. The subject of truth is what the next chapter will cover. This work will not explore the name of the real as condition.

GENERIC: THE NAME OF THE REAL AS TRUTH

Infinity has always been philosophy's holy grail. From its inception as an idea, it has anticipated the Cantorian revolution, and all previous interrogations into the relation between discrete numbers and the geometrical continuum should be read in the spirit of that anticipation. We can say now that the abyss which such interrogations have always faced, between the two orders of infinites, is none other than the difference between ω_0 and $P(\omega_0)$, the latter being the set that is capable of fixing points in space and the former of ordinary arithmetic. No doubt there is a mystery in this abyss. This mystery has been expressed in innumerably different concepts and metaphors, but never so clearly as in the ontological difference between the orders of infinity, which is a decisive step in the magnificent speculative development of the dialectic of discrete and continuous, from Plato to Husserl, via the inexorable path of Hegel's *Logic*.

This abyss, as we saw in the previous sections, comes into full focus for thought only when we have a way to think about being, not only qualitatively, but also quantitatively. It is only with respect to the quantification of being that the gap of different orders of infinity, and thereby their potential qualitative differences, opens up.

This gap, what Badiou refers to as ontological errancy, is thereby quantitative: the gap that the continuous (that is the quantitative aspect of its being) $P(\omega_0)$ has with respect to the denumerable ω_0, about which ontology cannot make a decision. The antinomy of quantity is thereby an enigma of Being that disturbs ontology. The unsayable of Being, what defies the comprehension of Being qua Being and thereby what is not expressible by knowledge, lies in the quantitative gap between presentation and representation.

Thought's dissatisfaction of the unmooring of being from knowledge—the errancy of representation—is manifested since the Greeks, whose efforts time and again to overcome the unsayable of being have failed. It is thought's tendency to avoid the errancy that causes the recurrence of those orientations to either mystify the errancy or make it comprehensible to, and thereby expressible in, knowledge. One predominant orientation is to think of infinity as that which transcends such errancy. This form, as we have examined previously, is no longer the dominant form of idealism of infinity, having been banished after Romanticism. The other tendency present in our earlier discussions, which can be captured as the measurable or continuous infinity, is dubbed linguistic idealism, which is the current dominant orientation of thought. But it is only now, through Cantor's famous Continuum Hypothesis (CH), that we understand what is truly at stake in both of these idealisms. Only by employing axiomatic thinking, ZFC, and the theory of large cardinals—what has enabled us to think infinity quantitatively—do we have an adequate framework for the event, at whose site knowledge falters. We formerly presented the general form of this problem as the General Continuum Hypothesis (GCH), which is a superset of Cantor's CH.

CH: $P(\omega_0) = \omega_1$, GCH: $P(\omega_n) = \omega_{s(n)}$

In this section we will cover two orientations, both supported by the framework we described, which result in opposite reactions to this faltering of knowledge: one exonerates knowledge by suturing it at the evental site as the cause, resulting in the foreclosure of the real as truth; the other de-sutures knowledge from the evental site and allows new consistencies to be drawn from it. Respectively, these two orientations are constructivism and generic.

Constructivism

The ontological moniker for linguistic idealism is the axiom of Constructibility. The ontological orientation that espouses this axiom is constructivism. Constructivism musters a response to the errancy inherent in ontology by eliminating this errancy altogether. It boils down to the following assertion: "that which is not susceptible to being classified within a knowledge is not."[48]

The constructivist orientation is an unavoidable nemesis due to the proximity of knowledge to the situation and its state. Knowledge is the ordinary regime of the relation to being wherever a new temporal shift is not possible, and when the prospects of fidelity have somewhat waned. Constructivism does not have an aggressive agenda. Its latent philosophy is to preserve the status quo and declare the only progress as that of knowledge and language. Knowledge hides the excess intrinsic to being by attributing this excess to itself, namely in the falling short of language/knowledge. The excess of being, however, is not a lack in knowledge. They are not the same, as the latter is predictable and the former is irruptive and unpredictable. The latter is manifested as weakness whereas the former as excess. The latter as a tamed and structural progression and the former as a tectonic shift, irruption of the inconsistent, as revolution.

Following this mantra, in relation to CH, constructivism only constructs those parts of a set that follow the strategies below:

1) The part is recognizable according to a rule that is expressible in first order logic.
2) The rule quantifies over variables from domains that are limited to presentation alone—that is, the domain of the rule is a set itself.

Take a set α, with the set of its parts $P(\alpha)$: everything that is included in α. That is the beginning of excess.

The general strategy is this: all existential and universal quantifiers select over set α only. This means that all free variables in the quantifiers will refer to an element of α. All parameters, that is, those parts of the formulae that refer to a fixed multiple also belong to α. This means that everything that we express about multiples, we express about the multiples of α.

The set of "definable parts" is defined in the above sense of quantifiers and parameters. Let's designate this subset of α as $D(\alpha)$. It is obvious that $D(\alpha)$ is a subset of $P(\alpha)$. The set of definable parts is the subset of the power set of α but filtered by a certain language in such a way that it separates the elements of α using a set of formulae that refers only to those elements. That's why we can call $D(\alpha)$ the set of constructible parts of α.

Using this method, we can define a constructible universe L which is stratified into levels. The first level $L_0 = \Phi$. Then, $L_1 = \{\Phi\}$ and $L_2 = \{\Phi, \{\Phi\}\}$. In general, for any successor ordinal we have:

$$\boldsymbol{L}_{s(\alpha)} = D(\boldsymbol{L}\alpha), \text{ or } \boldsymbol{L}\alpha = \cup_{\beta \in \alpha} \boldsymbol{L} \text{ (if } \alpha \text{ is a limit ordinal)}$$

So, the universe $\boldsymbol{L}$ is constructed such that

> the only multiples which are admitted into existence are those extracted from the inferior level by means of constructions which can be articulated in the formal language, and not 'all' the parts, including the undifferentiated, the unnameable and the indeterminate.[49]

We call a multiple γ constructible if it belongs to one of the levels, or hierarchies, of L, say $L\alpha$ where α is a successor ordinal. Let's assume $\alpha = s(\beta)$, which means $L\alpha = D(L\beta)$. That means there is a formula f in $L\beta$ that can separate elements of this hierarchy into the constructible set γ which belongs to $L\alpha$. All multiples at the hierarchy of the successor ordinal are separated from the previous ordinal via a formula such as f.

Recalling what we said about the von Neumann universe V as the general model of ZFC, it should be clear that L is a subset of V: $L \subset V$.

The Axiom of Constructibility

The question here is, if we suppose that our hierarchies are built as such, are there any multiples that are left out? In other words, are there any multiples that are not constructible and thereby not included in the successor hierarchy? The difficulty is this: there is no way to show that a multiple is left out, because once we can specify such a multiple, it is automatically constructible. This impasse of proof, as the case is so often in sciences, becomes a positive axiomatic statement, in this case being the axiom of Constructibility: For every multiple γ, there exists a level of the constructible hierarchy to which it belongs; that is, $(\forall\gamma)\ (\exists\alpha)\ [\gamma \in L\alpha]$, where α is an ordinal. Simply reformulated $L = V$.[50]

There is a double move embedded in the conceptual framework of the axiom of Constructibility that is paramount to its success involving an exclusion and an inclusion. The exclusive move consists of restricting the domain of input to only those multiples from an already programmatically constructed domain. It builds a model for all (relevant) axioms that comprises only a constructible universe. The inclusive move is then the proof that every multiple in such a model is constructible. It is important to realize this: it is solely based on the "only" that "every" will work in this proof. As we shall see later, Cohen adopted the exclusiveness of the model, and showed that not "every" multiple is constructible. We can "condition" a multiple in a way such that it cannot belong to the constructible ground model.

Within the whole approach of constructivism, there is a conscious attempt to relativize the quantifiers to a constructible hierarchy *only*. As such, if we have a formula with a free variable such as $f(\beta)$, we are saying one of the following:

$(\exists\beta)[\boldsymbol{L}(\beta)\ \&\ \ldots]$
$(\forall\beta)[\boldsymbol{L}(\beta)\ \&\ \ldots]$

This means exactly what we said before, we are relativizing, that is, restricting, to the domain of the function f to only select from the constructible hierarchies. That being said, the second (inclusive) point in the constructivist approach is to prove that every multiple in the constructible universe is constructible.

The Restriction of Being and the Restriction of Thought

We said earlier that the goal for both orientations of ontology—constructivist and generic—is to render the inability of presentation to measure the excess of representation intelligible, that is, thinkable. The first thing to realize in this relation is that it was never the excess itself that was the troubling issue, but its immeasurability: the fact that we do not know whether different infinities grow at a lock step or not. To put it differently, as captured by the Easton theorem, the notion of succession is undecidable when it comes to infinite cardinals: it is impossible to say whether a cardinal is a successor of another cardinal; or given a cardinal, what its successor cardinal is.

Although there are several ways to build higher size classes of ordinals, it is not possible to show whether a given construction yields a successor cardinal.

It is this immeasurability that gave the impetus to move toward the reign of the language. The idea here is this: if the cardinality of a given constructible multiple S is $\omega\alpha$ for some ordinal α, then the cardinality of $P(S)$ is $\omega_{s(\alpha)}$. This means that the cardinality of the power set of S is measured by the next limit ordinal. The growth of $|P(S)|$ over $|S|$ is at lock steps with α and $s(\alpha)$. In other words $|P(S)|$ is measured by the next limit ordinal—there is no other limit ordinal between $\omega\alpha$ and $\omega_{s(\alpha)}$. This is the general form of CH.

In the spirit of saving thought from the immeasurable, constructivism intervenes by inserting language between belonging and inclusion. What licenses this effort is the tendency of thought to avoid errancy. That is why nominalism is constructivism's spontaneous philosophy, because it avoids what it cannot name, and what it does not name does not exist. It is to declare being only under the label of discernible. We are living (if not more so than any other time) in the era of the dominance of nominalism. Linguistic idealism, as its ideological form, is the dominant philosophy of our time. But the outcome does not so much pertain to being as it does to thought. Constructivism comes to restrict being only by restricting thought to what language can do or admit. Language possesses infinite power, but it is no match for the denumerable. Therefore, we sacrifice in order to remain in language, and it seems impossible to give measure of this sacrifice. We can only say that the demanded sacrifice "as the price of measure and order is both intuitively enormous and rationally incalculable."[51]

In 1971 Frederick Rowbottom published the following result: if there exists a Ramsey cardinal, then for every cardinal smaller than it, the set of constructible parts of this cardinal has a power equal to this cardinal. In other words: if π is a Ramsey cardinal, and if $\omega\alpha < \pi$ we have:

$$|P(\omega\alpha)|L = \omega\alpha$$

This result proved that the price we pay for constructibility is even more than what we had intuited. Constructibility does not restrict the excess, it simply eliminates it. This is simply too big of a price to pay.

Even so, the above is not the only cost of the axiom of Constructibility. This axiom also forecloses the event. We have seen how ontology forecloses the event through a decision that is manifested as the axiom of Foundation. With the axiom of Constructibility, the foreclosure of event is no longer a decision, but a theorem—we can now prove that there can be no self-belonging multiple in the constructible universe. In ZFC, the existence of events is anticipated, but disallowed—and it is disallowed precisely because it is anticipated. This is the price that ontology pays to ground a consistent science of presentation of multiplicities that does not unravel at the site of paradoxes of self-belonging. In <ZF+Constructibility> on the other hand, events cannot exist—because they cannot be constructed. That means ontology as a historical multiple itself, what we dubbed a discipline, will not be possible.

The axiom of Choice meets the same fate as the axiom of Foundation. The choice function becomes a property of the constructible universe. AC was an axiom of ontology because, at the height of disorder, we needed a guarantee that we can produce a minimal order, which is the ground for building multiples that have rich enough properties (such as well-ordering) such that they render themselves to investigations. But without the possibility of reaching the height of disorder, there is no longer any use for an axiom for ordering. Of course, for Badiou, foregoing AC means the foregoing of intervention. But, that is also coherent with the fact that events are nonexistent in the constructible universe.

The foregoing of event and intervention is too much of a price to pay. If what we have established earlier holds—that the mode of thinking through axioms is necessarily evental—then foregoing the event and intervention means the foreclosure of axiomatic thinking itself. The axiom of Constructibility defaults the axiomatic thinking, and in particular, it defaults the very evental consistency prodigal to ZF as the result of the axiom of Infinity.

This is accomplished by making a choice at the evental site named the Continuum Hypothesis. There, the decision to restrict the excess sutures thought to the existing representation—the void of the site acts as a cause to suture the knowledge that is punctured around the site of the event. This choice excludes that which could transpire from the evental site as new consistencies, which thought would be able to explore.

Generic

The concept of generic is the culmination of the axiomatic orientation and a consequence of the axiom of Infinity. For us, generic is the name of the real as truth. That is, generic is the real of the situation and its encyclopedia of knowledge. Using the concept of generic, Paul Cohen proved that <ZFC+¬CH> is consistent in (a model based on) ***L***.

In the previous section, we showed that the axiom of Constructibility embeds two moments in its construction. In the first moment (exclusion), the axiom restricts the domain of input to exclude anything that is not syntactically correct (quantifiers and predicates). Using this moment, the axiom builds a model for all (relevant) axioms that comprises *only* a constructible universe. In the second moment (inclusion), the axiom then proves that *every* multiple in such a model is constructible. It is solely based on the "only" that "every" will work in this construction. The axiom of Constructibility provides a syntax according to which we can both build a universe, and at the same time, posit that the universe is complete. Constructivism, in short, is the dominance of the formalism of language over the materialism of multiples, in a way which it eventually eliminates the excess of being over language, the excess that was originally hinted at by the immeasurability of inclusion over belonging. What is then obvious is that a strategy to prove the opposite, which is the sheer dominance of being over the language, cannot be through a proof—because any proof is formalization. In other words, the general strategy is not to rely on the proof of ¬CH but to show that if ZFC is consistent then <ZFC+¬CH> is consistent in some constructible model. We saw earlier that in mathematics the consistency proofs use the model theory. That means proving that a set of axioms is consistent is equivalent to showing that there exists a model for those axioms. So, the general strategy can be summarized as follows:

ZFC has a model → <ZFC+¬CH> has a model

We saw earlier that ZFC cannot be shown to consist in $V\omega$ due to Gödel's incompleteness theorem. Therefore, if ZFC is consistent, it must be consistent in a different model $V\kappa$ for another inaccessible cardinal κ, $\kappa > \omega$. While $V\kappa$ is a completely veridical universe for ZFC, Cohen's proof takes place in a much smaller model that is semi-veridical or quasi-complete wherein a certain subset of ZFC axioms are veridical (as we shall see shortly). Badiou calls this model *S*. If we denote as $S[G]$ (read it as the generic extension of *S*)[52] the model in which <ZFC+¬CH> consists then we can state the strategy of Cohen's proof in the following way:

$S \rightarrow S[G]$

Generic and Event: Disposition and Composition

Badiou provides a masterful demonstration of Cohen's proof. But from our perspective, his discussion includes an important nuance that separates it from an otherwise normal account of Cohen's generic and forcing. In general, there are two sequences that are in play in Badiou's presentation. These sequences have to do with the being and existence of truths. Let's formulate the first sequence in the following way:

1. We have a definition of the constructivist orientation in ontology. According to this orientation what is, is constructible in language, and conversely what doesn't have such construction has no existence.
2. We have a definition of knowledge according to the constructivist orientation. Knowledge is the encyclopedia of determinants. Determinants

are properties of a given situation. Properties are collectibles through discerning multiple parts of the situation.

3. Truth cannot be part of the knowledge. In this sense, we must break away from the Aristotelian and notion of truth as adequation and espouse the notion of truth as the knowledge's impasse. Therefore, truth must be indiscernible—as knowledge by definition is finally the outcome of the discernment of properties.
4. The indiscernibility of truth results in the infinity of truth as every finite subset is discernible.
5. But the infinity condition is not enough as discernible subsets of an infinite situation could also be infinite. So, for truth to be indiscernible it must not only be infinite but also every one of its elements must defy at least one property of the encyclopedia. Indiscernibility is a stronger condition than infinity, which requires a more pervasive application of the axiom of Infinity—the extent of which constructivist program hopes to avoid.
6. The general strategy for "conditioning" an indiscernible subset is to collect terms from the situation, which has nothing in common that could be remarked, save the belonging to the situation. As such, the indiscernible part solely possesses the properties of any part whatsoever, and thereby it is referred to as the generic. The only "property" that is sayable of the indiscernible part is of the belonging of its elements, which is the supreme genre, the genre of being of the situation as such.
7. The universality of this indiscernible part also lies in its generic constitution, as it rejoins the properties of the entire situation without ever committing to any one of them and thusly representing the entire situation through exhibiting the very being of what belongs insofar as it belongs, which is the very being of the situation, in contrast to what language carves out therein by the way of discernible properties.
8. How do we know the indiscernible even exists? We know because mathematics is able to back it up. Cohen's proof shows that a semi-veridical denumerable ground model can harbor indiscernible subsets.

The above sequence illustrates a rough schema of what is sayable of truths and their beings: we have a definition (truth is indiscernible in the situation), we have a categorization (truth is universal) and we have a modality (truth is apodictic or we have a guarantee for its existence).

In the second sequence, Badiou wants to show one more thing. He wants to show that the infinite subset of those enquiries that have a positive relation to the name of an event is a generic, in the above sense. His illustration goes like this:

1. Let p be a determinant of the language L of the situation S. Let x and y be two elements of S such that x is determined by p and y is not. Then

the finite subset $w = \{x,y\}$ is not determined by p, since it has at least one element that is not determined by p. We say w avoids p.

2. Now let's consider the sequence of $x_i(+)$ where each x in this sequence is an element of S that is positively enquired in relation to an event that is named. If within such infinite sequence we find one x that is not determined by p then we can say that the entire sequence avoids p.
3. Now, we say

> "*if* a faithful procedure contains, for every determinant of the encyclopedia, an enquiry which avoids it, *then* the positive result of this procedure will not coincide with *any* part subsumable under a determinant. As such, the class of multiples which are connected to the event will not be determined by any of the properties which can be formulated in the language of the situation. It will thus be *indiscernible and unclassifiable* for knowledge. In this case, truth would be irreducible to veridicity."[53]

4. This is not an unreasonable assumption that such a procedure exists. But we would still need a proof of its existence (which is Cohen's proof).

Badiou's presentation in relation to these sequences may appear disjoint, but seeing ontology as a discipline, we can understand them precisely in this way: the latter sequence is what the thought of generic *composes* in ontology, while the former is what the thought of generic *disposes* in ontology.

In the first sequence, ontology disposes of the concept of the generic and the guarantee of its existence in any situation that fits a certain set of requirements. That is the discussion of the type: ontology disposes of the thought of generic as something any such situation possesses the techniques to name, and such that an extended model could be constructed containing the generic subset. In the second sequence, however, we compose a particular subset in reference to an event. Therefore, the second sequence takes up the discussion of the token (not the type): there is a certain procedure according to which we could construct a subset that has the characteristic of an indiscernible. The former sequence is the way ontology disposes of the thought of the generic in any situation, and the latter is the way in which a discipline (including ontology) composes toward a generic in a situation, with event marking the difference between the two. In this case, event marks the shift of the perspective between composition of the token and disposition of the type. While the second sequence does not speak of the first, it supports it. The generic is the thought that ontology supports, or in our terminology, the discipline of ontology disposes of the thought of generic. The generic, being an ontological category, is thereby applicable to any situation of being, including to that historical situation that constitutes the interiority of the discipline of ontology itself. The thought of event in contrast is the thought that belongs to the theory of discipline—as we have discussed in the previous chapter. The indiscernible in a given discipline, while supported ontologically, is the matter of composition and depends on faithful procedures and the figure of the subject. The theory of discipline speaks of the composition of indiscernibles and the categories of the event, subject, and truth that matter to the operations of composition. The dual operations of composition and disposition also clarify the incommensurability of ontology and truth, what Badiou emphasizes so often. Ontology does not dispose of truth, because truth is a category that belongs to the theory of discipline, not ontology.

This begets another incommensurability between being and truth. Along with saying truth has no meaning in ontology as a disciplinary disposition, we should also abandon the project of collocating being and truth. Instead, the thought of being is the thought that is disposed by axioms—what can be extracted or derived from thinking through them. The thought of truth, in contrast, is not disposed by axioms, but that does not mean truths do not exist. The fact that the indiscernible has a being does not endow the role it performs in a historical situation as the supplement. The being of what will have been the truth becomes the being of truth through the disciplinary composition, which is the work of the discipline as the subject—as the subjective aspect of the discipline, which is the topic of our next chapter.

Therefore, ontology can think the being of the indiscernible and can devise a method for how to name it, as well as the modification required in the situation of being such that the indiscernible can be supplemented in the situation's dictionary. But in its disposition, ontology does not name the event, the subject, or truth. However, as we see with the second sequence, what ontology allows in the first sequence must be composed in *every case* through a procedure that is thinkable as a disciplinary composition.

This difference in the modalities of a discipline helps us to understand Badiou's gesture toward the differentiation between the being of the truth as what is thinkable by ontology, and the existence of a truth as something unthinkable by ontology. The generic is part of the situation but not specifiable by the knowledge supplied by the situation. But the being of truth is ontologically thinkable. This is the meaning of the compatibility of truth with ontology, but the existence of a truth, that there is a truth in a given situation, is not something of which ontology disposes, as ontology does not speak of a given situation. So, while the being of truth is thinkable in ontology, its existence is not, as one belongs to the disposition of ontology and the other to the compositional operations in any discipline.

Generic: The General Strategy

We said that Cohen's proof takes place with a model S within which ZFC is semi-veridical. The general strategy is to show that there is a model for

<ZFC+¬CH>. This model, say *S*[*G*], is constructed by adding a subset of *S*, which is *G*, to *S* in a way that ZFC still holds but CH becomes untrue. We cannot construct *G*, we can only condition it. But during the conditioning of *G* we get enough information about *G* such that, when we add it to *S*, we can show the resulting model is such that CH is untrue in the resulting model. *G* is such that it provides a complete paradigm for veracity in *S*[*G*], the generic extension of *S*. Because of this peculiar structure, Cohen called this set a generic filter, as opposed to a set: while filter does not allow us to clearly see inside of *G* (as it is non-constructible) we can get just enough information about what is inside so that we are able to prove things about the new model. Cohen's strategy is for the construction of a new model in general and this method can be used for many results other than CH. The general strategy can be formulated as the following: given a particular proposition *p* that has veracity in a model *S*, we can build another model from *S* such that we are able to force ¬ *p*. This is performed by creating a generic filter *G*. But the nature of this generic filter will depend on the nature of the proposition that we are refuting. Cohen built one of these filters, which is specific to the proposition of CH.

The first thing to notice is that *S* cannot be a finite set: it would neither support the operations nor the "quantity" that these operations presuppose. *S* must be such that the Ideas of the multiple, the axioms, are veridical within it. Another aspect that is important to us is that there is a differentiation between two perspectives: the perspective of an ontologist and the perspective of an inhabitant of the situation *S*. This distinction harkens back to everything that we have developed from "Mark and Lack" and *Theory of the Subject* in relation to the dialectic of the structure and the meta-structure, and how this dialectic can be seen as a regionalization of a more "global" logic—the logic of discipline. At this juncture we can ask: How does one think beyond a given model? The answer is: by thinking axiomatically. We have seen that the axiom of Infinity itself is already the archetype for this sort of thinking: thinking beyond the given model. The thinking that established the model $V\omega_0$ which enabled us to theorize arithmetic as a closed and consistent field, is now inviting us to think the generic. In contrast to this orientation, we have the orientation of thinking that is informed by the knowledge alone, the type of thinking that saw infinity only as inaccessibility (pre-Cantorian mathematics), and immeasurability as inexistence (constructivism). In this sense, the division of ontologist and inhabitant is very much an anticipated division—it is rooted in the division of two orientations of thinking.

Accordingly, we can retrofit this division of perspectives in an even more general context that will assist us in our discussions of the subject. We have three moments of the dialectic of composing and disposing we spoke of earlier. We have the ontological moment, which is the distinction between inclusion and belonging—the inevitable excess of being over the presentation that registers itself in the very difficulties that ontology faces—the ontological impasses we have formerly discussed. Then we have the moment of mediation when this distinction will give rise to two perspectives. There is a figure for whom the excess of part over elements is thinkable, whom we call the ontologist, and the figure for whom this distinction is unthinkable, whom we call the inhabitant. Then there is a third moment in which the distinction itself is projected unto the inhabitant. This is literally a theory of incarnation that begins with being, which composes and disposes of itself in the form of structure and meta-structure. This gives us two positions: a structured position of the inhabitant and the pure meta-structural position of the ontologist. This distinction itself then is reflected back in the structure: How do you construct a meta-structure that is immanent to the structure itself? The latter provides the framework within which we can ask the question of truth.

We shall recall that a part of the theory is absolute if it is independent of the underlying model—in this case, the theory holds for both the inhabitant and the ontologist equally. For an inhabitant of S, elements of S and elements of $P(S)$, when S is transitive, are both absolute. But the elements of the power set of an element of S are not absolute, because there are no resources in S that empower the inhabitant of S to have this recognition. That's a difference between an ontologist and an inhabitant. This is also important for the general strategy of discernibility: there is discernibility from the perspective of an inhabitant only, which means that when we say G is indiscernible, it is said for an inhabitant of S.[54]

By constructing $S[G]$ as a model for <ZFC+¬CH> Cohen went on to prove that CH cannot be derived from ZFC, even in a veridical universe such as L. From a model that belongs to L within which CH is veridical (it is a theorem) it is possible to construct a larger model (which supports the same paradigm of veracity as the original model) within which we can force the statement ¬CH to be veridical, whose veracity can be checked by an inhabitant of the smaller model. Therefore, a statement that is undecidable in one model can be veridical in a generic extension of the base model, which means that we have an ability to force undecidable statements (in the Gödelian sense) in one model to be veridical in another model that is the extension of the first model. This point of forcing is important for us for a different reason, which we will return to shortly: the techniques of forcing invented by Cohen and advanced by Martin and others force the undecidable at the point of the indiscernible. Therefore, as the foregoing attempted to show, there is a sort of coupling between indiscernible, undecidable, and what will have been decidable when it comes to the techniques of forcing. We shall exploit this point shortly toward the identification of a place where the thinking of the subject could be exercised.

Accordingly, we can retrofit this division of perspectives in an even more general context that will assist us in our discussions of the subject. We have three moments of the dialectic of composing and disposing we spoke of earlier. We have the ontological moment, which is the distinction between inclusion and belonging—the inevitable excess of being over the presentation that registers itself in the very difficulties that ontology faces—the ontological impasses we have formerly discussed. Then we have the moment of mediation when this distinction will give rise to two perspectives. There is a figure for whom the excess of part over elements is thinkable, whom we call the ontologist, and the figure for whom this distinction is unthinkable, whom we call the inhabitant. Then there is a third moment in which the distinction itself is projected unto the inhabitant. This is literally a theory of incarnation that begins with being, which composes and disposes of itself in the form of structure and meta-structure. This gives us two positions: a structured position of the inhabitant and the pure meta-structural position of the ontologist. This distinction itself then is reflected back in the structure: How do you construct a meta-structure that is immanent to the structure itself? The latter provides the framework within which we can ask the question of truth.

We should recall that a part of the theory is absolute if it is independent of the underlying model—in this case, the theory holds for both the inhabitant and the ontologist equally. For an inhabitant of S, elements of S and elements of P(S), when S is transitive, are both absolute. But the elements of the power set of an element of S are not absolute, because there are no resources in S that empower the inhabitant of S to have this recognition. That's a difference between an ontologist and an inhabitant. This is also important for the general strategy of discernibility: there is discernibility only from the perspective of an inhabitant, which means that when we say G is indiscernible, it is said for an inhabitant of S.

By constructing S[G] as a model for <ZFC+¬CH> Cohen went on to prove that CH cannot be derived from ZFC, even in a veridical universe such as *L*. From a model that belongs to *L* within which CH is veridical (it is a theorem) it is possible to construct a larger model (which supports the same paradigm of veracity as the original model) within which we can force the statement ¬CH to be veridical, whose veracity can be checked by an inhabitant of the smaller model. Therefore, a statement that is undecidable in one model can be veridical in a generic extension of the base model, which means that we have an ability to force undecidable statements (in the Gödelian sense) in one model to be veridical in another model that is the extension of the first model. This point of forcing is important for us for a different reason, which we will return to shortly: the techniques of forcing invented by Cohen and advanced by Martin and others force the undecidable at the point of the indiscernible. Therefore, as the foregoing attempted to show, there is a sort of coupling between indiscernible, undecidable and what will have been decidable when it comes to the techniques of forcing. We shall exploit this point shortly toward the identification of a place where the thinking of the subject could be exercised.

As we remarked earlier, the ability to force veracity in an extended model is a categorical enhancement made by Cohen to Hilbert's formalist and Gödel's constructivist paradigms. We cannot overemphasize the importance of this point: Cohen's work is in fidelity to the formalism of both paradigms. As we have clearly seen, Cohen's proof works from within a veridical universe and not from an epistemological point of view located outside. We refer the reader again to Badiou's criticism in "Mark and Lack" of the effort to "ground" mathematical logic in something outside of mathematics (e.g., the theory of discourse), or to find a different discourse wherein a transcendental meaning of what mathematics is doing is manifested (e.g. epistemology). Borrowing the terminology of the logic of discipline that we have developed so far, Cohen's advancement made Gödel's achievement a regionalization within a larger framework, in the same way Gödel's work made Hilbert's formalism a regionalization of a larger framework.

One more important note should be added here, which also harks back to our foregoing remarks related to the axiomatic nature of infinity and what proceeds from the axiom of Infinity. Cohen's proof relies heavily on the regularity of the cardinal ω_1 (next cardinal or limit ordinal after ω_0). This proof relies on the fact that ω_1 as a cardinal cannot be decomposed: or as a cardinal it cannot be reached from below. We have seen such a cardinal, ω_0, whose existence is decided by the axioms. This is an important aspect: the existence of a regular cardinal cannot be proven or derived, it has to be decided, but this decision is not a one-time decision, it is something that axiomatic thinking must encounter every time when it reaches the vicinity of a regular cardinal. Even after the first regular cardinal, ω_0, there is no way that through the set theory operations (von Neumann or power set) another regular cardinal could be obtained. Having the first limit ordinal and Peano arithmetic we can obtain successor ordinals but never a limit ordinal. The existence of any limit ordinal shares the same pedigree as the first limit ordinal: it is a matter of decision, not veracity.

THEORY OF THE DISCIPLINE: FROM INFINITY TO SUBJECTIVITY

Our theory from the start has been that Badiou's work from his early contributions to *Cahiers* have followed a more or less recognizable thread that we have labeled the thought (or the logic) of discipline. Our aim in this work has been to recuperate elements of Badiou's work and organize them around this guiding thread. We articulated three ways or aspects in which the idea of discipline has emerged in Badiou's work from the late 1960s, to the time of the publication of *Being and Event*; namely, interiority, novelty, and beginning. We also articulated that in the three episodes that we have highlighted in Badiou's intellectual development, each was occupied with one of these themes—"Mark and Lack" was centered on interiority, *Theory of the Subject* on novelty and *Being and Event* for the first time took on the difficult question of beginning (while rethinking interiority and novelty under the new commitments that pertain to the beginning). Despite this episodic focus, however, Badiou's intervention has always been in the vicinity of two major categories: the infinite and the subject. In other words, what characterizes his intervention in all three aspects has been the influence of the thought of actual infinites to sustain an understanding of the category of the subject.

In this chapter, our focus has been on the influence of the idea of infinite on the thought of beginning and the axiomatic method of thinking. We made the claim that to understand Badiou's Platonism, we need to return to the principles of axiomatic thinking best captured by Albert Lautman's work between the two world wars. In Lautman's view, axiomatic thought sees the progression of sciences through a network of ideas and not a hierarchy of concepts. Axiomatic thinking pushes for the maximality of thought through two of its distinct features: it is objectless and subversive. It is objectless since the real of its thinking and thereby the real of its formalism is the formalism itself: the real of mathematics is the very formalism it employs. It is subversive in that it can go beyond its own limits. The real of axioms, the consistency of the theories that are in comprehension of the generalities that the axioms sustain, is itself a generality: the generality sustained by the axiom of Infinity and the other axioms that it helps to engender. The axiom of Infinity manifests the very ability of the axiomatic thinking to go beyond its own limits, its very ability to create new consistencies. That is why we have claimed a special "status" for the axiom of Infinity among all other axioms of ZFC. While all other axioms are the outcome of the engagement of thought with the question of beginning in a mode where this beginning is not grounded, and thereby activates the thinking without "objects," *the axiom of Infinity is where the maximality of thinking is truly exercised, according to which new investigations open up to show the limits of ZFC itself.* Through the axiom of Infinity, ZFC has been able to transgress its own limits.[55]

In this sense, we claim that the idea of infinite is that which not only thinking disposes of, but also what plays a constitutive role in the composition of other Ideas, such as the idea of independence (independence of CH from ZFC) we just saw. In addition, the idea of infinite un-sutures thought from several other pre-Cantorian principles that have dominated thought. In our view, this newfound commitment to the idea of the infinite lies at the center of Badiou's unveiling of a consistent and crucial theoretical position and his critical stance toward structuralism, dialectic, and the modern versions of idealism. We shall review some of these principles that, in the presence of the idea of infinite, will come undone.

The first and the most important principle, as has been numerously seen in this study, is the principle of finitude. We now know that the opposition of finite and infinite is far from being trivial, and its dialectic in some sense underwrites many other major philosophical categories. Badiou's works in some sense have been aimed toward shifting this dialectic in ways that were not available before. The orientation that openly declares the existential seal of infinity as an axiom of all thinking detaches thought from the foundational status of finitude (everything that is, is finite), and renders finitude as a regionalization of infinity.[56] One may measure the consequences of this detachment by honing into the conception of interiority as a consequence of the primacy of the finite. The interiority of a discipline should not be thought of as the ability of the discipline to "fold" upon itself. The model of interiority according to folding is a model that has its foundation in finitude. Folding is the general model that ultimately maintains the illusory effect of infinity for a being that is finite; something that is much celebrated by German Idealists Kant, Fichte, and Hegel—the model that informs Kant's conception of life and Fichte's conception of rationality, and we have spoken extensively about Hegel's conception of Affirmative Infinity. In contrast, and as early as the first chapter of this study, we have been dealing with another model of interiority: the model that sustains the interiority of a discipline through the ability for the discipline to infinitely stratify itself. We can now establish that this infinite stratification is underwritten by the existential seal of actual infinities. In the presence of the idea of infinite, the interiority is no longer thought as an effect of some other operation, but as an immanent state of the being of discipline itself. As such, discipline, as a region of being that is infinite in size, is not reflective, and its progression is not established through the dialectic of self-realization. In addition, the concept of reflectivity which is at the center of the concept of folding is ultimately bound to the thought of the One: the One of discipline (or discipline counted as one), that must remain reflective so that it could enjoy the illusory experience of infinite immanence. In contrast to interiority founded on the operation of folding, interiority in the presence of the axiom of Infinity is non-operational; it is immanent, material, and tied to the being of the discipline itself.

The axiomatic infinity of discipline supersedes the One and the thought of finitude both in terms of composition (how it is composed: its genesis and progression) and disposition (what it disposes of: the thought it donates) of disciplines—they are both axiomatic.

In the infinite region of a discipline, the progression is thereby best understood not by folding, or by dialectic of self-annihilation and self-realization, but through breakage and assembly. We observed these modalities at length in Badiou's brilliant essay "Mark and Lack" in the way in which a science circumscribed by Bachelardian principles is bound to encounter and overcome impasses that are inherent in its being. It is guaranteed that they encounter impasses, because what constitutes the discipline is of the order of the infinite. They are guaranteed to overcome these impasses because they are infinitely stratifiable, that is you can absorb the impasse into the constitution of the discipline and toward new consistencies.

As this study has hinted at earlier, Lazarus has made a similar attempt for the discipline of politics.[57] First, a discipline in Lazarus' case is a region of thinking that thinks about its own thinking from within. Hence, both the thinking and the thought belong to the same "act" (or the same region). This for Lazarus is what differentiates disciplines from "objective" sciences (or disciplines from discourses, using Badiou's and Miller's terms of debate) in which what is being stated, investigated, and theorized refers to certain representations, properly outside of both the relation of the real as well as the activity of thinking. After Lautman, we now know that the sort of thinking that thinks immanently is a modality of thinking activated by axioms. Politics, Lazarus argues, is a discipline with just such an interiority: it is a thinking that thinks its own thought. The two axioms of this discipline are: "People think," and "Thought is a relation of the real." These two statements—understood as axioms for the discipline of politics, what Lazarus dubbed "anthropology of the name"—circumscribes an interiority for politics that activates the subjective modality of thinking, a modality of thinking that has no "objects":

> The interiority designates the attempt to elucidate the subject from inside the subjective and not in *exteriority*, that is, not by calling on objectivist or positivist referents. With respect to the statement *People think*, the approach in interiority will examine *What is thought in the thought of people*. *Anthropology* designates an investigation into forms of thought, which requires study.[58]

Lazarus stresses that the statement "people think" is an identifier of the thought, and not a reflexive statement.

> What makes this anthropology original is not only that there is no external system of reference but also that the relation of people and their thought, instead of being reflexive, is one of confrontation and it emerges from interrogating the thought of people about what people think.[59]

In our view what makes "people think" an identifier of thought is precisely its axiomatic nature: it carves out a region of thinking/being that has its own interiority. This interiority is thus not sustained by making politics reflective, but by allowing politics to have its own immanent and infinite region; the immanent region of thought that is of the order of the real (according to Lazarus second statement, which is the other "axiom" of politics).

> The moment when the question of thought as a relation of the real emerges is the moment that forges thought as a category in its own right. . . . The movement of thought is then very precisely that which is opened by the application of the foundational statement to its own field. . . . What is thought in what people think is a relation of the real.[60]

One can clearly see the parallel between Lazarus and Badiou with respect to politics and science (in *Anthropology of the Name* and "Mark and Lack" respectively).

There are several other pre-Cantorian paradigms heavily weaved through the classical conception of infinity. A particular one, used even by us, is the inadequate paradigm of potentiality and actuality. Despite having used the qualifier "actual" many times in our study so far, we must admit that this qualifier burdens us with a false dichotomy. While our usage of the qualifier "actual" has been to emphasize the status of the idea of infinite in set theory, which is different from its sacralized conceptions, we must make this clear that the standard Aristotelian modality inferred by this term imports the paradigm of potentiality-actuality which we believe is false. The idea of infinite does not infer the actuality that stands in relation to some other idea of infinite that is in potentiality. There is only one idea of infinite, articulated in set theory, and that idea ratifies the existence of an infinite limit ordinal—the first limit ordinal that contains all the natural integers. This ratification of the existence of such limit ordinal is the sense in which we used the term "actual."

Limit is another principle that has been interwoven with infinite, that somehow what happens "at the limit" is more complex, and also more obscure, than that which is in play in a succession. The principle of limit has wretched away the "sublime" conception of infinity from the "mundane" task of counting, succession, and technique. But in fact, as we have already discussed in the relation of infinity and real,[61] the profound mystery is not what is at the limit, but what is the consequent of the succession: as without having the resoluteness that the category of the real earns from the idea of infinite, we are forever mired in the impotence in relating that which we commence to that which we conclude. Limit is already in the composition of succession, and it is the invisible seal that allows the succession to succeed. The principle of limit is none other than disposing of what is already in the composition. The thought that the limit is ontologically incommensurable with successor is pre-Cantorian. With Cantor, for the first time we discovered that this presumed incommensurability is incorrect, and in fact what we call a limit has been within the finite numbers all along. For example, in an algebraic field we can construct a limit that is finite yet is inaccessible using a definitive operator. This means inaccessibility is a relative notion, which means there is no ontological being for the limit: something is a limit only relative to a particular configuration of numbers and operators. Consequently, when we turn the limit into a number we are not transmogrifying an entity from a certain quality to another quality, we just assert something that has been its quality all along. Therefore, Kronecker's intuitionist intervention, which questioned Cantor's theory to turn the set of natural numbers into a number, completely missed what numbers were from the beginning. Cantor's idea of infinite was a total destruction of the ontological heterogeneity of the limit and succession: ontologically speaking, they are both numbers. The limit-successor distinction only pertains to particular roles that certain beings which are ontologically equivalent (they are all numbers) fulfill according to a certain operator (i.e., belonging).

Here, we again turn to our framework informed by the logic of discipline to assist in this disentanglement. We call upon that which we showed to be present in Badiou's discussion, despite not being made explicit: the disciplinary operations of composition and disposition. We have clearly seen these operations both in two sequences pertaining to the concept of generic as well as the two roles of the inhabitant and the ontologist. This is what the logic of the discipline helps us to realize. The ontological principle of infinity points us to a different paradigm: the paradigm that the disposition of the idea of infinite is no different than this idea in composition; infinity is at the same time an existent as well as a vector of information, a commitment to maximality, and a method of organization. So, for example, in the classical pre-Cantorian paradigm, we take the concept of the infinite series of succession of integers, and the elusive limit we never reach, and from this we fold it upon the succession of integers. This means that we need to first positivize the limit and then imagine that it is ontologically commensurable with the series, which allows us to say it can follow the series. But as we have seen in this chapter, this is not what the idea of infinite involves. Composition and disposition, in this case, makes more sense because rather than positing a disruption of the sequence and the limit, we posit a shift of perspective: that which we compose is that of which we could also dispose. This paradigm does not ask us to go to the limit and come back in order to be able to dispose of the idea of infinite. This idea is constitutive of the very process of the thought of Ideas in general. Composition and disposition are not two separate phases in the lifecycle of the discipline: at every moment, what is in composition could at the same time be disposed. Disposition is not the culmination, the end result, the ending of the process of composition. It is the re-rendering of what is being composed in a different "perspective." In other words, to loosely borrow the

Hegelian famous phrase, the idea of infinite is not only the "substance" (in disposition) but also the "subject" (in composition).

This paradigm shift—from the crystallization of infinite as an absolute Other, or that which is out of reach, or that which is the outcome of the folding, to what discipline allows us to think as the concurrent dual character of composition and disposition—pertains to the category of subject. The lesson we learn from the theory of discipline is that the axiom of Infinity is an outcome of axiomatic thinking but at the same time this thinking implies a prior commitment to the idea of infinite. That is, while axiomatic thinking disposes of the axiom of Infinity, it is the idea of infinite that is inherent in the composition of any axiom and the axiomatic orientation in general: the underlying intuition sustained by the idea of infinite is that when we add a totally arbitrary axiom, we are not bringing a visitor from a strange land into an otherwise homogenous zone, because all other self-evident axioms are equally strangers. This ability to experimentally devise and test axioms owes its allegiance to the underlying commitment to infinity that thought attributes to being and thinking.

It is under this prior commitment that we would like to understand the category of the subject and the subjective character that is in the disciplinary composition. In that sense, we would like to claim this: the salient properties of axiomatic thinking, objectless and subversive, can be supplemented by two conditions, one material and the other formal: while infinity is the material condition of the discipline whereby the discipline in its composition and disposition is conditioned by it being infinite in matter, the subject is the formal condition of the discipline in its composition alone. Thinking in this framework, the following chapter will try to establish that the theory of discipline allows us to think the subject without having to rely on the classical categories of the Other, the object and the other subject, which have yielded us the theory of recognition, the theory of experience, and the theory of otherness respectively. When we are allowed to think the subject as being woven out of the idea of infinite, then we suddenly no longer need the standard mediations of hitherto theories of the subject, which have always thought the subject in relation to something outside. To speak of the subject has meant to simultaneously speak of some sort of operative to allow for reflexivity: through experience, or through inter-subjectivity, or metaphysical otherness, which at the end always amounts to having the subject mediated through the limit—through what is outside. This means, in the grammar we have developed here, the subject is traditionally thought through the categories of finitude and limit—categories emblematic of the pre-Cantorian thinking.

With the grammar of discipline at our disposal—the compositional and dispositional empowered by the idea of infinite—we are all of the sudden able to think the subject without having to posit external categories and markers—to think subject, as Lazarus has mentioned, in a non-relational context.

Establishing this conception of the subject, in accordance to the idea of infinite, is the focus of our next chapter.

NOTES

1. Badiou, *Being and Event*, 227.

2. Alexander Koyré, *From the Closed World to the Infinite Universe*, Baltimore: The Johns Hopkins Press, 1957.

3. See "Meditation 10," Badiou, *Being and Event*, 112–120.

4. It is important to recall that the void is the mark of excess not lack. It marks the excess of being over presentation or appearing: excess appears as the void in the structured field of presentation and the knowledge of it.

5. Benedict de Spinoza, "The Ethics," in *A Spinoza Reader*, 85–265. Princeton: Princeton University Press 1994, 103–104.

6. See Edwin Curly, *Behind the Geometrical Method – A Reading of Spinoza's Ethics*, Princeton University Press, 1998.

7. See Don Garrett, "Spinoza's Theory of Metaphysical Individuation" in *Individuation and Identity in Early Modern Philosophy, Descartes to Kant*, Kenneth F. Barber and Jorge G. E. Garcia (eds.), Albany: State University of New York Press 1994.

8. Badiou, *Being and Event*, 162.

9. This is affirmed by the fact that there are orientations in mathematical logic that work with ZF minus the axiom of Infinity, such as the intuitionist logic. Those branches of mathematics built upon ZF minus the axiom of Infinity lose large portions of the actual mathematical theorems and constructions, but nevertheless able to think consistently about multiplicities. We will come back to this point in this chapter and the next.

10. The axiom of Infinity, as we shall see shortly, states the existence of an infinite set, but does not make any stipulation beyond that.

11. Badiou, *Being and Event*, 163.

12. Badiou, *Being and Event*, 164.

13. This harkens back to one of the senses of the void that we enumerated in *Being and Event* as the operation itself. (cf. chapter 6).

14. Badiou, *Being and Event*, 165.

15. Hegel, *Logic*, 138–139.

16. Hegel, *Logic*, 139.

17. Hegel, *Logic*, 143.

18. Isn't this splitting Hegel thrice (we observed two splits earlier in chapter 2)?

19. Badiou, *Being and Event*, 165.

20. Badiou, *Being and Event*, 166.

21. Jeremy Gray, *Plato's Ghost: The Modern Transformation in Mathematics*, Princeton: Princeton University Press 2008, 18–38.

22. As the famous quote from the mathematician Stanislaw Ulam attests to: "The infinite we shall do right away. The finite may take a little longer" (John Stillwell, *Yearning for the Impossible: The Surprising Truths of Mathematics*, Natick: A K Peters Ltd. 2006, 198).

23. Stillwell brings the example of the infinite series $1-\frac{1}{3}+\frac{1}{5}-\frac{1}{7}+\ldots=\frac{\pi}{4}$ whereas the arbitrary series. : $1-\frac{1}{3}+\frac{1}{5}-\frac{1}{7}+\ldots+\frac{1}{1000001}$ does not have a readily available. answer. (Stillwell, *Yearning for the Impossible*, 198).

24. Alain Badiou, *Number and Numbers*, Cambridge: Polity Press 2008, 32.

25. Gauss was one of those geniuses who was among such mathematicians as shown in the following quote from a letter he wrote to H.C. Schumacher on July 12, 1831, "As to your proof, I must protest most vehemently against your use of the infinite as something consummated, as this is never permitted in mathematics. The infinite is but a figure of speech: an abridged form for the statement that limits exist which certain ratios may approach as closely as we desire, while other magnitudes may be permitted to grow beyond all bounds" (Stillwell, *Yearning for the Impossible*, 199).

26. Badiou, *Number and Numbers*, 44.

27. See John Stillwell, *Roads to Infinity: The Mathematics of Truths and Proofs*, Natick: A K Peters, Ltd 2010, 168–182.

28. For example, set *V*4 contains 24 = 16 elements. *V*5 contains 216 = 65536 elements. And the set *V*6 contains 265536 elements, which very substantially exceeds the number of atoms in the known universe. *V*7 contains $2^{2^{65536}}$. So, the fin. ite stages of the cumulative hierarchy cannot be written down explicitly after stage 5 (see Wikipedia https://en.wikipedia.org/wiki/Von_Neumann_universe, 2018-01-03). The set $V\omega$ has the same cardinality as ω. The set $V\omega+1$ has the same cardinality as the set of real numbers (as per Cantor's proof $|\mathbf{R}| = |P(\omega)| = 2\omega$, and as we saw above $|V\omega+1| = 2|V\omega| = 2\omega$, therefore, $|\mathbf{R}| = |V\omega+1|$).

29. The sets asserted to exist in such axioms—namely, sets larger than any whose existence can be proved from the ZF axioms—are often known simply as large cardinals.

30. See https://en.wikipedia.org/wiki/Axiom_of_choice, 2018-01-03.

31. Gregory H. Moore, *Zermelo's Axiom of Choice, Its origins, Development, & Influence*, New York: Dover Publications, Inc. 2013, 3.

32. Stillwell, *Roads to Infinity*, 171.

33. Stillwell, *Roads to Infinity*, 173.

34. Stillwell, *Roads to Infinity*, 176.

35. Stillwell, *Roads to Infinity*, 176.

36. Stillwell, *Roads to Infinity*, 181.

37. Badiou calls this "impotence of the rule" (Badiou, *Being and Event*, 146).

38. Badiou, *Theory of the Subject*, 226.

39. Jacques Lacan, *Le Seminaire XXII de Juacques Lacan R.S.I. Annee 1974–1975*, Translated by Jack W. Stone, xx.

40. Lacan, *Le Seminaire XXII*, xx.

41. See "Meditation Thirteen," Badiou, *Being and Event*, 142–149.

42. Badiou, *Being and Event*, 147.

43. Badiou, *Being and Event*, 148.

44. For reasons that will become clearer shortly *Being and Event* re-posits and resolve all three questions through axioms and the concept of generic.

45. See the preliminary remarks in chapter 6.

46. As discussed in the last chapter, as a logic not only structuralism but also dialectic is a regionalization of the discipline by foregoing the requirement for the beginning and only retaining the requirements of interiority and novelty.

47. In particular we consider W to be an ordinal in von Neumann's sense, that is, a transitive set all of whose elements are transitive. W is thus a set whose elements are all of the elements of W and W itself.

48. Badiou, *Being and Event*, 293.

49. Badiou, *Being and Event*, 299.

50. The majority of working mathematicians don't really care if $\boldsymbol{L} = \boldsymbol{V}$, because their everyday work is likely independent of this problem—and the nice aspect of mathematics is precisely this "regional" property that we talked about before—the acceptance or refusal of one axiom doesn't "ripple" to all other areas of math, so mathematicians don't need to care about every axiom.

51. Badiou, *Being and Event*, 310.

52. We prefer the usage of G for the generic over Badiou's usage of symbol ♀.

53. Badiou, *Being and Event*, 338.

54. In appendix 5 of *Being and Event* Badiou clearly states the absolute terms are those terms that are true for both inhabitant and ontologist at the same time. As Badiou mentions, the ontologist has the vantage point of being "outside" the situation. Accordingly, it can be inferred that there are a number of terms that may be veridical for an ontologist and not veridical for an inhabitant of the situation. This is an important property in the subsequent discussions. As we have discussed in this work, this sense of "outside" is precisely from the inhabitant's frame of reference: ontologist can see outside through a "prophetic vision." But, as we have seen in detail, ontology is that whose practice is of absolute interiority, and as such the ontologist is the one who is not outside of anything, because from the standpoint of ontology there is no "outside." In other words, ontology is a situation that has no state: the relation between mathematics and meta-mathematics is a mathematical relationship. The modeling of a mathematical universe is itself a mathematical object. In the terminology that we have developed here, ontology is the discipline whose composing and disposing consist of the same domain (same data and operation). This is an important distinction. The vantage point of ontologist is not that she transcends the situation: that she is within a larger model, that is, $P(S)$ instead of S. The vantage point of the ontologist is that she thinks the beginning axiomatically, and that mode of thinking gives her access to a sort of maximality according to which she can construct models that are bigger than what is given. In contrast, the inhabitant thinks according to what is allowed by the knowledge of the situation and what it allows her to access. To the inhabitant the ontologist sounds prophetic and transcendental because what is "absolute" for the inhabitant (the situation, the knowledge and its real) is "relative" to the ontologist.

55. Although the debate between Badiou and Heidegger has not been a focus of this study, we have hinted at this debate often enough to warrant us to add one more

thought. The idea of infinite encapsulates the difference between the two philosophical systems: in our time the asymptotical oneness of being and logos, what lies at the center of the new emerging materialism, is what the idea of infinite helps to capture.

56. The same goes for the foundational status of thought of the One (everything that is, is one). One may also argue that finitude furnishes thought with the orientation that takes the conception of the One as foundational or operational for all thinking. Kant's conception of transcendental unity of apperception may be instrumental in this regard. The one of the finite "I" as the foundation for all apperception may be the psychological ground for the de facto, that is uncritically acclaimed, status of both finitude and the One. Lacan captures the "symptomatic" nature of the One in a different way in his *Seminar XIX*: "That there should be Oneness does not go without saying. It *seems* to go without saying because, for example, there are living beings, and each of you here, lined up in such orderly fashion, looks in every respect to be quite independent from the others and to constitute what in this day and age is called an *organic reality*, to hold together as an individual. It was here that entire first philosophy derived a certain support" (Jacques Lacan, *. . . or Worse, The Seminar of Jacques Lacan Book XIX, 1971–1972*, Cambridge: Polity Press 2018, 120). We won't attempt such a derivation here. But, although the concept of the One may also be thought in transfinite context (the One of Cosmos, the One of God, the One of Nature, the One of Life), such a kinship between finitude and the One, at which we are hinting herein, may explain why the model of infinity through reflectivity is identical for both the infinite One and the finite one.

57. See the introductory remarks to chapter 2 for a summary of Lazarus' program.

58. Lazarus, *Anthropology of the Name*, 48.

59. Lazarus, *Anthropology of the Name*, 50.

60. Lazarus, *Anthropology of the Name*, 60.

61. See the section "Senses of Infinity" in this chapter.

Chapter 8

Subject

Before we elaborate on the ways in which the idea of the infinite reconstructs the category of the subject, it is useful to understand why this category has such a central place in Badiou's thought. Indeed, why do we need a new thinking of the subject in the first place? Badiou's answer is that the subject is that operator of philosophy from which, in our time, the *compossibility* of Badiou's four generic procedures is drawn.

We recall from *Manifesto for Philosophy*[1] that, for Badiou, philosophy is the *space* for the compossibility of all four generic procedures: it is where these generic procedures together form a philosophical *vision* compatible with each particular condition. This means the following three things:

1. That there is an operation in philosophy that brings together the four conditions and helps us comprehend a certain *philosophical period.*
2. Such comprehension enables us to examine the philosophical period we are in, whose name, according to Badiou, is *modern philosophy.*
3. The period we call *modernity* has not ended, and thereby modern philosophy continues.

Philosophy has *conceptual operators* by which it *configures* its conditions. This configuration is what orients thought in each epoch, and it is the way philosophy thinks about a given epoch, as associated with a particular configuration. To understand these *operators of philosophy*, the first thing to note is that one of these conditions, or generic procedures, is always closest to an evental site and serves as the main referent for the deployment of the compossibility of the conditions.[2]

Such was the case for Plato, Badiou proclaims. There were two events within the situation relevant to Plato: the emergence of the city-state and the shift in the theory of sizes from the Pythagoreans, based on arithmetic, to the geometrical and continuous method of Eudoxus. Plato invented this configuration capability of philosophy and thereby created, for the first time, a space for both conditions: the operator of this configuration was the theory of Forms—operable across both politics and mathematics. At the same time, this operator bestowed a suspicious status on poetry and made love an indiscernible—that which is neither discourse (*logos*) nor knowledge (*epistémé*).

A philosophical period is one in which

> a certain configuration, specified by a dominant condition, persists. Throughout such a period, the operators of compossibility depend on this specification. A period creates a nexus out of the four generic procedures, in the singular post-evental state in which a generic procedure is inscribed into the space of thinking and circulation that philosophically serves as the determination of the time.[3]

In the case of Plato, Forms are under the domination of the matheme (and thus mathematics acts as the dominating condition over politics). Plato asserts this in his Republic by having the governor educated in arithmetic and geometry. At the same time, imitative poetry is kept at bay, as that form of art that demonstrates the flexibility of language to sustain sophistry.

The question for us is now whether there is a *modern* period of philosophy. The *postmodern* declaration in some way corroborates this: the modern period did exist and is now over, according to the postmodern philosophers. If so, then the question is: What was (or still is) the operator of the configuration of the modern period? To understand that, Badiou first enumerates three distinct episodes that should fall in such a period:

1. Europe's classical age, of Descartes and Leibniz, in which the mathematical condition was dominant under the Galilean event, whose intervention introduced the infinite into the matheme.
2. Europe's romantic age, of Rousseau and Hegel, in which the historic-political condition was dominant under the event of the French Revolution.
3. Europe's post-romantic age, of Nietzsche and Heidegger, in which art, with poetry at its heart, was the dominant condition. Badiou does not name an event for this episode, but mentions that the return to art was "through an anti-Platonic retroaction, in the operators by which philosophy designated our time as that of a forgetful nihilism."[4]

In this temporal sequence, we observe a movement of principle, from which the compossibility of the generic procedure is drawn, from the three conditions involved in these episodes: science dominant in the first, politics in the second, and art in the third. But within this displacement, there is still an invariant, from Descartes up to Nietzsche, Freud, Husserl, and Lacan. This invariant is the theme of the subject. This theme endured devastating destruction by Heidegger and his followers, but was subsequently recast within Marxism (the subject of revolution) and psychoanalysis (the subject of the clinic).

This means, for Badiou, that the category of the subject stands for the operator of the configuration that designates the period that we call *modernity*.

According to Badiou, such is the question of the fate of philosophy: Is our time up to the task of upholding and reinserting the category of the subject back into philosophical discourse, now dominated by linguistic and analytical traditions? Our thesis points to the new orientation of thought constituted by the Cantorian invention and axiomatic thought as that which overhauls the thinking of the subject, which we sketch by closely tracking it through the theory of discipline.

As much as the Cantorian revolution has caused a total reorientation of thought, in Badiou's eyes, another disorientation had already been at the essence of what he names the Age of Poets, whose central figures, from Hölderlin to Celan, began to think outside the categories of the object and the subject. But, there is a difference: while the Age of Poets did disorient thought by playing down the role of knowledge in the thinking of truths, today we need a reorientation. Therein lies the difference between Heidegger and Badiou. While for Heidegger the destitution of metaphysics, and the opposition of knowledge and truth, on their own, inform an orientation of thought according to which the philosopher will only need to harken to the poet, for Badiou there still remains *one more step*; a step that our poets were unable—or perhaps lacked the conviction—to take. This step consists of an effort to rethink in the lineage of Cartesian meditation. So, while a disorientation of thought had already been accomplished in at least continental Europe, the reorientation is something that remains a task for philosophy. In summary, philosophy must affirm that "Disorientation can be *conceptualized*."[5] Badiou maintains that the category of the subject is the key to this conceptualization and reorientation.

That is why, throughout his career, Badiou's interventions have always taken place around the question of the subject: What can philosophy say (or not say) about the subject? The category of the subject has been a (if not the) central category of Badiou's philosophy. Early in his career, and in reaction to the proposal that grounds science in general, and the foundation of mathematics in particular, on the repressed figure of the subject, in "Mark and Lack" Badiou affirms that science has no subject because science does not have a lack that it cannot fill itself. In other words, Miller is wrong to look for some place within the signifying chain in order to metaphysically or metaphorically establish the primitive of the Lacanian psychoanalytical subject. The discipline of science does not rely on something exterior to the discipline.

Having banished the subject from the discipline (of science), in *Theory of the Subject* Badiou ruminates on another *aporia*: How can we think novelty in a structure? This thinking took us from the analysis of Hegel's *Logic* and the relation of Something and Other to the notion of periodization. We saw that while Hegel's *Logic* contains resources to think novelty beyond mere circularity, and hence according to the schema of periodization, he does not have an explicit enough categorization of the two dialectics, which Badiou denoted as structural dialectic (SD) and historical dialectic (HD). Badiou finds a more advanced theorization in Lacan. In this regard, he thinks that the great dialectician after Hegel was not Marx, Lenin, Sartre, or Althusser, but Lacan. According to late Lacan, an understanding of jouissance and the signifier requires a logic similar to what Badiou calls HD, in that the relation of these two is no longer that of a situation (language) and the situated (speech), which is captured through the structural logic of whole-part. Instead, language itself is situated and a source of exception. Singularity and regularity are both generated through the practice of language. Therefore, the relation between jouissance and signifier is better captured through the relation of forces, which is the central grammar of the historical dialectic.

While Lacan's thought can make irregularity thinkable, its model is based on the clinic. Badiou takes this theory and transposes it to the model that supports politics: the State, masses, and classes. While this theory justifies irregularity as a force with the same legitimacy and potency as the force of structure, in the life of an individual subject in the clinic, the analysis must finally find a cure in the structure of one's symbolic constitution. By changing the model from a correlate of analysis to a correlate of politics, Badiou expands the scope in which the concept of consistency emanates and operates. In the Lacanian model, consistency belongs to the imaginary register: while the knot is in the real, its effect is registered in the imaginary as consistency. When the knot is cut, this effect of consistency is also destroyed, as the knot is the cause (in the real) and the consistency is the effect (in the imaginary). There is an inconsistency here: from the way the singularity of the jouissance comes to cohabit the very constitution of the regularity of the language—this regularity being in fact the retroactive effect of singularity—we come back to the primacy of regularity again, which is protected by the knot. But why such a compromise? The answer, in my view, is that the model operative in analysis is insufficient to sustain the radicalism of which the theory is capable. Using the model for politics, Badiou shows that the real of the masses is not only thinkable, it is the source of consistency, the source of a newfound regularity. What Lacan's thought was capable of—the relation of singularity and regularity—had to eventually recoil in the face of the weak model to which he had to apply it—a model that was too limited to sustain the consequences of what was possible to think. And here was precisely where the subject as the black sheep of materialism came into the picture: the real of the (political) subject (i.e., the masses) is the provider and guarantor of consistency.

What does the move from interiority to novelty tell us about the category of the subject? The logic of the signifier borrowed its conception of the subject from outside of the logic of structure. In other words, it performed an ideological construction of this category. Badiou dismissed this conception of the subject by maintaining that the structure is an absolute interiority whose consistency is not dependent on anything from outside. The logic of force, however, reinstitutes the subject, but this time within the constitution of the structure itself. This way, while maintaining the absolute interiority of structure, a gap is opened up within the structure itself which harbors the subject—the subject is anchored in the very gap between the surface of the law and the real of the forces that the law "in-discerns," but in whose qualitative distinctions it depends. Subjectivization constitutes the moment when this qualitative indiscernible erupts to the surface, and the subjective process is that which keeps this eruption at the level of the ongoing operations of the structure. Novelty is the consistency brought forth by the subjective process, and thereby belongs to the category of the subject. From this, we now understand why Badiou's salient philosophical undertaking during this phase of his career is captured within a book entitled *Theory of the Subject*: novelty is a category that requires a theory about the embedded subject. The subject is that which fastens the interiority and novelty within the boundary of a discipline.

Hence, for Badiou, the Age of Poets establishes that access to truth presupposes the destitution of the category of object as what presents the being: object, in this view, is a correlate of knowledge and not of being or truth. For Badiou, the task of philosophy is to sharpen the achievement of poets from Hölderlin to Celan, whose insights take philosophy in an opposite direction from what the philosophical avatar of that age, Heidegger, had pronounced. In fact, contrary to what Heidegger had surmised from the poets, the destitution of the object does not mean the destitution of the subject. Rather, it enjoins us to think the subject independently, without a vis-à-vis. For Badiou, the ability to think the subject without object consists of a possible renaissance for philosophy.[6] While Heidegger and the post-structural milieu were pronouncing the end of the subject, Marxism and psychoanalysis proposed its restoration. But both schools, ironically similar to Heidegger, missed the great insight of the poets and restored the subject at the expense of also restoring a form of object and objectivity (the proletariat and the object of desire). So, while their return to the category of the subject was warranted, their construction still owes too much to objectivity. For Badiou then, the task of philosophy involves a complete commitment to the withdrawal of the object, while traversing those theories that either abandon the subject along with the object, or buttress the subject while still being unable to forgo the object.

Below, we will examine one of these traversals, and it should come as no surprise that the philosopher whose thoughts are material to his task is none other than Lacan. Nor should it come as a surprise that, in our view, the main coordinates of this traversal are two pivotal thoughts: Badiou's commitment to the maximality of thinking and the idea of infinity.

Accordingly, we picked two encounters with Lacan involving the category of the subject, both definitive of what Badiou intends for this category. The first encounter is the very last meditation of *Being and Event*,[7] and the second is "Subject and Infinity."[8] Our reading of these texts will take place through our main theoretical apparatus: the discipline at whose center is the idea of infinity. Through this, it will become evident, during the first encounter, that Lacan's formulation of the subject of the unconscious owes too much to the wrong side of Descartes: it owes too much to the *I* in *I think*. The proposal of this chapter begins by showing that Lacan's theory of the subject relies too heavily on the count-as-one of an individual, and thereby it is not capable of overcoming the Sartrean reflexive subject. Disciplinary thinking shifts the emphasis from *I* to *think*, or more precisely, "being thinks." The real, which is the *it*, is still a derivative of Descartes', rather than a recomposition of thought from the ground up. In this sense, we will investigate the thesis that the true motivation of Descartes' *cogito ergo sum* is disciplinary thinking, in which the new category of the subject will arise.

The second encounter takes place around Lacan's treatment of infinity. Lacan actually had engaged with Cantor in *Seminar XIX*, but makes it very clear that, for him, actual infinities are imaginary. However, Cantor is still useful to Lacan: the concept of actual multiplicity, understood as a set, authorizes psychoanalysis to speak of what the latter calls a *unary trait*. The theory of sets thinks multiples without totalization; it only thinks the *unicity* of each multiple. Lacan points out how psychoanalysis is still stuck with Freudian doctrine, because it needs a representation of otherness. However, this requires a unifying otherness. So, we can never operate in the model of the psychic apparatus with the idea that something of otherness is captured by the structure. In Badiou's terminology, nothing of the outplace is preserved as an outplace by the place. We are authorized to move further and no longer think identification and representation only in these terms, because Cantorian mathematics has shown us that multiplicity without unity, and only with unicity, is perfectly thinkable. We can distinguish between multiplicities, without unintentionally unifying them. So, the extensive treatment of multiplicities authorizes us to think otherness without subsuming it as a finite multiple under a common property. Therefore, for Lacan, the gift of Cantor is the power to think unicity, which preserves otherness. This is what Lacan means by the unary trait, which states that something of the otherness is preserved by a trait. But, in *Seminar XX*, as we shall see shortly while reading Badiou's intervention, Lacan still maintains that the infinite must be thought of as inaccessible, something fictitious. The fact that we can think consistently about actual infinities and derive positive results from such thinking is totally out of Lacan's reach, not because it is unthinkable, but because Lacan has no name for that which is both inaccessible and consistent. This includes the fact that infinity is totally mis-measured by finitude, and is totally indiscernible from the standpoint of countable situations, but is still thinkable. These two positions, in successive seminars of Lacan, are, of course, inconsistent. Lacan accepts that it is possible to think the unicity of the multiplicity in extension (that is, in their existence), while claiming that we cannot accept actual infinity.

For this study, the Cartesian and Cantorian frameworks are combined: there is a way to think of the *cogito* and infinity within the same framework, which we call discipline. We claim that a rejuvenated understanding of the subject, in line with Badiou's work, involves a reconstruction of the subject using the idea of the infinite, and through this reconstruction we find that it is not the *I* or *it* that thinks, but the discipline itself.

Let us now elaborate on these points.

THE SUBJECT AND THE COGITO

For Lacan, the imperative to return to Freud was doubled by the directive to also return to Descartes. How can these two imperatives work together?, Badiou asks. He provides the answer as follows:

> The key to the matter resides in the statement that the subject of psychoanalysis is none other than the subject of science. This identity, however, can only be grasped by attempting to think the subject in its place. What localizes the subject is the point at which Freud can only be understood within the heritage of the Cartesian gesture, and at which he subverts, via dislocation, the latter's pure coincidence with self, its reflexive transparency.[9]

Understanding the subject in its place is the key to understanding how the subject of the unconscious can only be understood as the *cogito*. How does Lacan link the Cartesian subject and the subject of psychoanalysis? What does the term "subject" mean for Lacan? He provides a thesis: the subject is the Cartesian subject, that is, the subject of the *cogito* or the statement "I think." Lacan makes this point explicit in his *Seminar XI*:[10]

> I dare to state as a truth that the Freudian field was possible only a certain time after the emergence of the Cartesian subject, in so far as modern science began only after Descartes made his inaugural step.
>
> It is on this step that depends the fact that one can call upon the subject to re-enter himself in the unconscious—for, after all, it is important to know *who* one is calling. It is not the soul, either mortal or immortal, which has been with us for so long, nor some shade, some double, some phantom, nor even some supposed psycho-spherical shell, the locus of the defenses and other such simplified notions. It is the subject who is called—there is only he, therefore, who can be chosen.
>
> In order to understand the Freudian concepts, one must set out on the basis that it is the subject who is called—the subject of Cartesian origin.[11]

Let us first make four quick remarks about the *cogito* that we think are salient to the present discussion.

First, *cogito* is the subject of thought. It is only because the subject thinks that it is certain about itself. In addition, *cogito* is a subject of certainty. It is certain about its own existence, not about its essential being, but its existence as it presents itself in thinking: the mere fact of thinking is what the subject as *cogito* is certain about.

Second, the subject of *cogito*, which is the subject of thought and certainty, is not a subject of truth. The *cogito* suspends any considerations of truth: the subject's thoughts can be true or false, they can be hallucinations, dreams, or mistakes. Regardless, the truth status of these thoughts does not change the operation "I think."

Third, the statement "I think therefore I am" rips away any particularity of what a subject is. A subject *is*, insofar as it *thinks*, and that eliminates all qualities and marks of empirical or substantial individuality, whether corporeal or non-corporeal. This emptiness at the core of the subject, marked by "I think therefore I am," is what makes it possible for it to engage with/in the world through a process that is devoid of any presuppositions—the process that we call the modern scientific process. In this sense, *cogito* is the true subject of science.

Fourth, the statement "I think therefore I am" enunciates the fact that, insofar as I cannot doubt that the process of thinking is occurring, I am certain that I am. But the existence that is being affirmed is not the existence of a substance, but a process, since at this point, Descartes has not yet deduced a substance form in "I am." The Cartesian subject is thus the subject of the process of thought (the subject of enunciation) rather than the subject in the Aristotelian sense (the subject *I* in the statement "I am," or the subject of a statement) that is backed by a substance. The subject is *in* the enunciation, not *in* the statement.

What then are the grounds for Lacan claiming that the subject of psychoanalysis is the same as the subject of science, which is the Cartesian subject? In what sense is Freud Cartesian?

At the outset, we observe that the subject of psychoanalysis, what the practice refers to as the analysand—the person who asks for therapy—seems to be very different from the subject of the *cogito*. An analysand is firstly the one who suffers: if there is a *cogito* for the analysand, it seems to be more like: "I suffer therefore I am"! It appears that the subject of psychoanalysis is the subject of affects, not thoughts. In addition, the analysand is not the subject of certainty. On the contrary, an analysand is someone who mostly doubts: she does not know what is happening to her. She suffers but does not know why. An analysand, one who says, "I suffer therefore I am," is therefore also a subject of doubt. The analysand who suffers and doubts about the cause of her suffering also wants to know about the true reasons behind this suffering. An analysand is therefore not ambivalent to the truth.

What does the Freudian process do for such an analysand? It transforms the subject of suffering into a subject of thought. This process is called *free association*, through which the analysand is to speak her thoughts; to say what she is otherwise silently thinking. Free association transforms the subject of suffering into a subject of thinking, which is a step closer to the subject of *cogito*.

But an asymmetry exists herein: as the therapy brings more of the subject's affects into spoken words, and transforms the subject of affect into the subject of thoughts, it does not necessarily bring about certainty. In fact, just the opposite! To the analysand, the thoughts that she speaks about during free association may remain incoherent, inconsistent, insignificant, and meaningless. The analysand doubts the truth of her thoughts, and here lies the asymmetry between Descartes and Freud: for psychoanalysis, the moment of doubt is the moment that the subject of the unconscious *comes to be*. This was the case with Freud's discovery of the unconscious itself: it was his doubts about his dream thoughts that first suggested to him the existence and efficacy of the unconscious. Lacan mentions this point, and the parallel between Descartes' and Freud's experience, in the following way:

> In a precisely similar way, Freud, when he doubts—for they are his dreams, and it is he who, at the outset, doubts—is assured that a thought is there, which is unconscious, which means that it reveals itself as absent. As soon as he comes to deal with others, it is to this place that he summons the I think through which the subject will reveal himself. In short, he is sure that this thought is there alone with all his I am, if I may put it like this, provided, and this is someone thinks in his place.
>
> It is here that the dissymmetry between Freud and Descartes is revealed. It is not in the initial method of certainty grounded on the subject. It stems from the fact

> that the subject is "at home" in this field of the unconscious. It is because Freud declares the certainty of the unconscious that the progress by which he changed the world for us was made.[12]

The dissymmetry pointed out by Lacan pertains to certainty: in psychoanalysis, the certainty is not found in the subject of thought, that is, in the analysand, but is situated in the Other, or the analyst. It is the analyst who is responsible for the presence of the unconscious.[13] Therefore, the difference between Descartes and Freud, according to Lacan, can be summarized as follows:

> We know, thanks to Freud, that the subject of the unconscious manifests itself, that it thinks before it attains certainty.[14]

In the end, the Cartesian subject is a conscious subject defined by its mastery over the process of thought. On the contrary, the psychoanalytic subject is subjected to language and the signifier. Both subjects exist insofar as they think, or more precisely, speak; both of their existences depend on speech and are in relation to the signifier, but these relations to the signifier are inverted with respect to each other. The relation of *cogito* is that of the mastery of the signifier, and the relation of the subject as unconscious is that of slavery to the signifier. In that sense, one may say that Freud is Cartesian insofar as he adopts the notion of the Cartesian subject, but it is a subverted subject.

There is another way in which the subject of the unconscious is a subversion of the subject of *cogito*: the existence of a substantial being becomes the existence of a process—the subversion of the *home* of the subject from the statement to the enunciation. This introduces a measure of ex-centricity, a divide, which Freud sought to exhibit in the relation of the subject to itself and to the whole experience of therapy. In *Écrits*,[15] Lacan highlights the difference between where I am presented by my thought (the subject of statement) versus where my thought is (the subject of enunciation), which is captured in his re-transcription of "cogito ergo sum": *ubi cogito, ibi sum.*[16]

The subject of psychoanalysis, therefore, looks like a generalization of the Cartesian subject: first, the two moments of thinking/doubting and being/certainty are not at the same *place* in therapy (as we shall shortly see in Lacan's usage of topological surfaces). Second, the analysand is not the only constituent: the object (or the analyst), the couch, and free association are also constituents of the subject. Third, the relation of the analysand to analyst changes as the process evolves. So, although psychoanalysis still works within the overall framework of the Cartesian subject, it makes this architecture more complex.

So, when Badiou says that Freud can only be understood (according to Lacan) "within the heritage of the Cartesian gesture, and at which he subverts, via dislocation, the latter's pure coincidence with self," we understand this sense of *within* according to the above structure that not only *subverts* but at the same time *generalizes* what is meant by the category of the subject.

> What renders the cogito irrefutable is the form, that one may give it, in which the "where" insists: "Cogito ergo sum" ubi cogito, ibi sum. The point of the subject is that there where it is thought that thinking it must be, it is. The connection between being and place founds the radical existence of [the] enunciation of [the] subject.[17]

Badiou continues:

> The subject thus finds itself ex-centered from the place of transparency in which it pronounces itself to be: yet one is not obliged to read into this a complete rupture with Descartes. Lacan signals that he "does not misrecognize" that the conscious certitude of existence, at the center of the cogito, is not immanent, but rather transcendent. "Transcendent" because the subject cannot coincide with the line of identification proposed to it by this certitude. The subject is rather that latter's *empty* waste.[18]

The moment, or place, of certitude is where thought coincides with being. But that is not an immanent operation, Badiou claims. How does this stand with Lacan?

Lacan, throughout his early carrier after discovering Freud, was in search of a framework for articulating Freud's discoveries on a solid basis that would be scientific, while also non-humanist and non-essentialist. This leads Lacan to the concept of the *split subject*. As Badiou remarks, Lacan's "Science and Truth"[19] is the primary source of this theoretical innovation and serves to highlight the important topological characteristics of the split subject. It is in this text that Lacan examines the status of the split subject as the true conceptual heir to the Freudian discovery of the unconscious, and how the rigor of the Freudian legacy might be maintained through this particular advancement. This, of course, amounts to a different reading of Freud's corpus, in particular from ego psychology, popular at the time and recognized by many of Freud's followers as where his legacy should reside.

Lacan proposes that Freud's work should be read in anachronological order (from later to earlier): from *Ichsplatung*, to the articles on fetishism (1927) and the loss of reality (1924). This allows the reader to grasp how the theory of the divided subject is the groundwork for the topography introduced through the terms (which Lacan opposes): *Ich*, *Über-Ich*, and *Es*. Lacan introduces the concept/term of "subject" to assist with the navigation of concepts introduced by Freud that, due to their lack of clarity, contributed to the *aporia* of depth psychology. Lacan professes that he borrowed the term "subject" from structuralism so that he could avoid the tendency to substantialize, or to confuse it with brain, mind, ego, or similar substance-like constructs. He asserts that Freud's second topography (ego, superego, and id) is best understood by thinking it through the concept of *subject.*

Freud's scienticism, although not completely adopted by Lacan, has very important consequences for the fate of psychoanalysis, one that should be preserved and continued. Lacan says that it is impossible to imagine psychoanalysis as a practice and the Freudian unconscious as a discovery without the birth of science in the seventeenth century, the century of geniuses. Lacan seems to allude to a science which could be applied to psychoanalysis, but one with a notion of subject (as a split between knowledge and truth and as the "two sides" of the Möbius strip). This notion will render irrelevant all humanist references in psychoanalysis.

As a practice whose main focus is human suffering, psychoanalysis should be fraught with humanist references, distinguishing it from the *hard* sciences. But Lacan does not like this proximal notion, and thereby stresses that "humanist references become superfluous in science, the subject cutting them short."[20] This portrays the tightrope that Lacan walks between humanist and scientific domains, and, at least in this particular text, in relation to the Freudian legacy as well as reasons that are his own, he approximates psychoanalysis more toward science than humanism. As evidence, his references to two historical anecdotes show how a scientific posture for psychoanalysis (which Freud was also in favor of) helped psychoanalysis gain popular acceptance, and furthermore, inspired Freud to look for a scientific formulation of human behavior, which led him to the discovery of the unconscious. Lacan sees the root of the latter inspiration in the scientific tendency to formulate all physiology, both mental and physical, by the laws of thermodynamics. Lacan persists on a certain path that, in his view, helped Freud investigate the unconscious with scientific rigor. This is the same path that set Freud apart from Jung, and from ego psychology, which was popular among American Freudians at his time. Lacan finds it important to draw a line with these tendencies, and suggests an alternative formulation from another field of study: linguistics. Borrowing from the structuralist approach, Lacan introduces the subject, around which the unconscious can be elaborated and further developed, with a rigor analogous to scientific method, and without recourse to an *object*, which is the way in which both scientific (physics, chemistry, etc.) and non-scientific (Jung, ego psychology, etc.) approaches tended to conduct their studies.

Lacan specifically distinguishes the subject of enunciation from the subject of science, something only discernible from within the context of linguistics, and this is why he promotes this as a model for psychoanalysis. How can this similarity be mapped? Here is a short suggestion by Lacan:

> It is in the realm of logic that the theory's various refractive indices appear in relation to the subject of science. They differ as regards the lexicon, syntactic morphemes, and sentential syntax.[21]

In other words, in linguistics, there is recognition at the level of the basic operators of discourse (lexicon, syntax, etc.) with relation to the subject of enunciation and the subject of science (which we take to be the subject within the statement).

It is also important to point out that a science can be understood finally as a text about a particular subject. The subject appears in this text, the same way that *I* appears in "I think therefore I am." The *I* in this text is "the subject of science" partly because it is what a text is about—that is, when a physics text discusses objects in motion, for example. Science is ambivalent to enunciation; what is important is the text itself. Compared to the sciences, the position of psychoanalysis is peculiar, as it primarily engages with the textual material of speech while its proper subject is the analysand. As such, while the speech (or text) is where the subject of other sciences will solely reside, the subject of psychoanalysis is where the speech is enunciated from.

Lacan's ingenious proposal is that there is nothing but speech, so if there is a *where* for the subject of enunciation, it is speech itself (so there is no depth psychology, no mind, soul, etc.). The subject of enunciation can only be found in what has been enunciated. This sounds enigmatic: How can that which enunciates be in the enunciated? Here is where Lacan's genius strikes again, by suggesting that the structure of speech should be viewed as a Möbius strip: speech is structured such that the two subjects (the subject of the statement and the subject of speech) constitute the two sides of this strip, and the split subject is the very split of the two surfaces. Psychoanalysis is the practice that is capable of analyzing the two-sidedness of speech, so that the subject of enunciation can be traced there.

How does psychoanalysis do this? This is where another of Lacan's famous formulations enters. Lacan declares that the unconscious is structured like language: that is, the unconscious is subject to linguistic operations. Therefore, the *analysis* in psychoanalysis is the practice of decoding those linguistic operations that allow us to recognize the specific discourse produced by the analysand's unconscious, which is incidentally intermixed with the analysand's conscious speech. That is why, as we pointed out in the foregoing, psychoanalysis employs the method of free association. In this method, the analysand speaks freely of thoughts as they come to her mind, and through the course of analysis, the analyst begins to recognize those patterns that express unconscious thoughts.

We finally come back to this: both conscious and unconscious thoughts, both subjects, are found in speech, in the same text that is the vocation of the sciences, as the *I* appears in "I think therefore I am." This is perhaps yet another way to understand why the subject of psychoanalysis is finally the same subject of the sciences. And this is how, for psychoanalysis, truth and knowledge are finally sutured within the same medium. Science can never suture knowledge and truth: this much is certain following Gödel's proof. In Lacan's model, truth and knowledge are seen as two sides of the same strip, traversable through speech, and no longer localized within the same *domain* and thus subject to Gödel's incompleteness theorem: topology sutures what arithmetic could not. This is what Badiou alludes to in the quote above: certitude, knowledge about a subject can never be immanent to *where* the truth of the subject is—this truth cannot be obtained through or culminate in the former certitude—they belong to two sides of the strip only sutured through speech: certainty is never the vocation of the subject of unconscious.

The Lacanian subject is, therefore, a topological structure, where the truth of the subject is a void subtracted of all knowledge, yet maintained in a structure that is intertwined with the same conscious subject. This is what a split subject is, the split between (the subject of) truth and (the subject of) knowledge (if you cannot unite them, as per Gödel, keep them separate, but within one structure, aka a Möbius strip).

Lacan, in his *Seminar XIV* and *Seminar XV*, developed a schema to describe the relation of thinking and existing, as in the statement "I think therefore I am."[22] In this schema, the two sides of the statement, that is, *I think* and *I am*, have no intersection: no part of one belongs to the other; in other words, the thinking being is without existence and the existent being does not think.[23] It is impossible to have both thinking and existing at the same time. This schema is a way, in a logical form, to account for the torsion within the structure of the split subject, which became thinkable through the topology of the Möbius strip. It is easy to see how the split subject supports the modalities of *I am*, underwriting conscious knowledge and the certitude about who the subject (of the statement) is, and *I think*, underwriting the signifier, the unconscious thought, and the truth of the subject (of enunciation) at the same time. In fact, the only issue that remains concerns *therefore*, there is no relation between the two, other than both belonging to the same structural torsion that Lacan calls the split subject. This sense of *therefore* will not find any place in this logical interpretation of the Cartesian subject. Below, we will discover that the elimination of *therefore* proves to be an important step in positioning the torsion of thinking and being, not inside the subject, but inside the discipline.

Now, when Badiou emphasizes the localization of the void, he absolutely targets the center of this debate, because Lacan's project is precisely an attempt to suture the truth of the Cartesian subject (i.e., the void of knowledge, as we said before, not merely in "I think," but the mere process of thinking, that is, the pure act of enunciation) with the knowledge of *I am*, which is the conscious subject. In this sense, Lacan's project of split-subject could be understood as the coming to maturity of the Cartesian project in a pure formal structure. Why is it important for Lacan to suture the void and knowledge into a subjective structure? The answer seems obvious: because of the alternative results in various kinds of depth psychology; the void is placed either in a chemical/biological brain, or in transcendental substances such as the mind, the soul, pure practical reason, and common consciousness.

At this point then, we are faced with a decision as to the location of thinking. Descartes originally attributed this to the Cartesian ego as the human subject, which is entirely transparent and transcendental. With Freud, we find that the thinking *thing* is not precisely in the same place as that to which the statement of certainty refers. So, we at least have two places: a place of thinking (enunciation, doubt) and a place of being (statement, certainty). Lacan provided an architecture for this *spaltung* based on the structuralist theory of language. Speech is where both certainty and doubt, being and thinking, are woven together, whose topological paradigm could be thought in terms of a Möbius strip. In this architecture, the subject is the very split between thinking and being as the two surfaces of the strip. The two sides share nothing with each other: there is an empty lacuna that acts as the mere differentiation between being and thinking, which Lacan dubs the subject.

HOW DOES THIS MODEL MEASURE UP TO OUR DISCIPLINARY REQUIREMENTS?

What Lacan accomplishes is extraordinary as he avoided the metaphysical and humanist presumptions looming behind the subject from Descartes to Sartre, and at the same time he was able to extend structural causality by distinguishing the law from the cause. This opened up a gap within the structure between the automatic unfolding of the signifier (*automaton*) and the agency that evades formalization (*Tuchè*)—which we could name subjective causality, to distinguish it from structural causality. Despite this, however, Lacan's notion of the subject, similar to his forerunners, still remained sutured to the I of I think. Speech, which is ultimately the place of the subject, is ineluctably one's speech: for Descartes, as well as for Freud, Sartre, Husserl, and Lacan, thinking is ultimately sutured to the human individual. Miller, as seen at the very beginning of this study, had realized this and attempted to generalize the category of the subject such that it becomes the latent and repressed core of the logic of structure. According to the latter thematization, and inspired by Lacan's innovations, Miller argued that not just one's speech but any discourse, such as science or politics, is structured around a repressed kernel. Structure itself is not-all. Every structure forecloses or bars an element that paradoxically makes the structure consistent. Badiou uses this important discovery by Lacan: structure is not the place of consistency and stability; it is precisely the place of instability. This instability is the main reason for the metonymy of signifiers—inaugurated by a traumatic loss of jouissance. Accordingly, the subject is not constitutive of just an un-symbolizable excrement of the structure, but as Žižek puts it,

> "subject" designates the contingency of an Act that sustains the very ontological order of being. "Subject" does not open up a hole in the full order of Being: "subject" is the contingent-excessive gesture that constitutes the very universal order of Being.[24]

We call Lacan's position *modest* because we think Lacan was justified in putting a disciplinary constraint on the reconstruction of the concept of cogito as a mere differentiation with a claim to universality. That is, ultimately, Lacan chooses to stick to *I think*, with the emphasis on the *I*, for the same reason that psychoanalysis is the only rigorous theoretical field where the statement "I love you" is relevant. The *I* in "I love you" connects to individuals. This move is justified precisely for the same reason that, in the procedure of love, the reference to the individual body still counts in a way that does not in any other procedure. So, Lacan is correct in asserting that if we care how people come to be desiring subjects, how they come to inhabit their bodies, and how they come to desire other bodies, as psychoanalysis does, we must remain faithful to the idea of the Cartesian subject. Because of this, the I is a relevant concept, and a more general construction of the concept should not erase it, as psychoanalysis in the end is concerned with how thinking and the body encounter each other in an individual. We call this a disciplinary constraint because it makes the practice of psychoanalysis possible. In other words, seen through the lens of the disciplinary theory in which we are engaged, psychoanalysis as a practice is that which subjectifies. It is not the case, however, that the subject in the clinic involves only the analysand. Rather, it involves the whole experience of the analysis, including the sessions, the free association, the analyst, the transference, the theory, as well as the analysand. In other words, thinking and being reside in a situation that we call analysis, or in our parlance, within the concrete discipline we call the psychoanalytical situation. The *psychoanalytic* cure is therefore not the end of the discipline per se, but rather it only eliminates the dependency of the disciplinary subject on the *external* facets of the analysis: the sessions and the analyst. This is where the analysand is now capable of participating in the subjective experience, which is the discipline itself. That is where the constraint put on the individual human body becomes material to the practice of psychoanalysis—it becomes the boundary of what will have been the elusive cure.

The issue that arises is thereby not the constraints set in order to circumscribe the disciplinary practice, but rather concerns the generalization of the constraints as a rule of thinking as such. In Badiou's parlance, it is the suturing of thought to a particular condition of truth. This suturing, although heavily suggested by Lacan himself throughout his work, was never transformed by him into a *program*, but, by his exceptional reader, his son-in-law-to-be, and the future heir to his theoretical legacy, Jacques-Alain Miller.

The other, and more important issue, regarding this generalization, which makes the conception of the subject in psychoanalysis unsuitable for us, is that the suturing of the subject to the chain of signifiers has the inaugural assumption of a lost or impossible object. Why is this important? Because the conception of the subject for both Lacan and Miller, as the case was for the philosophers before them, is effectively mediated through the object, and this does not bode well from either the perspective of axiomatic thought, or Badiou's perspective, which holds philosophy responsible for thinking our era through the category of an objectless subject.[25]

In fact, the conception of an objectless subject is precisely what the theory of discipline posits: the discipline is the cogito; it is the discipline that thinks, not the I. It is only within the movement of thought, whose coordinates traverse actors, artifacts, assumptions, theories, discourses, and experimentations spread across a disciplinary landscape, that subjective thinking can take place. The discipline thinks its own being through axioms, which means disciplines can think their beginnings. As we have discussed in detail in the previous section, the main difference between dialectical and disciplinary thinking is that a discipline can think its own beginning, but a dialectic cannot—as axioms are the only viable means of thinking the beginning, through a violent rupture from presumptions based on opinions. And it is precisely through the thinking of beginning that the thinking of truths finds legitimacy, as the latter goes beyond what the discipline could think through its beginning: thinking truths is precisely thinking beyond the point of the impasse of the beginning. But that is the other aspect of axiomatic thinking: that it can summon itself to carry out the task yet again. An example of this convocation under the event of infinity is Zermelo–Fraenkel set theory (ZF). ZF (which includes the axiom of Infinity) as a set of axioms could in fact recoil, as it did, in the face of the infinity event toward the foreclosure of infinity under the assumption of constructivism. This is the event operating as the cause. But, at the same time, the fidelity to the infinity event can orient thought toward the assumption of the generic, where event now operates as consistency. Ontology cannot decide between the two, itself falling short in the face of being's excess. This manifests the movement of axiomatic thought in two modalities: disposition and composition. When the discipline conveys a concept, such as the generic, it does so by positing a new generality that supplements the set of previously posited generalities that had carved the being of the discipline. In the compositional mode, however, the discipline experiments with the avowed generality using the material available to it in the situation of which it is a discipline. This we have dubbed the discussion of the token, or the material condition of a truth multiple.

For example, when the concept of the party in power, as a consequence of the Lenin event, recoiled to an organized State, as an abstraction of the socialist society after the October Revolution (event as cause), Mao offered a different conception of the socialist state and the relation of the Party to the masses in which the Party, as an abstraction of the State, was supplemented by the Cultural Revolution (event as consistency). At this level, the Cultural Revolution is a generality that merely supplemented the original set of axioms that defined the discipline of socialist revolutionary politics conveyed by Lenin. The Cultural Revolution is the conveyor of a new set of axioms (Lenin's party + the Cultural Revolution, akin to ZF + infinity). However, this generality was put to work in the real situation of Chinese society. This experimentation amounted to having every cadre tend to crops for six months. The implementation of the generality or idea is the compositional modality through which an instance of the truth is approximated within the discipline. The debate in the politburo, the organization of brigades, the new regiments for the cadres, were all part of the implementation of the supplemented generality toward the truth of the Leninist doctrine of the dictatorship of the proletariat, the creation of soviets, and the dismantling of the State.

On this basis, we propose to think of the objectless subject as the discipline in its compositional modality. In that sense, the cogito, the "I think," is the discipline when perceived compositionally. In contrast, the being of discipline is always its modality in conveyance; it is what the discipline conveys of itself. As a result, "I am" is the marker of the thinking of the discipline over its beginning, its thinking over its being. Using this grammar, the phrase "I think therefore I am" conjoins the two modalities of the discipline. This is, in my view, what Lazarus meant for the discipline of politics: politics thinks its own thought. This is also what Badiou said about the discipline of science when he proclaimed that there is no subject of science: there is no subject of science because primarily there is no object of science. Science is that thinking which aims for its own interiority—at least science in the Bachelardian sense. Badiou further claims that art and love are regions of thinking that manifest the same characteristics as politics and science.

Thought in this way, we construe Badiou's exposition of the subject in *Being and Event* within the above general disciplinary landscape. The relation of the subject to an event from one end, and to truth from the other, is best understood when we place that relation in the context of the dual modalities of the discipline. In that sense, Badiou has provided a more elaborate framework for how the compositional logic operates. In "Meditation 23," Badiou introduced the problem of *double origins* concerning procedures of fidelity. This problem referred to the naming of the event, which is the first step in an intervention. And the second is the operator of connection, the procedure of truth itself. The operator of fidelity emerges consecutively to an interventional nomination. Badiou calls this emergence of the operator of fidelity subjectivization. Thereby, subjectivization is an emergence of the Two: it is connected to the evental site through the intervention and it is connected to the situation through the operator of fidelity. Every intervention performs a naming, which puts into circulation a supernumerary name. But every such naming leads to a consecutive counting—a count that is different than the count-as-one of the presentation, and the count-as-one of the State. This counting is the second aspect of the subjectivization—the operator of fidelity. Subjectivization, designated by a proper name: St. Paul, Lenin, Cantor, is the coming together of a name, which names the event, and an operator of fidelity, which is the generic procedure of a simple encounter. However, subjectivization is neither only one nor the other, but something in excess. From the perspective of the situation and its state, the name has no significance. The insignificance of the proper name of the subject is an attestation to the fact that subjectivization is connected to the occurrence of the void. The Two, in our view, should be understood as a commitment to the maximality of thinking, which is *occasioned* by events and as that which invokes this maximality. This maximality is then *exercised* by experimentation in thinking. By being able to issue statements whose truth will go beyond the established system, thought demands and occasions its extension. Therefore, we see a very clear resemblance between Lautman's views with respect to the way that generalities are identified or named, and how new theories are developed based on the inauguration of those generalities. Here we see the clear role of subjectivization, from one end caused by an event (the cause) and from the other oriented toward a truth (consistency). Ontology does not speak of this relation. As we have repeated time and again, ontology deals with the being of the generic as an impasse of the knowledge of the situation: the being of truth in general, whereas subjectivation is where a singular truth comes to exist in a situation, in relation to the occurrence of an event. The being of truth is compatible with ontology: ontology does not speak of truth but does speak of the generic and its being.

But this much we knew from Lacan too: we already knew that the real has a being. If the generic is the name of the real as truth, what does the association of the generic and the real give us that we did not have before? How do we think more of the real by utilizing the name generic or indiscernible? In two ways: because the real is not just an impasse to the knowledge of a situation, it is the source of its novelty too—the situation could change and the real of the situation is the guarantor of the possibility of this change. Accordingly, the relation of the real and the interiority of the situation is not only of the category of cause, but also of the category of consistency. This attests to the connection of truth and consistency—whose inner relation belongs to the disciplinary relation of novelty and interiority. But, the generic, although providing a guarantee of the being of truth, does not speak of the existence of a truth in a given situation. Badiou's philosophical edifice, starting from the possibility of an event and the existence of its evental site, to the act of intervention and the procedures of fidelity, speaks to the existence of a truth associated with a particular evental site. We can also refer to the latter as the local possibility of a truth. So, when Badiou says the "local status of a generic procedure" has dependence on "a simple encounter," we understand the above sense of the locality of a truth, vs. the ontological globality of the being of the truth that pertains to every historical situation in which an event decrees its void.[26]

What constitutes a simple encounter? A finite set of enquiries, + or −, subsequent to the fixation of the name of an event, starting from its evental site. This procedure is so simple that it is non-prescriptive of the trajectory it should take: the encounter is completely aleatory. Although its effect is prescriptive (+/- results), its trajectory is not. That is how it is simple: it is an aleatory and unplanned encounter. The only thing that we can say about this encounter is that its trajectory starts from an evental site. Therefore, in its locality, we have a procedure whose genesis and progression are constituted by aleatory, non-prescriptive, and lawless encounters. Each encounter enriches the one or the other set of enquiries with the names of the elements according to whether or not they are affected by the name of the event.

From the perspective of this simple encounter, the truth (of that encounter) appears as an ideal: "a truth is the ideal assemblage of *all* the evaluations, it is a complete part of the situation."[27] We can perhaps say that the ontological being of truth as a *complete* generic subset of the situation appears to the locality of the above simple operations as an unachievable ideal.

This leads to a thought: If the subject is the discipline in its compositional mode, and if discipline is infinite, then what does the association of the discipline and the subject tell us about the subject? Is the subject finite or infinite? Here is where the conception of the locality is important. The subject is related to the locality of an event and the locality of a procedure. As everything local is finite, the locality of what determines the subject would mean that it is also a finite multiple. Using this conception of locality, Badiou provides a definition of the subject in which the subjection is *qualifiable* and *unqualifiable* at the same time:

> Thought in its operation, the subject is qualifiable, despite being singular: it can be resolved into a name (e_x) and an operator (□). Thought in its multiple-being, that is, as the terms which appear with their indexes in effective enquiries, the subject is unqualifiable, insofar as these terms are arbitrary with regard to the double qualification which is its own.[28]

What does being qualifiable mean? To me, the term "qualifiable" shall mean that the term "subject" is defined by a formula or predicate. The name guarantees the singularity of the subject: the singular name of the event is the source from which the subject inherits its singularity in the situation. The qualification (and thereby the predicative aspect) of the subject comes from the operation of fidelity. Furthermore, the matter of the subject comprises the terms submitted to enquiry at a given moment. Due to the aforementioned locality of the enquiry, the matter of the subject is unqualifiable. That is, the matter of the subject has no definable relation to the operation that indexes the terms of the enquiry (as + or -). This means that the multiple terms that constitute the matter of the subject is arbitrary and lawless. But the locality and thereby arbitrary nature of the enquiries that constitute the matter of the subject make the subject necessarily finite: locality means finitude, and every finite multiple is discernible and thereby belongs to the realm of knowledge. This means that the matter of the subject is not co-extensive with truth, despite the fact that "a truth proceeds solely via the assemblage of those enquiries," it always surpasses the finite support of the subjectivization that makes the truth proceed.[29]

The finitude of the subject is not in contention with the infinitude of the truth. The local configuration of an instance of the truth is always finite, because of its locality. But the truth, as an idea that supplements the thought of the discipline, is infinite in its extension and in how the discipline is conveyed.

What is the importance of all this? We believe that the category of the subject whose vicissitudes are discussed in *Being and Event* gives us the framework of how the discipline operates in its compositional modality, whereas ontology, and in general axiomatic thinking, constitutes the working of the discipline in its modality of conveyance. The discipline is a region of thinking that has no object. Its ground refers to no objectivization. The important claim is thus this: understanding the subject as what belongs to a modality of the discipline as a whole should give us a concept of the subject with no object. A non-referential and non-objective category.

The journey is quite amazing here: as the Cartesian subject is the original template according to which the psychoanalytical subject is constituted, the roots of the latter are evidently in philosophy, matured and ripened in psychoanalytical theory. Now we have a philosophy that takes the concept of the subject from psychoanalysis back into philosophy, by keeping the same commitment to formalism in defining its constitution.

This periodization is none other than the great philosophical exchange through the history of modern thought, and more specifically in our era between philosophy and psychoanalysis. For both, there is a truth, but whereas for philosophy the being of truth is trans-individual and eternal, for psychoanalysis, truth is singularly placed in the subject's unconscious. In addition, we now have a new possibility for conceptualizing the subject that emerges from the concept of discipline, from which we can also ingest Badiou's theorization of this concept from *Being and Event*.

But what does this theory, based on the concept of discipline, give us that the theory of the subject in *Being and Event* did not? Why do we need a different formalization aside from, or in addition to, the formalization that Badiou provides? The concept of discipline speaks to three crucial conceptual apparatuses in Badiou's enterprise: first, it speaks to his Platonism through the commitment to the maximality of thinking; second, it speaks to his fidelity to the Cantor event through the commitment to the idea of infinity; and third, to his revolutionary politics through the commitment to the category of the subject. The discipline comes together in these three apparatuses, which gives a new logic that is more general than the dialectic. This generality allows us to finally see that what the category of the subject names at the historical level (the operator of compossibility in modernity), discipline names at the ontological level, that is, the trans-historical level (the conditions of thinking that make the thought of interiority, novelty, and the beginning compossible in general). In a way, we could even think of the Platonic form as a local name whose global instance is discipline. Therefore, "discipline thinks" is a trans-historical statement, whose name in modernity is "subject thinks" (and "form thinks" in antiquity).

Therefore, we are quickly arriving at a point where, with the concept of discipline, we are able to merge the Cartesian and Cantorian frameworks toward a revived understanding of the cogito, in the tradition of Parmenides as discipline in composition. And this, we believe, is the proper reconceptualization of the Cartesian meditation after the Age of Poets.

THE SUBJECT AND INFINITY

Lacan, in his theory of sexuation, develops two subjective logics: man and woman, each with its own determinations. This grounds a singular truth in psychoanalysis: there is no sexual relationship. This theory is of utmost importance to us for two reasons: first, in the position of woman, we can effectively recognize a *partially* objectless determination of the subject. Furthermore, Lacan proceeds to introduce infinity as the proper determination of this objectless subject. The focus of this section is precisely the encounter between Badiou and Lacan regarding the sense of infinity employed here. This encounter takes place in a surprisingly obscure text called "Subject and Infinity,"[30] in which Badiou mostly comments on a single passage from Lacan's *Seminar XX* (Chapter VIII, entitled "Knowledge and Truth").[31] The text Badiou quotes is as follows:[32]

> In that logic, on the basis of the fact that one can write "not-every [*pas-tout*] x is inscribed in Φx," one deduces by way of implication that there is an x that contradicts it. But that is true on one sole condition, which is that, in the whole or the not-whole in question, we are dealing with the finite. Regarding that which is finite, there is not simply an implication but a strict equivalence. It is enough for there to be one that contradicts the universalizing formula for us to have to abolish that formula and transform it into a particular. The not-whole becomes the equivalent of that which, in Aristotelian logic, is enunciated on the basis of the particular. There is an exception. But we could, on the contrary, be dealing with the infinite. Then it is no longer from the perspective of extension that we must take up the not-whole [*pas-toute*]. When I say that woman is not-whole and that that is why I cannot say Woman, it is precisely because I raise the question [*je mets en question*] of a jouissance that, with respect to everything that can be used in the function Φx, is in the realm of the infinite.
>
> Now, as soon as you are dealing with an infinite set, you cannot posit that the not-whole implies the existence of something that is produced on the basis of a negation or contradiction. You can, at a pinch, posit it as an indeterminate existence. But, as we know from the extension of mathematical logic, from that mathematical logic which is qualified as intuitionist, to posit a "there exists," one must also be able to construct it, that is, know how to find where that existence is.[33]

Lacan affirms the Aristotelian conclusion that the negation of a universal statement is equivalent to a particular negative statement. If I say "not all apples are red," I imply that there is *at least one* apple that is not red. In logical notation:

$$\neg\,(\forall x){:}\ f(x) = \exists x{:}\ \neg f(x)$$

But Lacan puts a proviso on this. He says that this equivalency only works when x belongs to a finite set. In a finite set, when I posit $\neg\,(\forall x){:}\ f(x)$, I must be able to find an x such that I could say $\exists x{:}\ \neg f(x)$. But Lacan thinks that this cannot be the case if x belongs to an infinite set. In an infinite set, there is no guarantee that I could find such an x. At most, I can say that such an x is probable but not determinate. But why does that make this equivalency less true in the case of the infinite? This is where Lacan appeals to intuitionist logic. In intuitionist logic, the existence of something is only guaranteed if that something can be constructed or demonstrated in a mathematical proof. This points to an explicit disparity in Lacan's text when dealing with two different logics, Aristotelian and intuitionist, which we will return to later in this section.

But why does Lacan use this logical jargon? The issue is that the function Φ, which stands for castration, is a universal function in psychoanalysis. Castration is a process through which a speaking (i.e., human) subject is made. As such, if x is a subject, in psychoanalysis, then (analytically) $\Phi(x)$ holds (see figure 8.1). The issue is that if $\Phi(x)$ is true, then how can we at the same time assert $\neg (\forall x), \Phi(x)$?

When Lacan says woman is not-all, and on that basis he concludes that there is no *Woman*, it suggests that, while there is a universal genus of *Man* that all particular men fall under, *Woman* is not such a genus, there is no generalization under which particular women fall, and thus there is no *Woman*.[34]

The passage we cited above is from the eighth session of Lacan's *Seminar XX*. The passage was produced after the main lecture was over (apparently there was a quarter of an hour still remaining and Lacan continued talking). Lacan, who was excited about a lecture delivered by François Recanati four months prior to that date (during the first session, entitled *On Jouissance*), provides some remarks regarding "the schema Recanati had to erase earlier," which apparently had to do with Aristotelian logic and the topic of *not-all* (*pas-toute*):[35]

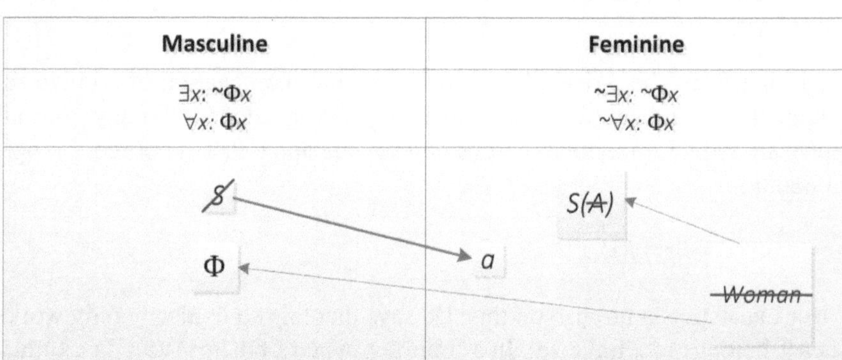

Figure 8.1 Lacan's Formulae of Sexuation. *Source:* Author created.

> This not-whole (*pas-toute*), in classical logic, seems to imply the existence of the One that constitutes (*fait*) an exception. Henceforth, it would be there that we would see the emergence in an abyss—and you will see why I qualify it thusly—of that existence, that at-least-one existence that, with regard to the function Φx, is inscribed in order to speak it (*s'inscrit pour la dire*). For the property of what is said is being, as I said earlier. But the property of the act of saying is to ex-sist in relation to any statement (*dit*) whatsoever.
>
> The question then arises whether, given a not-whole, an objection to the universal, something can result that would be enunciated as a particular that contradicts the universal—you can see that I am remaining here at the level of Aristotelian logic.[36]

It appears to us that Lacan's argument, comprised of both this and the previous quote, can be divided into three parts. First, he acknowledges that in Aristotelian logic, when one speaks of "not-all x satisfy $\Phi(x)$," one also implies that there exists an x that contradicts $\Phi(x)$. If I assume Φ is "is red," then the expression "not all apples are red" means that we can find an apple that is not red. Lacan then adds that this is only true if we are dealing with the finite. In the case of x belongs to a finite set, the existence of a particular x such that $\neg\Phi(x)$ is not only an implication, but also an equivalence. Such a condition is not necessary in predicate logic. In fact, as far as predicate logic is concerned, the equivalence of "not-every" and "there exists" is universally true regardless of the scope of x.[37]

In the next part of his argument, Lacan considers the possibility of x to select over an infinite domain, and he says the following:

> *then it is no longer from the perspective of extension that we must take up the not-whole* [*pas-toute*]. When I say that woman is not-whole and that that is why I cannot say *Woman*, it is precisely because I raise the question of a jouissance that, with respect to everything that can be used in the function Φx, is the realm of the infinite.[38]

Lacan makes a turn here. He says that, in dealing with an infinite, *not-whole* domain, the logic in question should no longer be approached from the perspective of *extension*. What does this mean? Here is a possible reading: When I say, "not all apples are red," I am referring to the sets of all apples. But when I say, "not all of an apple is red," the not-all does not refer extensionally; it refers to a single apple and says that not-all of that single apple is subject to the function *is-red*. It appears, solely based on this passage from Lacan, that the sense in which he wants to use the quantifier *not-all* suddenly shifts from an external scope (not every x) to an internal scope (not every part of x).

In the third part of the argument, Lacan is completely submerged in the discourse of the infinite: a point that pertains to the dialectic of existence and ex-sistence.[39] Lacan says this: if we say "not all of x is $\Phi(x)$" (based on an internal domain of x) and the domain is infinite (which is the case when Lacan speaks of woman's jouissance), we do not thereby conclude that there exists an x such that Φ is not true for it. In the context of jouissance, Lacan asserts that, unlike man's jouissance, not all of woman's jouissance is determined by the phallus: this is made obvious by the double arrow to the crossed-out *Woman* (Woman) in the diagram, which illustrates that ~~Woman~~ has two partners, the phallus and *S*(Ⱥ), the signifier of the inexistent/inconsistent Other. Therefore, the assertion that "not all x such that $\Phi(x)$," in my view, should mean that part of x, woman's jouissance, is determined by its relation to the signifier of the Other, but by writing *S*(Ⱥ), Lacan indicates that he does not affirm the existence of the Other. Intuitionist logic can afford that to Lacan: it allows us to keep *S*(Ⱥ) at the level of ex-sistence instead of existence. This reading is confirmed by the second passage we quoted from *Seminar XX*, which is just before the original passage quoted herein. A little before the passage we cited, Lacan says:

> Is there One or not? In other words, this not-whole (*pas-toute*), in classical logic, seems to imply the existence of the One that constitutes (*fait*) an exception.[40]

Therefore, when we say "not all x is $\Phi(x)$," the question is: are we confirming, as Aristotelian logic requires us to do, that "there exists an x such that $\neg\Phi(x)$"? We know that the Primal Father is the referent of the One in the above quote: there is a One who is not subject to the effect of castration. What is the existential status of the Primal Father? That is precisely what takes us to the next part of this quote:

> Henceforth, it would be there that we would see the emergence in an abyss—and you will see why I qualify it thusly—of that existence, that at-least-one existence that, with regard to the function Φx, is inscribed *in order to speak it* (*s'inscrit pour la dire*). *For the property of what is said is being, as I said earlier. But the property of the act of saying is to ex-sist in relation to any statement* (*dit*) whatsoever.[41]

What is implied by the not-all is precisely what allows us to speak of something that is not under Φ, but at the same time, this not-all does not force us to commit to "there exists," thanks to the intuitionist logic. In the end, we have a pure signifier for the Other, which has no existential commitment.

In his *Seminar XIX*,[42] Lacan had previously adopted a different strategy toward the same end, by elaborating on a proposition that has a similar construction: *Il y a de l'Un* ("There is something of the One," or "Sometimes there is One"), which he contracts as *Yad'lun*. After affirming the "Aristotelian angle" for the contradiction between "the statement that *for all x, x* fulfils in Φx the function of an argument, and the fact that there *some x* that can fulfil the place of argument only in the enunciation that is the exact negation of the former," he proceeds to say: "I'm daring to advance something that is plainly lacking in the aforementioned logic. I am able to do so to the extent that the term existence has quite certainly changed meaning since then." How has the sense of "existence" changed? When we say "*There exists an x, at-least-one*, lends a value that can be qualified as true to what is posited as a function." In this sense, we can say: "There is a distance between this and natural existence, which I will not name otherwise today, for want of a better word." This sense of existence is what Lacan attributes to the One, and it is why Lacan finds Cantor's invention of sets so useful: "the mainspring of set theory hinges entirely on how the One—that *there is*—of the set is distinct from the One of the element."[43] In Lacan's mind, set theory supplies a sense of one (or oneness) without having to ground it in the one of being, or the existential sense of oneness and sameness (as we have seen Plato had attributed to Parmenides). Instead, the commitment to the one is grounded in the being of multiples, and as in set theory, there are nothing but multiples:

> It is in this respect that it is, let's say, *inadequate* in the Platonic dialogue to make anything whatsoever of the *existent* participate in the realm of the *like-for-like*. Without the crossing-through whereby the One is initially constituted, the notion of *like-for-like* would not appear in any way whatsoever.[44]

So, when Lacan refers to "there exists," this does not refer to the sense of existence from Aristotelian predicate logic. Instead, borrowing from set theory, he is "daring to advance" the existential sense of the standard quantifiers.

This new "logic," proffering a sense of "existence" distinct from "natural existence," is the same sense that Lacan wishes to leverage when he speaks of the "infinite" field of woman's jouissance—the not-all. At the bottom part of the diagram, ~~Woman~~ is connected to Φ and $S(\not{A})$. $S(\not{A})$ is that not-all which, at the same time that it signifies the non-existing Other, plays a part in formulating feminine sexuality. Using the foregoing interpretation, the sexuality (jouissance) of every woman is subject to the effects of castration, but castration does not fully determine a woman's sexuality (jouissance): it is a necessary, but not sufficient, condition. This is obvious from the first formula on top of the right side of the above figure that says $\neg\exists x$: $\neg\Phi(x)$: there is no woman who is not affected by castration, which means all women are affected by castration, but that is not all that determines their sexuality (unlike men). The key to this interpretation is, as we have said, to posit an internally scoped sense of the quantifier. However, in the sense of not-all, "there exists" does not predicate the existence of a thing outside of the effect of castration. That is why Lacan takes recourse in intuitionist logic, since what he wants in the end is a logic where the affirmation of *not-all* does not result in a "there-exists":

> For an intuitionist, therefore, from the statement "not for every x Φ(x)" there is no reason to conclude that "there exists x such that not-Φ(x)." On this point intuitionism coincides perfectly with Lacan's aim.[45]

In intuitionist logic, the traditional concept of truth is qualified by the concept of constructive provability: a method of proof that works by demonstrating the existence of an object by creating or providing a method for creating it. Therefore, while predicate logic finds the proposition "not all x such that $\Phi(x)$" to be equivalent to the proposition "there exists an x such that $\neg\Phi(x)$," such that the truth of one will imply the truth of the other, intuitionist logic does not consider these equivalent. Therefore, the truth of the former proposition does not imply the same for the latter: the existential statement must be independently constructed. This is, as Badiou points out in his commentary on *Seminar XIX*, exactly the effect that Lacan is seeking: woman, the not-all, has a relation to the phallic function, and nothing more—somewhere means not everywhere; the somewhere—not everywhere—of the woman's position is expressed as not-whole.[46]

But is this a temporary alignment with intuitionist logic, or is Lacan an intuitionist? Badiou emphatically asserts the former. Lacan neither fully subscribes to intuitionist logic (e.g., the renunciation of *reductio ad absurdum* as per his approach in the interpretation of symptoms) nor renounces the existence of actual infinites (as per his usage of alephs). He even, in the first passage we quoted from *Seminar XX*, makes use of the Aristotelian logic of finite sets, something that an intuitionist would not agree with. The key that Lacan mistakenly sought in intuitionist logic does in fact lie on the side of the infinite. This confusing relationship with intuitionism points to another important matter as well. While intuitionism is not a branch of logic to which Gödel, Cohen, and Badiou subscribe, it is nevertheless one that many mathematicians have worked fruitfully within. This makes the debate between intuitionists and realists in mathematics an interdisciplinary debate, one that Lacan is not party to. Recall Badiou's detailed analysis of the ideological dispositions that are developed in relation to the sciences. As early as "Mark and Lack," Badiou had recognized two different representative stances, one transcendental to the sciences, such as epistemological theories, and the other immanent to a given science. The latter stance represents what a science thinks about itself, and despite being ideological, is part of that science—furthermore, science already contains the means of eradicating such internal ideological representations. Hilbert's program is evoked as such an ideological representation generated within the practice of mathematics itself—subsequently addressed by Gödel's inconsistency theorems. In contrast, we have ideological instances that are produced outside of a given science, those that do not pertain to the practice of the science itself. Miller's representation of the logic of the signifier is an example of this transcendental-ideological stance. Likewise, we could say that intuitionism is a representative disposition inside the practice of mathematics, and Lacan's usage of intuitionism, like Miller's, is an ideological disposition that is transcendental to the discipline of mathematics.

But there is more. While Miller's ideological usage of Frege's logic (itself ideological, but nevertheless intra-mathematical) amounts to a consistent program, Lacan's usage here is not so. As per Lacan's articulation, we are dealing with an infinite field of woman's jouissance. Somewhere in that infinite field, there is also phallic jouissance. But does not the existence of this somewhere, by way of negation, also affirm the existence of the everywhere of the infinite, the not-whole? Lacan's answer is no: the not-whole, the everywhere that is infinite, is too indeterminate for the finite to be circumscribed by phallic jouissance. This thought of Lacan's is seemingly confirmed by mathematics: there is no dialectic between the finite and the infinite—the infinite is the absolute alterity and inaccessibility with respect to the finite. The paradigm of the absolute alterity of the infinite, as we have formerly seen, renders the Hegelian dialectic of the finite and the infinite impossible—the infinite is not the negation of the finite, nor its sublimation, but its inaccessible determination. According to the paradigm of the mathematical infinite, phallic jouissance functions similarly as the succession operator, such that, despite its unending insistence (an unending chain of desires), a circumscribed finite jouissance will never be commensurable to the infinite not-whole; the same way that no whole number is commensurable to ω:

> This is a crucial adjective, both in set theory and in the logic of the formulas of sexuation. This enjoyment of the feminine not-whole is properly the inaccessible infinity in which castrated enjoyment is determined.[47]

But in axiomatic set theory, ω is an actual number, infinite in size, but actual nevertheless—that is, this number exists in the same way that the number 2 exists. But actuality is not what Lacan is after. It is enough for him to say that the infinite is an inaccessible horizon for the finite. This gives us a good explanation as to why feminine enjoyment ultimately has the structure of a fiction: it is a fiction of the inaccessible. From here stems the organic relation between this enjoyment and God. This is precisely where Lacan drops his association with Cantor:

> Thus, set theory is designed to restore the status of number. What proves that it does indeed restore it, within the perspective that I have been laying out, is that, setting out as it does the grounding of the One, and making number lean on this as a class of equivalence, set theory thereby manages to highlight what it calls the *non-denumerable*, which is very straightforward, as you are about to

> see. It's readily accessible, but to translate it into my vocabulary, I call it, no the *non-denumerable*—an object I would unhesitatingly qualify as mythical—but the *impossible to denumerate*.[48]

Lacan takes from set theory what he deems appropriate so he can "advance a new logic" to explain the non-existence of the sexual relationship. But the passage above completely shows his non-axiomatic comprehension of the denumerable. The axioms (of Separation, Power Set, Union, Empty Set, etc.) allow for the construction of the whole numbers. But none of these foregoing axioms are responsible for the construction of the denumerable itself, because the denumerable is not constructible. It exists according to a separate axiom: the axiom of Infinity—using Lacan's parlance above, ω is readily accessible axiomatically. Asserting the status of the denumerable as mythical is what makes Lacan a pre-Cantorian (and in our parlance, pre-disciplinary) in the end: for him (and this is why intuitionism sounds appealing to Lacan), the infinite is ultimately the paradigm of inaccessibility.

The latter point is accentuated by the way Lacan mobilizes Cantor in his treatment of the theory of identification. As we mentioned earlier, Lacan made use of the concept of sets in his *Seminar IX*, on identification, where he wants to move beyond Freud's general theory of how representations associate themselves through an economy based on the quantity of excitement. At this point, Lacan took advantage of a method different than his usual one. Freud was already able to show that what is significant are associations wired in one's psychic apparatus. In contrast, Lacan wanted to show that, before this can happen, one needs to bring into the psychic economy something that has no signification, which can represent the lack of representation, that is, desire itself. Therefore, the question is: How can one have a representative with no representation? A difference with no signification? The problematic that this approach opens, according to usual philosophical arguments, is that of the representation of nothing. Lacan needed to show that a relation to some unity, which is also not a representation of unification, is thinkable. This was not strictly a psychoanalytical problem, but rather a necessary stepping stone in Lacan's argument. At that moment, Lacan found his solution in Cantor's theory of multiples—because a set provides exactly this minimal measure of unicity without having to unify what it contains. With this step, Lacan mobilized Cantor in the correct sense: mathematics authorized Lacan to continue thinking maximally. From this, he proceeded to develop the concept of a unary trait, as a potentially denumerable set, as an addendum and extension to Freud, and to the existing psychoanalytical theories of the time.

This anecdote shows the positive influence that Lacan received from mathematics, set theory, and axiomatic thought. Here, Lacan does not yet condemn Cantor for having mythical or imaginary commitments. It was not psychoanalysis that was interpreting mathematics (as per Miller), but mathematics that was interpreting psychoanalysis. But there resides another point in this anecdote. We now know that the paradigm that Lacan wanted to move beyond is constructivism, that is, he wanted to move to the paradigm according to which a representation of a set does not have to be necessarily constituted by a representative, or a formula. We also know now that this is only possible by a commitment to the axiom of Infinity that is deeper than what constructivism could afford. In other words, what Lacan wanted to achieve is possible not just by a commitment to the axiom of Infinity, but also by a commitment to the genericity of denumerable sets: the movement we have been pursuing from Gödel to Cohen.

This finally takes us to the question crucial to our inquiry: What would it mean for the subject (in the analytical sense a woman) to have a determination that is of the order of the infinite, when this order is not merely fictitious and inaccessible? Let us see how Badiou answers this question, which summarizes the crux of our endeavor:

> Since feminine enjoyment is of the order of the infinite, does it not turn out that, rather than having the structure of a fiction, it has the structure of an axiom? A woman then would, as the condition of her enjoyment, have to decide the inaccessible as regards its existence. This axiomatic character of secondary enjoyment in no way contradicts its unutterable character. Granted, the axiom does state something, but the decision on this statement, the gesture by which it is inscribed are [sic!] not all stated in the axiom itself. The axiom does not express its axiomatic dimension, and so the decision that it is remains tacit. Silently, in the infinite element of her enjoyment, a woman would have decided that with respect to primary or phallic enjoyment there exists an inaccessible point that supplements its effect, and determines her as not-whole with regard to the function Φ. It is properly this silent decision that would forever block the sexual relationship from existing. For, summoned to the place of enjoyment, the position "man" and the position "woman" would continue to be separated by that layer (épaisseur) without substance (*épaisseur*) not even of the axiom, but of the axiomatic, and thus instituting, dimension of this axiom.[49]

What is important about the above *paradigm shift* (or a figure-ground change of perspective)? That the real of thinking—as what was thought of as impossible and impassable—becomes a positive category, from which new possibilities will arise: it is the pass of the real, the real as consistency and as generic.

We can interpret this change of perspective, a disciplinary operator which we call regionalization: this operator lays bare the presuppositions embedded in a certain theoretical edifice. This operator touches upon the themes of beginning, novelty, and interiority—and sits at the center of the theory of discipline. This operator has been utilized by Badiou, time and again: in his debate with Miller regarding the theory of discourse, with Lacan regarding the operation of subjectivation in "Logical Time," regarding the presuppositions of the doctrines of foundation and ground, and finally with regard to the cogito, those presuppositions that tie us to an objective determination of the subject. It is therefore fitting that we complete our last construction with yet another application of this operator by our master.

If the subject has a determination that is of the order of the infinite, and if we abandon the pre-Cantorian presupposition of the infinite as inaccessible—if we espouse the idea of the infinite as a (or the) positive category of thought (in our time), apropos the Cantorian revolution in mathematics—then this infinite determination does not merely touch upon the real as mythical or mystical, but as generic. The essence of the subject is no longer unreachable, but comprises an infinite truth. Following our construction of the subject as a non-relational category—subject minus the object—we come to realize that the determination of this subject is infinite, and this determination under the idea of infinity and axiomatic thought is nothing but an infinite truth.

If we agree with Badiou that the category of the subject in philosophy has acted as modernity's operator of compossibility—a constant theme that has been present from Descartes (science), to Hegel (politics), to Heidegger (art), and to Lacan (love)—and if the return of philosophy will depend on the reworking of this category after its destitution in the hands of the linguistic turn, then, in my view, this reworking after Badiou must occur through the thought of discipline, in whose center we find the commitment to the maximality of thinking, the idea of infinity and its capacity for regionalization.

NOTES

1. Alain Badiou, *Manifesto for Philosophy*, Albany: State University of New York Press, 1999.
2. Badiou, *Manifesto*, 41.
3. Badiou, *Manifesto*, 42.
4. Badiou, *Manifesto*, 43.
5. Badiou, *Manifesto*, 74.
6. Badiou, *Manifesto*, 93.
7. "Meditation Thirty-Seven: Descartes/Lacan," in Badiou, *Being and Event*, 431–435.
8. Alain Badiou, "Subject and Infinity," in *Conditions*, London: Continuum, 2008, 211–227.
9. Badiou, *Being and Event*, 431.

10. Jacques Lacan, *The Four Fundamental Concepts of Psychoanalysis, The Seminar of Jacques Lacan, Book XI*, London: W. W. Norton & Company, Inc., 1998.

11. Lacan, *Seminar XI*, 47.

12. Lacan, *Seminar XI*, 36.

13. The role of the Other in acquiring the certainty is also emphasized by Lacan in the case of the Cartesian *cogito*: Descartes, although he affirms the presence of the thinking subject, has to assume that the truth of even this affirmation is dependent on an Other who is not deceptive.

14. Lacan, *Seminar XI*, 37.

15. Jacques Lacan, *Écrits, the First Complete Edition in English*, London, W. W. Norton & Company, Inc., 1999.

16. Lacan, *Écrits*, 429.

17. Badiou, *Being and Event*, 431.

18. Badiou, *Being and Event*, 431–432.

19. Jacques Lacan, "Science and Truth," *in Écrits, the First Complete Edition in English*, 726–745. New York: W. W. Norton & Company, Inc. 1999.

20. Lacan, "Science and Truth," 728.

21. Lacan, "Science and Truth," 731.

22. Here we use Bruce Fink's representation of Lacan's schema. See Bruce Fink, *The Lacanian Subject, Between Language and Jouissance*, Princeton: Princeton University Press, 1995, 44–48.

23. Fink, and perhaps Lacan too, contrasts thinking with being, but we chose to contrast thinking with existing instead, for reasons that should be obvious.

24. Slavoj Žižek, *The Ticklish Subject, The Absent Centre of Political Ontology*, London: Verso, 2000, 160.

25. This assertion is certainly true for the subject of castration, as the universal operator of subjectivation for Lacan. But, in his theory of sexuation, Lacan does develop a conception of the subject that is not wholly determined by castration alone. Properly speaking, that subject could be considered to be an objectless subject. We will examine that sense of subjectivity in Lacan in the next section.

26. Badiou, *Being and Event*, 324.

27. Badiou, *Being and Event*, 324.

28. Badiou, *Being and Event*, 395.

29. Badiou, *Being and Event*, 395.

30. The original French title of this essay is "The Position of the Infinite in the Split of the Subject," which is a far more crucial and interesting title than the one chosen for its English publication. This title almost summarizes the topic of this study.

31. Jacques Lacan, *The Seminar of Jacques Lacan, Book XX, 1972–1973*, London: W. W. Norton & Company, Inc., 1999.

32. The marginal page number of the quote is 94 (which is the page number Badiou cites).

33. Lacan, *Seminar XX*, 211–212.

34. One can only be a man if one assumes that there is an exception to castration in the set called Other. *Woman* is (one of the names of) that exception, and so is actually part of the masculine logic. It is important to highlight the dependence men have on *Woman*, and how this dependence does not go both ways.

35. We are using "not-all," which is what is widely used in the literature, and "not-whole," which is used in Badiou's "Subject and Infinity" and *Seminar XX* as the translation of *pas-toute*, interchangeably.

36. Lacan, *Seminar XX*, 102.

37. For example, see https://en.wikipedia.org/wiki/First-order_logic#Provable_identities.

38. Lacan, *Seminar XX*, 211–212.

39. For Lacan, existence is only possible in the symbolic register. We may refer to things that do not exist (i.e., they are not within the symbolic register), but this reference does not mean that they exist. Lacan's term for the sense of being of things that we can speak of but do not exist (i.e., they are not in the symbolic register) is "ex-sist." Lacan borrows this term from Heidegger.

40. Lacan, *Seminar XX*, 102.

41. Lacan, *Seminar XX*, 102 (our emphasis).

42. Jacques Lacan, . . . *or Worse, The Seminars of Jacques Lacan: Book XIX, 1971–1972*, Cambridge, Polity Press, 2018.

43. Lacan, *Seminar XIX*, 120–124.

44. Lacan, *Seminar XIX*, 125.

45. Badiou, *Conditions*, 215.

46. Badiou, *Conditions*, 215.

47. Badiou, *Conditions*, 218.

48. Lacan, *Seminar XIX*, 124.

49. Badiou, *Conditions*, 211.

Conclusion

Thinking Discipline

Badiou has been thinking and writing for nearly six decades, beginning since he and other young intellectuals supported Sartre in defying the Algerian war in the 1950s and 1960s. The third volume of his major work, *Being and Event*, is under publication in France as we speak (*The Immanence of Truths*). Therefore, his intellectual career covers a lot of ground. Badiou developed his philosophy through his debates with many great thinkers as well. We've examined several of these interlocutors, including Miller (and, through him, Frege), Althusser (and through him Bachelard), Hegel, Lacan (and through him, Descartes and Freud), Spinoza, Heidegger, and Cantor. Badiou's philosophy also heavily relies on other thinkers, but from the perspective of our study, there are two relatively unknown thinkers who have had the greatest influence: Lazarus and Lautman (and through him, Plato).

From this vast terrain of Badiou's works, we chose three, each belonging to distinct episodes of his life: "Mark and Lack" from his structuralist and logicist era, *Theory of the Subject* from his political and revolutionary era, and *Being and Event* from his mature era. These three works contain the best expressions of Badiou's commitments, captured as a distinct fragment of that time. However, by reading these works together, we have discovered, in fact, that there is an internal thread of thought that runs between them. We named this thread discipline in our project. In other words, discipline is the framework within which we can reconstruct Badiou's main ideas as part of a continuous work. While discipline was originally a way for us to read Badiou and to help us to reveal the internal coherence of his overall thought, over the course of our journey, this framework has proven its value beyond merely didactic.

Philosophically, none of the works that we've reviewed seem compatible with each other. But assuming we were successful in drawing out elements of continuity according to the concept of discipline, it would be possible to conclude that this theory must be distinguishable from the individual philosophical architectures of these works. From this perspective, discipline is the name of that which is continuous in Badiou's thinking. The saturation of logic, the dialectic of SD and HD, or the problem of the One and multiple are all concepts that are brought to bear on the question: what is a discipline? And accordingly: Is there only one discipline, are there many disciplines, and are they compossible? In fact, the detection of this single thread inside all of these major works was what signaled to us that we should think of this continuity as being independent of the philosophical architecture of Badiou's works.

We first recognized the logic of discipline in Badiou's critique of Miller in "Mark and Lack," whose main point was to defend the interiority proper to science, in defiance of Miller's project for a logic of all logic. From Badiou's structuralist works to *Theory of the Subject*, we showed that he started to think about the problem of structuralism in dialectical terms. That was the move which allowed him to think about the interiority of politics. The pertinent question for us here was: Where is it that Badiou can make this criticism? By pursuing this line, the theory of discipline was not yet spelled out. Yet, we soon discovered the answer to be the standpoint of dialectic, in particular the dialectic of Something and Other, that provided the resources to understand the principles of structure and to develop two strands of this logic: SD and HD, that which preserves and that which changes a structure as an interiority, respectively. In this way, *Theory of the Subject* provided the required means to think of a structure as an interiority, and the logic that allowed Badiou to do this was dialectic. In working through the *Theory of the Subject*, we were able to show that this interiority of the dialectic leads to a regionalization of the structures: If structures, seen under the light of the dialectic of SD and HD, do have the interiority that Badiou had intended, then where do these interiorities come from, in terms of their regionalized reals, the possibility of their multitude, their compossibility, and so on? It is one thing to show the inner dialectic of the interiorization of structures, and it is a different thing to show how this interiorization is supported. To provide a critique of structuralism, and to show that there is historicity to structures, Badiou "moved back" from structuralism to dialectic. And in order to provide the regionalization of the dialectic itself, Badiou had also to step back to a space that allowed the critique of regionalized dialectics: ontology—but not only ontology—because many of the aspects that enable the dialectical property of regions of being do not belong to ontology proper. So, there is a need for an additional theory to provide this. That theory, as we've tried to show in this work, is discipline. In that sense, discipline is what justifies the very division of *Being and Event* into ontology and that-which-is-not-being-qua-being. Though procedures of fidelity are defined within the framework of *Being and Event*, the very organization of this approach into a theory of situations (amongst them, historical situations) and a theory of procedures (historical situations + fidelity) is itself a derivative of Badiou's prior concern with allowing the interiority of a discipline to be self-determining (meaning: it could not be determined by the ontology of disciplines itself). So many decisions which seem unconditioned in *Being and Event* are clarified if we realize Badiou was finding new means to solve an old riddle.

Discipline is, therefore, the being/thinking that is "prior" to dialectic and the historicity of multiple regions of being, the same way that dialectic was the logic that allowed historicity in a single region of being (i.e., a structure). Discipline is how we can conceptually preserve the continuity from "Mark and Lack" to the *Theory of the Subject* and to *Being and Event*.

This sense of "prior-ness" of discipline to the dialectic that the latter is a specific mode of being/thinking of the former can be expressed in a different way. While the question of interiority was so central to Badiou's thought, so was the question of beginning. In our investigation of the thought of beginning, we discussed its historical suturing with two other thoughts: that of the One and that of the foundation. This suture enables the overdetermination of the thought of beginning either by the question of the ground (Heidegger) or the question of the One (Hegel). In our view, it was only in the introduction of actual infinity into the situation of being that this overdetermination became visible. Axiomatic thinking for us is the way to de-suture the thought of beginning from the One or the ground—yielding eventually to a region of being we named discipline. In this way, Badiou was able to address both of his concerns: interiority and beginning. We could also argue that dialectic is the name of immanence when a region of thought is bound together by the thought of the One and by the suspension of foundation (the suture to the beginning, in this case, is the assumption that the One is, cf. beginning of *Logic*), that is, the very immanence of this region. A primary example is speech, whose ground is in the other (signification) and whose beginning is always missing (cause of desire), but remains nonetheless a discrete unity (signifier): dialectics names the immanent "work" of speech, how it preserves what it negates, by becoming the very other of what was said, and so on. Dialectics is the "mode of existence" for certain regions of being, and, under the constraint it imposes on the question of the beginning, certain operations follow from the very being of what is at stake. This makes it clear that discipline is the name of another—broader—region. In some sense, it is a concept in "competition" with dialectics. It names a different sort of immanence, one that also suspends the commitment to the One. This leads to the thought that discipline is a condition for dialectics.

Now, discipline certainly has its own "drama," its own contradictions and operators. But the contradiction we've highlighted—between ontology and that-which-is-not-being-qua-being, that is, the procedures of fidelity—is not thinkable through dialectic. Why? Because one sequence is in rupture with the other. Badiou was only able to think of their continuity by being able to "step out" of one to arrive at the other. This rupture, which takes place within a region of being we call ontology, is thus not thinkable through a dialectical mode. It demands the ability to "reconfigure" an immanent domain, active fidelity to an encounter with an unspeakable, and the operators to reconstruct the region through the supplementation of a new being: the being of truth. These are operators available to discipline, not to dialectic. Our recognition of this new logic at the verge of a rupture is re-affirmed in the difference between composing and disposing of thinking (participating in the constitution of a consistent homogeneity versus being allowed to think some heterogeneous or even inconsistent being as a product of this homogeneous field we helped to compose). This is precisely how the discussion of events took place in *Being and Event*: ontology, in composing its own consistency, disposes of historical situations. This will lead us to a formal definition of what discipline is.

During our journey, we revealed the two major categories of Badiou's philosophy: being and thinking. For Badiou, thinking is ultimately axiomatic and being is infinite. We showed that these two categories find their best combination together in a single expression: the axiom of Infinity.

The axiom of infinity deals the most devastating blow to what Badiou calls the linguistic turn (or democratic materialism: "there is nothing but bodies and languages"). This is because it opens up access to a world, subtracted of sense and meaning, to the world of truths. Furthermore, this axiom declares that axiomatic thinking itself is the only valid mode in such a world—that, in view of the infinity of being, the only valid beginning is that of axioms: there is no ground and no metaphysical One. The infinity of being lends axiomatic thinking its legitimacy. In figure 9.1, the axiom of Infinity watches over the gap between thinking and being, the gap that is always there, but is now thinkable.

We can now define a discipline as a local configuration of being and thinking. The mode of thinking of that discipline, through a set of axioms, circumscribes that discipline as a regional being. This gives sense to the assertion: disciplines think their beings (or simply disciplines think). The being of a regional discipline is understood under the ontological schema of historical situations. That is, the being of a discipline contains evental singularities. This property allows the being of a discipline to gain the character of the real for that discipline. In other words, part of the being of a discipline is not expressible by it. And the encounter with this unsaid singularity of that discipline may trigger a generation of axiomatic thinking anew. We call this the second configuration of a discipline.

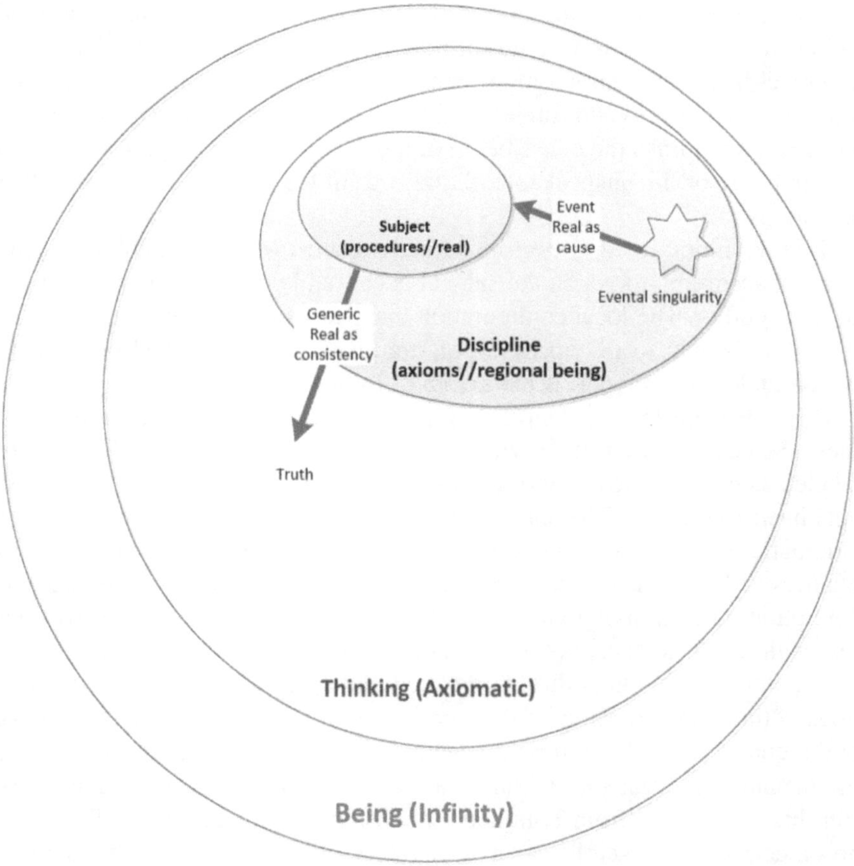

Figure 9.1 What Philosophy Says about Discipline. *Source:* Author created.

As a result of the encounter with the real, a new local configuration, within the regional configuration of being, will take place, which we call the subject. And this is the reason why, as Badiou says, the real is a category of the subject. This locality is exactly of a disciplinary configuration: that is, a configuration of formal (thinking) and material (being) conditions. The formal condition of this configuration is what Badiou calls the procedures of fidelity. From the perspective of this study, these are a subset of the operations of axiomatic thinking in general. The material condition, the being which a given configuration is drawing from, is the being of the singularity or the real at first. This is what Badiou, following Lacan, calls the real as the cause. But the formal procedures keep changing the material condition of the being of the subject. This material condition, under the procedure of truth (forcing), draws the being of the subject from the being of the generic toward the being of a truth. We call this second configuration, of a local being with the formal procedures of fidelity and forcing a subject. Therefore, a subject is a local configuration within a discipline that, this time, thinks a truth. In that capacity, the subject is a localized region of a discipline. While the discipline thinks the entire being of the discipline, the subject, motivated by the being of the unspeakable, thinks a truth for the real of the discipline it belongs to.

This definition juxtaposes itself directly against Miller's definition of the subject, according to which the subject is caused by repression or sinking in of the signifier. The local configuration that causes the subject in the case of discipline is akin to the operation of stratification we reviewed previously, whose underlying support is the axiom of Infinity.

Figure 9.1 illustrates a simple diagram that summarizes the relationship of these concepts to each other within the theory of the discipline. The sphere labeled as being is the being of consistent multiplicities. What of being presents itself to thought. The sphere that is labeled by thinking is the sphere of axiomatic thinking—the only type capable of thinking being, as opposed to realities and objects. The eclipse labeled by discipline depicts a regionalized configuration of thinking and being. We have seen many examples of such regionalization: science, politics, and ontology, for example.

The important point is this: axiomatic thinking creates regional configurations of thinking and being. In this configuration, what is of being in thinking is the consistency of the formal statements drawn from the axioms: this is the meaning of "realism" within axiomatic thinking. This is the great lesson that we learned from Lautman's philosophy of mathematics. There are no objects within disciplines—and therefore, there is no correlated subject either, because there is no subject-object dichotomy in the traditional sense. The domain of axiomatic thinking is objectless. This is the great lesson we learned from Lazarus' project of showing interiority and the immanence of politics. This does not mean that there are no objects referenced in disciplines or no knowledge of objects. What we mean is that disciplines are disciplines when they operate beyond their knowledge, which is when they apply the axiomatic thinking which has produced them. In that mode of thinking, disciplines do not think of objects—only what is beyond their point of impasse. This is a mode of thinking that is in operation through a local configuration of a discipline, the subject. But the architecture of this local configuration is no different than what, in the first place, made the being of discipline possible as a region of being: a regional configuration of thinking and being. So, if disciplines are possible, if disciplines are historical situations, if disciplines have singularities that are evental, then the being of the subject in a discipline is also possible and is defined in relation to a local configuration of thinking within the discipline, as well as the being that the general knowledge of the discipline has foreclosed. The subject is that which in a discipline composes consistency.

Seeing the architecture of the subject in continuity with the architecture of the discipline, in our view, makes the understanding of the former category a lot more robust. While a discipline thinks its being, the subject thinks a truth.

We further propose that what Badiou calls conditions of truths, politics, science, art, and love, should be reconstructed as different individual disciplines. That is ultimately why the truths of these disciplines, although eternal and inspiring to outside fields, solely belong to each of them. This is also why the subjects in each are different subjects. The regionality of discipline underwrites these effects.

The concept of discipline is an ontogenetic theory. It is the means to talk rationally about count-as-one and ultra-one. The theory of discipline tells us that subjects are not here from the beginning, even in the individual case. Certain ontological conditions must pre-exist them. In that sense, we can say that there is a "natural history" of the subject. We didn't have the tools to talk about this prior to Badiou's concept of discipline. The concept of discipline gives us the autogenesis of subjectivity. This is in contrast to the theory of events, which has often been read as something akin to miracles, leading to a judgment of Badiou as ultimately a theological thinker. We have clarified how this is not the case. We showed that his theory was always based on an idea of interiority, an interiority which thinks its own subjectivity.

Badiou's (unaccounted) concept of discipline, under the auspices of the actual infinities, and while still maintaining the real as a category of the subject, was able to do two things to the concept of the real: it regionalized the real to each situation, which allowed for multiple disciplines. It also allowed two types of the real: the real as the cause—the real of the event—and the real as consistency—the real of the generic. While the evental site does not need the axiom of Infinity (the axiom of Foundation ontologically guarantees its existence), the existence of generic in a historical situation does use this axiom. The inclusion of the event and the generic within an infinite historical situation, as well as the inclusion of the subject within the discipline as a local configuration, make the discipline not only the substance (discipline as a situation of being) but also the subject. Whereas in dialectics, we have substance and subject from the get-go, with the concept of discipline, we can ask about the conditions in which such an equation could be made. Furthermore, while in dialectic, the same situation of being is always dialectical, that is, it is always a substance and subject at the same time, in disciplines, this role is divided appropriately: a subset of the discipline is subjectivized as the result of the encounter with a local region of the discipline that exists but is not representable by the discipline. This change of conditions, in which discipline as the substance does not have to coincide with discipline as the subject, allows

Badiou to preserve knowledge separate from the real, and to finally show how truths are associated with the holes in knowledge (again, effects that were not possible in Hegel's system of logic, and although theorized by Lacan, are now given a proper ontological description). It is immanent to discipline to be substance and subject because the point of the real in a discipline is itself divided into substance and subject. It is both the irreducible point of the substance that resists formalization and the point at which one can split the situation to produce something new. This is proof that we are one step "prior" to the dialectic.

In our view, then, the definition of discipline as a configuration of thinking and being seems to be Badiou's absolute genius reworking of the Platonic theory of methexis, participation, now unbound from the constraint of objects: an object participates in the idea of which it is the expression—the disciplinary practice participates in the being of thinking which is the discipline itself, but this participation is not transparent; it's not "objective": the point of contact between the being in which the practice participates, and the thinking it is participating in, is the subject, and therefore is a negative point, with no transitivity implied. Furthermore, this configuration/participation happens at two levels in the "lifecycle" of a discipline: as a discipline and as a subject.

Coming back to our definition of composition and disposition, this can be stated in yet a different way. The logic of composing this thinking, and the disposition of its inner logic, must not share any properties! In other words: dialectic seemed to Plato the "highest" form of thought because he conceived the passage from mathematics to dialectic vertically, and it's true that dialectic comes "after" or "above" mathematical discipline, but in Badiou, this movement has become horizontalized, and now disciplines can be understood as the thinking of the conditions for a dialectical region—a Platonism of the multiple, indeed. This morphogenetic (in the strict sense) logic of participation seems to be the very being of what a discipline is, and even though the "inner logic" of the disposed-of thinking will be dialectical (due to being bound to the local One) and the inner logic of the composition will also be dialectical (due to being bound to the ultra-One of the event), the logic of participation itself, of composing-disposing, is not—since it has no dependence on the One, given that the point of contact of these two "levels" of a discipline is axioms and axiom-like statements.

At last, let us enumerate one more difference here. While the main operator of dialectic is negation, the main operator of the concept of discipline is regionalization: dialectical negation is conditioned by a regionalization that must have already circumscribed a place of interiority.

This brings us to the final thought in this conclusion.

In his essay published in 2014, Badiou mentions: "The fundamental problem in the philosophical field today is to find something like a new logic . . . or more precisely, a new dialectics." He refers to his other two masters, Marx and Plato, who were his forerunners in this task "to forge an explicit relation between revolutionary politics and a new dialectical framework." The new logic, in the wake of the failures of revolutionary politics, must possess new propositions adequate to "all forms of creative novelty," and in that regard, our problem is "the problem of negativity." The issue is that the principle of movement and creation in dialectic is that of negation. This is what we owe to Hegel. To demand the new is always to negate what is there, and that is true in any field, from politics to science to love, and so on. In demanding the new, "one stands in a relation of negation to the existing order."

But Badiou asserts that we are in a crisis of the power of negativity today. What, in his estimation, must happen is a sort of reinstalling of a logic whose affirmation comes before the negation, where one can "find a dialectical framework where something of the future comes before the negative present."[1] It is not a question of taking away opposition, struggle, and revolt. The point is that, prior to this, a minimum affirmative logic must have taken place.

The idea is simple, and now for us all too familiar. The negative conception of change is estranged from the power of infinity. It still works from inside the "known" configuration of the world, and when this configuration causes harm or discomfort or injustice, it demands a change—but with no proposal that goes beyond this demand for change. Badiou's proposal is an affirmation that is underwritten by infinity: if infinity is actual, then the new is possible. But that means "stepping out" of what is there. We saw this logic in discipline as the logic of the event, subjectivization and the procedures of truth. That is all good. Our problem with this gem of a text is rather the name "affirmative dialectic." In our view, and Badiou says as much, dialectic is not the logic for such an affirmation. The logic that Badiou is looking for is instead in his concept of discipline. First, because its main operator is not negation, but regionalization. Regionalization is an affirmative process by which a regional configuration of thinking and being is made possible: the consequence of axiomatic thinking (Lautman). The essence of axiomatic thinking is affirmation, not negation. In fact, not only the genesis but also the progression of axiomatic thinking is through affirmation. But what underwrites axiomatic thinking in the first place is the fact that we take being as infinite. To believe that actual infinities exist. If infinities do exist, then the only valid method of thinking the beginning is axiomatic, and not dialectical, thinking. All dialectics have some suture to the One, the same way that all phenomenological thinking has a suture to the foundation. Furthermore, there is no true affirmation of infinity in dialectical thinking. We saw this again in detail in Spinoza and Hegel. We saw the category of Affirmative Infinity for Hegel is the affirmation of the finite being: this affirmation fits the best in the infinity of the finite rationality—the infinity in a sort of reflexivity, which is a model for life and reason: the pedigree of German Idealism since Fichte that Hegel inherited and kept.

The only true "break" of this impasse of infinity was Cantor's axiom, which affirms that the mode of thinking that is adequate to the existence of actual infinity is, in fact, axiomatic and not dialectical.

So why are we repeating all this? Because the problem is not the operator of negation or only this operator. There is something else that gives the operator of negation its true power, its oomph, so to speak, and that, in our view, is the axiom of finitude as the fundamental axiom in all logic. The only way that we can think of the Other of A to be $\neg A$ is when we think of A as a finite being, or a finite situation, in Badiou's terms. If A stands, for example, for things such as free market economy, parliamentary politics, jobs, and education, then thinking the Other of these in terms dictated by the operator of negation is futile. The operator of negation fails, not because it is in crisis or has lost its potency, it is because these situations are not really finite, and it is not prudent or even practical to think of the Other of these situations in terms of negation alone. This is where the brilliance of Alain Badiou's proposal shines the brightest, and that is to defy the millennia-old rule of logic: in order for us to think the Other of our political and economic systems, justice and jobs, training and education, we should perhaps change our perspective and look at them as actual infinities, that is, situations with infinite sizes. Only then can we begin to apply the ontological models that mathematics has thought through regarding infinite sets. That is when we can begin to think about these situations as historical, find in them fragments which belong but are not included, speak of ground models for them, be able to develop generics for these situations and define their generic extension, and finally, their truths. By enacting the shift from finite to infinite, as the fundamental axiom of our thinking, we are suddenly capable of thinking about the Other in new ways, and we gain new operators not only to define but also to attain it.

We say this to illumine that what is at stake is not an operator of logic, but the entire founding assumption of the finitude of situations of being. This must be overturned if there is any dream of novelty in the world. To dream and speak rigorously of novelty in the context of disciplines, what Badiou calls conditions of truths, is to put this concept to work beyond what we have developed so far. It is apparent now that Badiou's entire project has been undertaken from a position afforded to him by a conception of discipline. This is the only concept that truly elaborates the consequences of the axiom of Infinity. And this axiom underlies Badiou's entire edifice: a theory for novelty.

NOTE

1. Alain Badiou, "From Logic to Anthropology: Affirmative Dialectics," in *Badiou and the Political Condition*, edited by Marios Constantinou, 45–55. Edinburgh: Edinburgh University Press 2014, 45.

References

Althusser, Louis. *For Marx*, London, New York: Verso 2005.

Althusser, Louis. *Lenin and Philosophy*, New York: Monthly Review Press 1971.

Althusser, Louis. *Reading Capital*, with Etienne Balibar, London, New York: Verso 1997.

Badiou, Alain. *Badiou: Theoretical Writings*, edited by Ray Brassier. London, New York: Continuum 2004.

Badiou, Alain. *Being and Event*, London, New York: Continuum 2007.

Badiou, Alain. *Conditions*, London, New York: Continuum 2008.

Badiou, Alain. "From Logic to Anthropology: Affirmative Dialectics" in *Badiou and the Political Condition*, edited Marios Constantinou, 45–55. Edinburgh: Edinburgh University Press 2014.

Badiou, Alain. *Infinite Thought: Truth and the Return to Philosophy*, London, New York: Continuum 2008.

Badiou, Alain. "Infinitesimal Subversion" in *Concept and Form Volume One*, edited by Peter Hallward and Knox Peden, 187–208. London, New York: Verso 2012.

Badiou, Alain. *Logics of Worlds: Being and Event 2*, London, New York: Continuum 2009.

Badiou, Alain. *Manifesto for Philosophy*, Albany: State University of New York Press 1999.

Badiou, Alain. "Mark and Lack: On Zero" in *Concept and Form Volume One*, edited by Peter Hallward and Knox Peden, 159–185. London, New York: Verso 2012.

Badiou, Alain. *Number and Numbers*, Cambridge: Polity Press 2008.

Badiou, Alain. *SAINT PAUL: The Foundation of Universalism*, Stanford: Stanford University Press 2003.

Badiou, Alain. *The Adventure of French Philosophy*, London, New York: Verso 2012.

Badiou, Alain. *The Century*, Cambridge: Polity Press 2007.

Badiou, Alain. *The Communist Hypothesis*, London, New York: Verso 2010.

Badiou, Alain. *The Concept of Model*, Melbourne: re.press 2007.

Badiou, Alain. *The Immanence of Truths, Being and Event III*, London: Bloomsbury Academic 2020.

Badiou, Alain. *The Rational Kernel of the Hegelian Dialectic*, Melbourne: re.press 2011.

Badiou, Alain. "Theory from Structure to Subject: An Interview with Alain Badiou" in *Concept and Form Volume Two*, edited by Peter Hallward and Knox Peden, 273–290. London, New York: Verso 2012.

Badiou, Alain. *Theory of the Subject*, London, New York: Continuum 2009.

Badiou, Alain. "We Need a Popular Discipline, Contemporary Politics and the Crisis of the Negative" in *Alain Badiou-Bibliography/Contemporary Politics and the Crisis of the Negative/Lacan Dot Com*, originally published in *Critical Inquiry* 645, 2008, http://www.lacan.com/baddiscipline.html.

Baki, Burhanuddin. *Badiou's Being and Event and the Mathematics of Set Theory*, London: Bloomsbury Academic 2015.

Bartlett, A.J. *Badiou and Plato, An Education*, Edinburgh: Edinburgh University Press 2015.

Belhaj Kacem, Mehdi. *Événement et répétition*, Auch: Tristram 2004. Also Belhaj Kacem, M., *L'Affect*, Auch: Tristram, 2004.

Belhaj Kacem, Mehdi. *Transgression and the Inexistent, A Philosophical Vocabulary*, London: Bloomsbury Academic 2014.

Bensaïd, Daniel. "Alain Badiou and the Miracle of the Event" in *Think Again, Alain Badiou and the Future of Philosophy*, edited by Peter Hallwad, 94–105. London: Continuum 2004.

Blanchette, Patricia. "The Frege-Hilbert Controversy" in *The Stanford Encyclopedia of Philosophy*, Spring 2014 Edition, Edward N. Zalta (ed.), URL = https://plato.stanford.edu/archives/spr2014/entries/frege-hilbert/.

Bosteels, Bruno. *Badiou and Poloitics*, Durham&London: Duke University Press 2011.

Carlson, David Gray. *A Commentary to Hegel's Science of Logic*, New York: Palgrave MacMillan 2007.

Clemens, Justin. *Badiou: Key Concepts*, with A.J. Bartlett (ed.), Oxfordshire: Routledge 2014.

Crockett, Clayton. *Deleuze Beyond Badiou*, New York: Columbia Press 2013.

Curly, Edwin. *Behind the Geometrical Method: A Reading of Spinoza's Ethics*, Princeton: Princeton University Press, 1998.

Dean, Jodi. *The Communist Horizon*, London, New York: Verso 2012.

Deleuze, Gilles. *What is Philosophy?* with Félix Guattari, New York: Columbia University Press 1994.

Desanti, Jean-Toussaint. "Some Remarks on Intrinsic Ontology of Alain Badiou" in *Think Again, Alain Badiou and the Future of Philosophy*, edited by Peter Hallward, 59–66. London: Continuum 2004.

Descartes, René. "Principles of Philosophy" in *The Philosophical Writings, Volume 1*, Cambridge: Cambridge University Press 2006.

Eyers Tom. *Post-Rationalism: Psychoanalysis, Epistemology, and Marxism in Post-War France*, London: Bloomsbury Academic 2013.

Faber, Roland. *Event and Decision: Ontology and Politics in Badiou, Deleuze and Whitehead*, with Henry Krips, Daniel Pettus (eds.), Cambridge: Cambridge University Press 2010.

Feltham, Oliver. *Alain Badiou: Live Theory*, London, New York: Continuum 2008.

Fink, Bruce. "Logical Time and Precipitation of Subjectivity" in *Reading Seminars I and II Lacan's Return to Freud*, 356–386. Albany: State University of New York Press 1996.

Fink, Bruce. *The Lacanian Subject, Between Language and Jouissance*, Princeton: Princeton University Press 1995.

Frazer, Zachary Luke. "The Category of Formalization: From Epistemological Break to Truth Procedure" in Alain Badiou's *The Concept of Model*, Melbourne: re.pre ss 2007.

Frege, Gottlob. *The Foundations of Arithmetic, A Logico-mathematical Enquiry into the Concept of Number*, New York: Harper & Brothers 1953.

Garrett, Don. "Spinoza's Theory of Metaphysical Individuation" in *Individuation and Identity in Early Modern Philosophy, Descartes to Kant*, edited Kenneth F. Barber and Jorge G. E. Garcia, Albany: State University of New York Press 1994.

Gillespie, Sam. *The Mathematics of Novelty*, Melbourne: re-press 2008.

Gray, Jeremy. *Plato's Ghost: The Modern Transformation in Mathematics*, Princeton: Princeton University Press 2008.

Hallward, Peter. "Introduction to Volume One: 'Theoretical Training,'" in *Concept and Form Volume One*, edited by Peter Hallword and Knox Peden, 1–56. London, New York: Verso 2012.

Hallward, Peter. *Badiou, a Subject to Truth*, Minneapolis: University of Minnesota Press 2003.

Hawking, Stephen. *God Created the Integers,* Philadelphia: Running Press 2007.

Hegel, George Wilhelm Friedrich. *Philosophy of Nature, Part Two of the Encyclopaedia of the Philosophical Sciences (1830)*, Oxford: Oxford University Press 2007.

Hegel, George Wilhelm Friedric. *Science of Logic*, Amhurst: Humanity Books 1969.

Heidegger, Martin. *Being and Time*, Toronto: Harper & Row Publishers Incorporated 1962.

Hilbert, David. *The Foundations of Geometry*, La Salle, Illinois: Open Court, 1965.

Houlgate, Stephen. *The Opening of Hegel's Logic, From Being to Infinity*, West Lafayette: Purdue University 2006.

Johnston, Adrian. *Badiou, Žižek and Political Transformations, The Cadence of Change*, Evanston: Northwestern University Press 2009.

Kirshner, Lewis A. *Applying the Work of Winnicott and Lacan*, in *Between Winnicott and Lacan: A Clinical Engagement*, Oxfordshire: Routledge 2011.

Koestler, Arthur. *Sleepwalkers*: *A History of Man's Changing Vision of the Universe*, New York: The MacMillan Company 1959.

Koyré, Alexander. *From the Closed World to the Infinite Universe*, Baltimore: The Johns Hopkins Press, 1957.

Kulstad, Mark. "Leibniz's Philosophy of Mind" with Laurence Carlin in *The Stanford Encyclopedia of Philosophy*, Winter 2013 Edition, edited by Edward N. Zalta, URL = https://plato.stanford.edu/archives/win2013/entries/leibniz-mind/.

Lacan, Jacques. "British Psychiatry and the War" in *Psychoanalytical Notebooks of the London Circle: Vol. 4, Psychiatry and Psychoanalysis: Spring 2000*, 9–34.
Lacan, Jacques. *Écrits, the First Complete Edition in English*, New York: W. W. Norton & Company, Inc. 1999.
Lacan, Jacques. "On a Question Prior to Any Possible Treatment of Psychosis" in *Écrits, the First Complete Edition in English*, 445–488. New York: W. W. Norton & Company, Inc. 1999.
Lacan, Jacques. *On Feminine Sexuality, The Limits of Love and Knowledge, Encore, The Seminar of Jacques Lacan Book XX, 1972–1973*, New York: W. W. Norton & Company, Inc. 1999.
Lacan, Jacques. . . . *or Worse, The Seminar of Jacques Lacan Book XIX, 1971–1972*, Cambridge: Polity Press 2018.
Lacan, Jacques. "Position of the Unconscious" in *Écrits, the First Complete Edition in English*, 703–721. New York: W. W. Norton & Company, Inc. 1999.
Lacan, Jacques. "Seminar on 'The Purloined Letter'" in *Écrits, the First Complete Edition in English*, 6–48. New York: W. W. Norton & Company, Inc. 1999.
Lacan, Jacques. "The Direction of the Treatment and the Principles of Its Power" in *Écrits, the First Complete Edition in English*, 489–542. New York: W. W. Norton & Company, Inc. 1999.
Lacan, Jacques. "The Number Thirteen and the Logical Form of Suspicion" *Autres Écrits*. Paris: Seuil 2001.
Lacan, Jacques. *The Ethics of Psychoanalysis 1959–1960, The Seminar of Jacques Lacan Book VII*, New York: W. W. Norton & Company, Inc. 1997.
Lacan, Jacques. *The Four Fundamental Concepts of Psychoanalysis, The Seminar of Jacques Lacan Book XI*, New York: W. W. Norton & Company, Inc. 1998.
Lacan, Jacques. *The Other Side of Psychoanalysis, The Seminar of Jacques Lacan Book XVII*, New York: W. W. Norton & Company, Inc. 2007.
Lacan, Jacques. *The Psychoses 1955–1956, The Seminars of Jacques Lacan Book III*, New York: W. W. Norton & Company, Inc. 1997.
Lacan, Jacques. *Le Seminaire XXII de Juacques Lacan R.S.I. Annee 1974–1975*, translated by Jack W. Stone.
Lacan, Jacques. *Transference, The Seminars of Jacques Lacan Book VIII*, Cambridge: Polity Press 2015.
Lacan, Jacques. "Science and Truth" in *Écrits, the First Complete Edition in English*, 726–745. New York: W. W. Norton & Company, Inc. 1999.
Lautman, Albert. *Mathematics, Ideals, and The Physical Real*, London, New York: Continuum International Publishing Group 2011.
Lawlor, Leonard. *Derrida and Husserl, The Basic Problem of Phenomenology*, Bloomington: Indiana University Press 2002.
Lazarus, Sylvain. *Anthropology of the Name*, Kolkata: Seagull Books 2015.
Lionnais, F.L. *Great Currents of Mathematical Thought, Vol. I.: Mathematics: Concepts and Development*, edited by F.L. Lionnais. New York: Dover Publications, 2004.
Livingston, Paul. M. *The Politics of Logic, Badiou, Wittgenstein, and the Consequences of Formalism*, Oxfordshire: Routledge 2014.

Lyotard, Jean-François. "Alain Badiou" in *Cahiers du Collège International de Philosophie* n.8, 1989, 211–225.

Maniglier, Patrice. *Acting Out the Structure*, in *Concept and Form Volume Two*, edited by Peter Hallward and Knox Peden, 25–46. London, New York: Verso 2012.

Meillassoux, Quentin. *After Finitude, An Essay on the Necessity of Contingency*, London: Continuum 2008.

Meillassoux, Quentin. "History and Event in Alain Badiou" *PARRHESIA* n.12, 2010, 1–11.

Miller, Jacques-Alain. "Action of the Structure" *Concept and Form Volume One*, edited by Peter Hallward and Knox Peden, 69–84. London, New York: Verso 2012.

Miller, Jacques-Alain. "An Introduction to Seminars I and II – Lacan's Orientation Prior to 1953 (Parts 1 to 3)" *Reading Seminars I and II Lacan's Return to Freud*, 3–35. Albany: State University of New York Press Albany 1996.

Miller, Jacques-Alain. "Matrix" *The Symptom 13*, Summer 2012, edited by Jacques-Alain Miller.

Miller, Jacques-Alain. "Paradigms of Jouissance" in *Lacanian Ink 17*, 10–46, Wooster: The Wooster Press 2000.

Miller, Jacques-Alain. "Suture (Elements of the Logic of the Signifier)" *Concept and Form Volume One*, edited by Peter Hallward and Knox Peden, 91–101. London, New York: Verso 2012.

Moore, Gregory H. *Zermelo's Axiom of Choice, Its origins, Development, & Influence*, New York: Dover Publications, Inc. 2013.

Nadler, Steven. *Spinoza's Heresy: Immortality and the Jewish Mind*, Oxford: Oxford University Press, 2004.

Norris, Christopher. *Being and Event: A Reader's Guide*, London, New York: Continuum, 2009.

Pfeiffer, Geoff. *The New Materialism: Althusser, Badiou, Žižek*, Oxfordshire: Routledge 2015.

Power, Nina. *On Beckett*, with Alberto Toscano (ed.), Manchester: Clinamen Press 2004.

Ramond, Charles. *Alain Badiou: Penser le multiple*, L'Hartman, 2002.

Regnault, Francois. *Dialectic of Epistemologies*, in *Concept and Form Volume One*, edited by Peter Hallward and Knox Peden, 119–150. London, New York: Verso 2012.

Roudinesco, Elisabeth. *Jacques Lacan*, Cambridge: Polity Press 2005.

Roffe, Jon. *Badiou's Deleuze,* Oxfordshire: Routledge 2014.

Ruda, Frank. *For Badiou: Idealism without Idealism*, Evanston: Northwestern University Press 2015.

Sartre, Jean-Paul. *Transcendence of the Ego*, New York: Hill and Wang 1991.

Spinoza, Benedict de. *The Ethics*, in *A Spinoza Reader*, 85–265. Princeton: Princeton University Press 1994.

Stillwell, John. *Roads to Infinity: The Mathematics of Truths and Proofs*, Natick: A K Peters Ltd 2010.

Stillwell, John. *Yearning for the Impossible: The Surprising Truths of Mathematics*, Natick: A K Peters Ltd 2006.

Tarby, Fabian. *Autour de Badiou*, with Isabelle Vodoz (ed.), Meaux: Germina, 2011.

Zalamea, Fernando. "Albert Lautman and the Creative Dialectic of Modern Mathematics" in *Mathematics, Ideals, and The Physical Real*, in Albert Lautman's *Mathematics, Ideas, and the Physical Real*, xxiii–xxxviii. London, New York: Continuum International Publishing Group 2011.

Žižek, Slavoj. "The Case of Alain Badiou" in *South Atlantic Quarterly*, 1998 - available here: https://www.lacan.com/zizek-badiou.htm.

Žižek, Slavoj. *The Ticklish Subject, The Absent Centre of Political Ontology*, London, New York: Verso 2000.

Index

About the Author

Mohammad Reza Naderi is a computer scientist and practitioner with a PhD in philosophy (Alain Badiou supervised his dissertation). His field of research includes logic, mathematics, and philosophy. He has published on the topics of fuzzy and mathematical logic. Reza is a member of the research collective Subset of Theoretical Practice (STP), focusing on linguistics and logic and their interplay with psychoanalysis and political theory.

www.ingramcontent.com/pod-product-compliance
Lightning Source LLC
Chambersburg PA
CBHW051256200926
51612CB00002B/59
* 9 7 8 1 6 6 6 9 3 1 0 4 4 *